PROGRESS IN MONOMERS, OLIGOMERS, POLYMERS, COMPOSITES AND NANOCOMPOSITES

PROGRESS IN MONOMERS, OLIGOMERS, POLYMERS, COMPOSITES AND NANOCOMPOSITES

RICHARD A. PETHRICK,
G. E. ZAIKOV
AND
J. PIELICHOWSKI
EDITORS

Nova Science Publishers, Inc.
New York

For permission to use material from this book please contact us:
Telephone 631-231-7269; Fax 631-231-8175
Web Site: http://www.novapublishers.com

LIBRARY OF CONGRESS CATALOGING-IN-PUBLICATION DATA

Progress in monomers, oligomers, polymers, composites, and nanocomposites / editors, Richard A. Pethrick, G.E. Zaikov, J. Pielichowski.
p. cm.
ISBN 978-1-60456-878-3 (hardcover)
1. Polymers. 2. Composite materials. 3. Nanostructured materials. I. Pethrick, R. A. (Richard Arthur), 1942- II. Zaikov, Gennadii Efremovich. III. Pielichowski, Jan.
TA455.P58P766 2009
547'.7--dc22 2009003876

Published by Nova Science Publishers, Inc. ✢ New York

CONTENTS

PREFACE

Knowledge is the foundation for mankind's most successful ventures. Today there is a visible shift in science, where people have shifted from chemistry to biology, and from biology to medicine. Everybody wishes to be wealthy and healthy. Jonathan Swift once said, "Everybody wants to live longer but nobody wants to be old."

Still, the role of polymer chemistry (pure and applied sciences) is very prominent in the world of science today, but it is heading away from polymers and polymer blends towards composites and nanocomposites. It allows for the creation of new materials with unique properties and new possibilities. This is a second edition on this subject.

If we measure the world's production of materials by volume and not by weight, then we find that it equals the production of iron, cast, steel, and colored metals together. This volume equals 250-270 million cubic metric of material per year. It is important to note that the rate of production of polymers is overcoming the rate of production of metals by 30-40%. Mankind used to live in the Stone Age, then the Iron Age, then the Bronze Age, and now it has come to be the Age of Polymers (leaning heavily towards composites and nanocomposites.) There are reviews and essays, according to the opinions of editors, that are helping to further develop polymer science and assist in solving practical applications (new materials with improved properties).

Chapter 1 - The methodology for estimating activation energy of diffusion processes in bio-membranes was developed based on spatial-energy notions (P-parameter). It is shown that transport relative penetrability in bio-membranes numerically equals the degree of conformation interactions calculated via the relative difference of corresponding P-parameters. Calculation criteria are in accordance with reference and experimental data.

Chapter 2 - The notion of spatial-energy parameter (P-parameter), a complex characteristic of the most important atomic values, was introduced for relative motion of two interacting material points based on the modified Lagrangian equation. The physical sense of P-parameter as an averaged energy of valence electrons in atom responsible for inter-atom interactions and having a direct connection with electron density in atom is demonstrated. Wave properties of P-parameter are found, its wave equation formally analogous to the equation of Ψ-function is given.

Chapter 3 - Now in connection with intensive development chemical, oil refining and other industries all rises a problem of preservation of the environment more sharply.

Millions tons of harmful gaseous substances are thrown out annually in an atmosphere. Creation of high-power plants results in necessity of application of clearing constructions of high efficiency [1].

One of perspective directions of gas purification are separation the devices using effect of action of a field of centrifugal forces, allowing most full to realize advantages new power - intensive technologies.

Centrifugal devices are characterized by high efficiency, simplicity of a design and low metal-intensive. Application of such equipment allows also essentially to intensify process mass transfer due to increase of speed of movement of phases [2].

As efficiency of separation is defined by requirements of concrete process, the big value gets correct calculation of speeds of movement of phases at which the high degree of clearing without occurrence of secondary ablation is achieved.

Without the decision of these, and also of some other questions, the further optimization processes of gas purification becomes difficult a problem.

Chapter 4 - The results of carbo-, organo-, glass-, hybrid- and stratified plastic composites are generalized in the paper. A number of new phenomena and effects conditioned by the interaction and influence upon each other of polymer matrix and armiring fibre in a composite are detected and described.

New methods to evaluate the structural characteristics of interphase strata and polymer matrix are suggested. Scientific bases of creating polymer composites possessing high relaxation ability on the basis of separate coating technology of components are worked up.

Chapter 5 - The modifying influence of polyazomethines on main physicomechanical properties of polybutyleneterephthalate has been investigated. The improvement of operation characteristics has been found out. It has been shown that the main reason of changing properties of polymeric mould is plastifying influence of polymeric add.

Chapter 6 - The processes of swelling and diffusion of solvents into the structure of tetraflurethylene-propylene coolymer can not be described quantatively using only one characteristic of a solvent. In all cases a molar volume of liquid has the determining effect, hampering its penetration into the structure of polymer. However, the effects of other factors – both solvation and density of cohesion energy of solvents – are significant. Adequate quantitative generalization of abovementioned processes can be obtained on the basis of free energies linearity concept by using of linear multiparameter equations taking into account the effects of different factors.

Chapter 7 - Physical chemical characteristics (composition, energetic stability and temperature of destruction) of crystallohydrates of cobalt(II)phthalocyanines were obtained by thermogravimetric analysis method.

Chapter 8 - The opportunity of the modification of the cellulose using biocidal polyelectrolite on the basis of metacrylate guanidine was investigated. Optimum contitions of modification, such as concentration and duration of modifying are found, dependence of the bactericidal properties on the maintenance of guanidine groups is investigated.

Chapter 9 - The analysis of literary data for last 17 years on synthesis, properties and application poliethersulphons (PES) and polietherketons (PEK) is lead. Polyarilatsulphons synthesize in two perspective ways. On the basis of poliethersulphons heat-resistant shock-resistant materials have been received. It is synthesized polyarylenetherketone with lateral carboxylate group in each part. Introduction of this group is.

Chapter 10 - The given results of the investigation make it possible to come to the following conclusions. Firstly, the expediency of receiving PBT with different antioxidants is shown. Secondly, an effective systems of stabilizers are received, the usage of which enable to synthesize PBT which has high thermal characteristics.

Chapter 11 - The work presented nowadays, is devoted to results of creation of compositions on the basis of polyethylene and the starch, capable to keep initial physic mechanical and physical and chemical properties and biodegradable under action of microorganisms of ground and the UV-irradiation after expiry of the term of operation. As rheological characteristics values of a parameter of fluidity melt (PFM), the basic physic mechanical properties - on an example of a pressure and deformation of destruction were investigated at single-axle a stretching, and in some cases - the module of elasticity.

Chapter 12 - The influence of the novel biocide guanidine – containing polyelectrolytes on the morphological properties of the St.Aureus bacteria is studied. The possible mechanism of the biocide influence of the polyelectrolyte is given. The process of the lysis of the bacterial cell caused by the guanidine polyacrylate has been detected by the atomic – force microscope Solver Pro – 47 in the Laboratory of the Nanoprobe Investigations of the Kabardino – Balkarian State University.

Chapter 13 - New phosphorus-containing polymers (polyproteinylmethenephosphinic acids) on base of peptides were synthesized through reaction phosphonomethylation by phosphinic acid and formaldehyde to the peptides with template ions. The product was characterized by IR spectroscopy and its ion-exchange properties toward Cu^{2+} cation was investigated. The material thermal behavior was studied by differential thermal analysis. The characterization results show that the templated phosphorus-containing polypeptides have improved capacity toward Cu^{2+} cation (2,9 mmol/g) and fireproof ability.

Chapter 14 - Physical and mechanical properties of the filled polymer composite materials (PCM) in dependence on the extent of filling, the rate of deformation were investigated. It was found out that structural properties of the filled composite materials are determined with the nature of polymer matrix, filling degree, nature of the fillers, structural organization of FCM, that is being formed in the process of receiving of the composite materials, and conditions of tests.

Chapter 15 - Quantum-chemical calculation of isoolefines: isobutylene, 2-methyllbutene-1, 2-methylbutene-2, 2-methylpentene-1, 2-ethylbuten-1, 2,3-dimethylbutene-1, 2,3,3-trimethylbutene-1, 2,4,4-trimethylpentene-1, 2-isopropyl-3-butene-1, 2-tret.-butyl-3,3-dimethylbutene-1 was made by classical method CNDO/2 in parametrization of Santri-Poppla-Segala. Electronic and geometric structure of these isoolefiines was calculated. Acid power of these compounds was determined. All under study isoolefines possess low acid power and pertain to class very weak H-acids (pKa >14). Exists the row an isoolefiines more capable to cation polymerization, than isobutylene

Chapter 16 - In the present research radical copolymerization of styrene with 2-methyl-2-vinyl-1,1-dichlorocyclopropane has been investigated. It is shown that the copolymer has a unit structure and is enriched with styrene units.

Chapter 17 - The comparative analysis of the basic physical and chemical properties of not stabilized polyalkilenetherephtalates and their compositions with polymers containing oximates groups is submitted. It is shown, that use of new polyformaloximates (PFO) and polyarylateoximates (PAO) as modifying additives allows improving operational properties and technical characteristics of polyalkilenetherephtalate compositions.

Chapter 18 - The purpose of the present work studying interaction of components in the heterogeneous environment in systems of type Me_2WO_4 - WO_3-MeCl. (Me-Li, Na, K, Rb, Cs) and development on the basis of received is stated at, it of the given electrolytes for synthesis oxides tungsten bronze alkaline metals with adjustable physical characteristics.

Study thermal stability of low-melting structures, enriched with tungsten and acidic tungsten (VI)

Chapter 19 - The principally new methods of electrochemical synthesis of the tungsten carbide nanopowders and carbon nanotubes have been developed in the centre of the collective use "The X-ray diagnosis of the materials" of the Kabardino – Balkarian State University. The research is a part of the priority project "The industry of the nanosystems and nanomaterials".

The electrolysis of the oxide and oxide – halogenide ionic solutions is employed for obtaining the tungsten carbide nanopowders. The basis of the electrochemical synthesis of the tungsten carbide is the simultaneous electrodeposition of atoms of both the tungsten and carbon on the cathode which may later join to form the nanoparticles.

Chapter 20 - The influence of the surface – active agent on the thermoplastic polymer adsorbed on the surface of the magnetic powder is studied.

Chapter 21 - Nanomaterials find wider application every day. The most promising ones are the carbon nanotubes (CNT) and nanofibres (CNF) and composites filled with them. The increased attention to these materials is caused by both the extra properties of ideal mono- or bi-layer CNTs (mechanical strength, heat and electroconductivity, density of the passing current) and the diversity in the morphology and properties of such nanoobjects.

Chapter 22 - The ozone application is based on its powerful oxidative action. The ozonolysis of chemical and biochemical compounds, as a rule, takes place with high rates at low temperatures and activation energies [1]. This provides a basis for the development of novel and improved technologies, which find wide application in ecology [2], chemical, pharmaceutical and perfume industries, cosmetics [3, 4], cellulose, paper and sugar industries, flotation, microelectronics, veterinary and human medicine, agriculture, foodstuff industry and many others [5-8]. At present it is very difficult to imagine a number of high technologies without using ozone. Here should be mentioned the technologies for purification of waste gases, water and soils, manufacture of organic, polymer and inorganic materials, disinfecting and cleaning of drinking and process water, disinfecting of plant and animal products, sterilization of medical rooms and instruments, resolving cosmetics problems, sterilization and therapy in veterinary and human medicine, surface cleaning and functionalization in the manufacture of polymer and microelectronics articles, flotation, deodorization and decolourization of gases, liquids and solid substances, oxidation in the intermediate stages of various technologies, etc.

Chapter 23 - In this chapter, a detailed review on processing of biodegradable and biocompatible natural nanofibers is presented.

Chapter 24 - The analytic review of recent advances in the field of synthesis and study of the structures and properties of the polymeric nanocomposites on the basis of the organomodified layered silicates is given

Chapter 25 - It was shown, that amorphous copper phthalocyanin (CuPc) forms molecular solutions in heptane, water and in polymeric matrixes (polythene, cellulose triacetate, polycaproamide). Is established, that the spectrum of individual molecules CuPc essentially differs from spectra of blue-green pigments CuPc: in a spectrum of individual molecules CuPc are absent chromogenes bands. The intensive electronic-oscillatory UV-bands of an acene type are characteristic for a spectrum of individual molecules. These bands correspond to three electronic $\pi \rightarrow \pi^*$ transitions. In this spectrum there are also weak $n \rightarrow \pi^*$ bands of p-electrons transitions of nitrogen atoms of aromatic macrocircle. From the received results follows, that the accepted literature treatment of an phthalocyanines chromogenity origin is incorrect. According to this treatment a known spectral Q-band (λ = 670 nm), belongs to individual molecules. Actually Q-band determining pigments colors, belongs to supramolecular units of CuPc.

Chapter 26 - Numerous experimental data on kinetics of photoinitiated stationary and nonstationary polymerization of mono- and di(meth)acrylates to high conversions show the possibility of the change of chain termination mechanism in reaction zones, where translation mobility of radicals (and, respectively, the rate of bimolecular chain termination) abruptly decreases. In this case linear from the formal stoichiometric point of view chain termination mechanism, which appears as an act of self-burial of active radicals at their propagation in polymer matrix, becomes determinative. Suggested earlier kinetic models with taking into consideration possible role of linear chain termination are in satisfactory agreement with all experimental data on homopolymerization, but need to be extended on copolymerization processes. In present article we propose kinetic model of copolymerization at linear chain termination, which appears as an act of self-burial of active radical in reaction of its propagation.

Chapter 27 - The contributors gave information about extrusion heads used in manufacturing polymeric casings for preventing pipes from corrosion. It was shown the high efficiency of this new design.

Chapter 28 - The contributors gave information about plant used for manufacturing building tile on the basis of polymer materials. Given plant allows to produce qualitative building tile

Chapter 29 - The contributors of this paper give information about constructions of multi-jet extrusion heads for manufacturing profile articles. It was shown high efficiency of these constructions.

Chapter 30 - The results of the present paper have shown the existence in polymer particulate-filled nanocomposites nanoadhesion effect, which consists in enhanced level of polymeric matrix-nanofiller interaction. This effect has purely dimensional origin and was realized for nanofiller particles with diameter less than 80 nm. The nanoadhesion effect influences strongly on macroscopic properties of those nanocomposites, in which it was realized. So, essential thermal stability enhancement and considerable thermal expansion coefficient reduction of nanocomposites on the base of phenylone at introduction 1 mass. % aerosil only are due to the nanoadhesion effect. Appreciation of nanoadhesion effect allows qualitatively and quantitatively to describe exactly nanocomposites reinforcement degree.

Chapter 31 - Perspectives in the field of creation highly effective anticancerogenic preparations have been evaluated. For their creation it is offered a new regio-selective method of glycosylation alkylurea in conditions nucleophilic catalysis with the following nitrozing of glycosyl carbamides of the D- and L-rows.

This method opens principal new possibilities of modification of compounds by means of glycosylamides bond allowing to get the preparations, possessing small toxicity and selectivity of the action.

Chapter 32 - The introduction of 0.035 wt% 1.1.5-trihydroperfluoropentanol decreases the rate of the photochemical destruction of polycaproamide fibres. The simultaneous modification of PCA with microquantity of 1.1.5-trihydroperfluoropentanol and 1.3-benzendicarboxamide, N.N'-bis(2.2.6.6-tetramethyl-4-piperidinyl) leads to the synergetic photostabilizing effect.

Changes in X-ray diffraction patterns of unoriented films of poly-ε-caproamide cast from its solutions in formic acid due to the introduction of 1,1,5-trihydroperfluoropentanol are analyzed. The plastification increases the fraction of polymer molecules in the planar all-trans conformation at the costs of those in the gauche conformation that constitutes the amorphous phase. This promotes the formation of hydrogen-bonded layers oriented perpendicular to the film plane as structural elements.

In: Progress in Monomers, Oligomers, Polymers, Composites ... ISBN 978-1-60456-878-3
Editors: R. A. Pethrick, G.E. Zaikov et al.

Chapter 1

ACTIVATION ENERGY OF MOLECULE AND ATOM DIFFUSION IN BIO-SYSTEMS

G.A. Korablev[1,*] and G.E. Zaikov[2,†]

[1]Basic Research-Educational Center of Chemical Physics and Mesoscopy, Udmurt Research Center, Ural Division, RAS, Izhevsk, Russian Federation

[2]Institute of biochemical physics after N.M. Emanuel, RAS, Moscow, Russian Federation

ABSTRACT

The methodology for estimating activation energy of diffusion processes in bio-membranes was developed based on spatial-energy notions (P-parameter). It is shown that transport relative penetrability in bio-membranes numerically equals the degree of conformation interactions calculated via the relative difference of corresponding P-parameters. Calculation criteria are in accordance with reference and experimental data.

Key words: Spatial-energy parameter, diffusion, activation energy, bio-structures, relative penetrability.

INTRODUCTION

Transport (transition) of molecules and atoms via bio-membranes is usually classified by energy-dependence [1]. Active transport occurs at the consumption of chemical energy due to hydrolysis of ATP and, strictly speaking, is not a diffusion process. Passive transport – the transition of a substance by the gradient of electrochemical potential and is not connected with the consumption of chemical energy by the cell: this is a diffusion process to the direction of the least electrochemical potential. The availability of trans-membrane gradient of molecule and atom concentration is its electromotive force.

[*] Correspondence to: E-mail: korablev@udm.net.

[†] Correspondence to: Institute of biochemical physics after N.M. Emanuel, RAS, Moscow, Russian Federation

The following types of passive transport are distinguished:

1. Simple diffusion of molecules conforming to the well-known Fick's law.
2. Diffusion through the pores (channels), selective penetrability of which can be limited by the solubility of substances in lipids.
3. Transport with the help of molecules-carriers that can be executed in two variants:
   a) a) mobile carrier – diffusion of molecules takes place alongside with a substance in the membrane;
   b) forwarding of a substance from one molecule to another.

Besides, the diffusion can proceed not only through the membrane, but also through movable water layers adjacent to the membrane from both sides, for example, oxygen diffusion into the cell. In these cases, the total system penetrability is usually estimated based on the principle of adding inverse values of penetrability in each subsystem:

$$\frac{1}{\gamma}=\frac{1}{\gamma_M}+\frac{1}{\gamma_W'}+\frac{1}{\gamma_W''} \tag{1}$$

The effective penetrability (γ) is proportional [2] to the diffusion coefficient (*Д*), membrane square (*S*), channel radius (*r*):

$$\gamma=\pi r^2 S\frac{Д}{l} \tag{1a}$$

where l – thickness of the membrane.

In a more modern model of limited diffusion [3,4] the protein fragment is considered as Brownian oscillator with great attenuation (K.V. Shaytan, A.B. Rubin). Its motion is described by two parameters: conformation potential and diffusion coefficient depending upon the coordinate *x*, which results in the general equation of the fragment movement in dense medium (Langevene equation).

In electro-diffusion processes the energy of ion penetration into the non-polar phase (*W*) is defined by Born formula [4]:

$$W=\left[\frac{Z^2e^2}{2r}\right]\left(\frac{1}{\varepsilon_W}-\frac{1}{\varepsilon_M}\right) \tag{2}$$

where: Z – electron valence, e – elementary charge, r – ion radius, ε_W and ε_M – dielectric penetrability of water and membrane.

It is interesting to find the direct dependence of main parameters of membrane diffusion with initial spatial-energy characteristics of free atoms. We showed such dependence before

4 Kosygina St., E-mail: chembio@sky.chph.ras.ru.

on the examples of diffusion and self-diffusion of atoms in many simple and complex structures [5,6].

P-PARAMETER

The analysis of kinetics of different physical and chemical processes shows that in many cases inverse values of velocities, kinetic or energy characteristics of corresponding interactions are added.

Some examples: ambipolar diffusion, total rate of topochemical reaction, change in the light velocity during the transition from vacuum into the given medium.

In particular, such supposition is proved by the aforesaid equations (1) and (2), and also by the formula of probability of electron transport process (W_∞) due to the overlapping of functions 1 and 2 (in stationary state) during electron-conformation interactions:

$$W_\infty = \frac{1}{2}\frac{W_1 W_2}{(W_1 + W_2)} \quad \text{or} \quad \frac{1}{2W_\infty} = \frac{1}{W_1} + \frac{1}{W_2} \tag{3}$$

Equation (3) is used [3] when estimating the characteristics of diffusion processes followed by radiationless transitions of electrons in proteins.

Lagrange equation can serve as an example. For relative motion of isolated system of two interacting material points with masses m_1 and m_2 in coordinate x it looks as follows:

$$m_{eq}x'' = -\frac{\partial U}{\partial x} \quad \text{where} \quad \frac{1}{m_{eq}} = \frac{1}{m_1} + \frac{1}{m_2}$$

here U – mutual potential energy of points; m_{eq} – equivalent mass. At the same time $x''=a$ (characteristic of system acceleration). For elementary sections of interactions Δx can be accepted as:

$$\frac{\partial U}{\partial x} \approx \frac{\Delta U}{\Delta x} \quad \text{i.e.:} \quad m_{np}a\Delta x = -\Delta U \text{. Then:}$$

$$\frac{1}{1/(a\Delta x)}\frac{1}{(1/m_1 + 1/m_2)} \approx -\Delta U \; ; \quad \frac{1}{1/(m_1 a\Delta x) + 1/(m_2 a\Delta x)} \approx -\Delta U$$

$$\text{or:} \quad \frac{1}{\Delta U} \approx \frac{1}{\Delta U1} + \frac{1}{\Delta U2} \tag{4}$$

where ΔU_1 and ΔU_2 – potential energies of material points on elementary section of interactions, ΔU – resultant (mutual) potential energy of these interactions.

The atom system is formed from differently charged masses of nucleus and electrons. In this system energy characteristics of subsystems are as follows: orbital energy of electrons and effective energy of nucleus taking into account screening effects (by Clement).

Therefore, assuming that the resultant interaction energy of the system orbital-nucleus (responsible for interatomic interactions) can be calculated by the principle of adding inverse values of some initial energy components, we substantiate the introduction of P-parameter [6] as the averaged characteristic of valence orbitals based on the following equations:

$$\frac{1}{q^2/r_i}+\frac{1}{W_i n_i}=\frac{1}{P_E} \tag{5}$$

$$P_E=\frac{P_0}{r_i} \tag{6}$$

$$\frac{1}{P_0}=\frac{1}{q^2}+\frac{1}{(Wrn)_i} \tag{7}$$

$$\mathrm{q}=\frac{\mathrm{Z}^*}{\mathrm{n}^*} \tag{8}$$

where: W_i – bond energy of electrons [7]; or – ionization energy of atom (E_i) – [8]; r_i – orbital radius i–orbital [9]; n_i – number of electrons of given orbital, Z^* and n^* – effective charge of nucleus and effective main quantum number [10].

P_0 will be called spatial-energy parameter, and P_E – effective P-parameter. Effective P_E-parameter has a physical sense similar to some averaged energy of valence electrons inside the atom and is measured in energy units, for example, in electron-volts (eV).

Now we will briefly discuss the reliability of such approach. According to the calculations [6] the values of P_E-parameters numerically equal (in the range of 2%) the total energy of valence electrons (U) by statistic model of the atom. Using a well-known relation between electron density (β) and interatomic potential, we can obtain the direct dependence of P_E-parameter upon the electron density at the distance r_i from the nucleus by statistic model of the atom:

$$\beta_i^{2/3}=\frac{AP_0}{r_i}=AP_E$$

where A – a constant.

The rationality of such an equation was proved by the calculation of electron density using wave functions by Clementi [10] and its comparison with the value of electron density calculated through the value of P_E-parameter.

The correlations of modules of maximum values of radial part of Ψ-function with the values of P_0-parameter were carried out and linear dependence between these values was found. Using some properties of wave function as applied to P-parameter, we obtain the wave equation of P-parameter formally analogous to the equation of Ψ-function.

Based on the calculations and comparisons, two principles of adding spatial-energy criteria depending upon wave properties of P-parameter and systemic character of interactions and charges of particles are substantiated:

1. Interaction of differently charged (heterogeneous) systems consisting of I, II, III, ... grades of atoms is satisfactorily described by the principle of adding the corresponding inverse energy values based on the equations (5-8), thus corresponding to the minimum attenuation of oscillations taking place in anti-phase;
2. During the interaction of similarly charged (homogeneous) subsystems the principle of algebraic addition of their P-parameters by the following equations is fulfilled:

$$\sum_{i=1}^{m} P_0 = P_0^{'} + P_0^{''} + \ldots + P_0^{m} \tag{9}$$

$$\sum P_Э = \frac{\sum P_0}{R} \tag{6a}$$

where R – dimensional characteristic of atom (or chemical bond) – this corresponds to the maximum intensification of oscillations taking place in the phase.

Applying the methodology of P-parameter based on the experimental data of about a thousand of different systems, spatial-energy principles of isomorphic replacement were found:

1. Complete (100-percent) isomorphic replacement and structural interaction occur at approximate equality of P-parameters of valence orbitals of interacting atoms: $P_E^{'} \approx P_E^{''}$
2. P-parameters of valence orbital of the atom of the least value define the orbital that is mainly responsible for the isomorphism and structural interactions.

Modifying the rules of adding inverse values of energy magnitudes of subsystems as applied to complex structures, we can propose the equation for calculating the P_C-parameter of a complex structure [6]:

$$\frac{1}{P_S} = \left(\frac{1}{NP_E}\right)_1 + \left(\frac{1}{NP_E}\right)_2 + \ldots \tag{10}$$

where N_1 and N_2 – number of homogeneous atoms in subsystems.

During the formation of solution and other structural interactions in the areas of the contact of atoms-components, the same electron density has to be established. This process is followed by the redistribution of electron density between valence areas of both particles and transition of the part of electrons from external spheres into the neighboring ones.

Apparently, with the nearness of electron densities in free atoms-components, the transition processes between boundary atoms of particles will be minimal, thus favoring the formation of a new structure. Thus, the task for estimating the degree of structural interactions, in many cases, comes to the comparative evaluation of electron density of valence electrons in free atoms (on the averaged orbitals) participating in the process based on the following equation:

$$\alpha = \frac{P_Э^{'} - P_Э^{''}}{P_Э^{'} + P_Э^{''}} 200\% \quad (11)$$

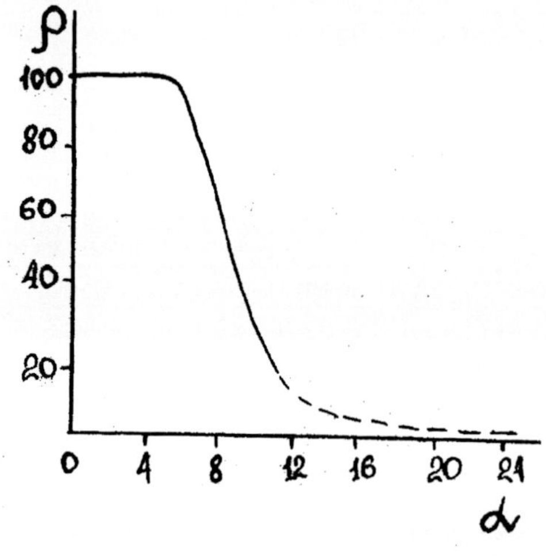

Figure 1. The nomogram of the dependence of the degree of structural interaction (ρ) upon the coefficient α – for P-parameters calculated via the ionization energy (E_i).

Following this methodology the degree of structural interactions (ρ) was evaluated in many (over a thousand) of simple and complex systems [6]. The nomogram of dependence of ρ upon the coefficient of structural interaction (α) – Fig. 1 (for P-parameters calculated via the ionization energy) was obtained.

Isomorphism as a phenomenon is usually considered as applied to crystalline structures. But obviously, the analogous processes can also take place between molecular compounds where their role and importance are not worse than of purely coulomb interactions. Diffusion processes, in particular, the replacement processes, apparently, can also be evaluated through the methodology of P-parameter

METHODOLOGY OF CALCULATIONS

It is found out that during the self-diffusion the activation energy often equals the total of enthalpies of formation and transition of vacancies. Apparently, in any actual case, different diffusion mechanisms can perform, but activation energy has to be always defined by the specifics and degree of interatomic interactions in structures.

Thus, the task for estimating the activation energy, in many cases, comes to defining the actual energy of paired interatomic interaction of diffusion atom and atoms of diffusion medium – for each specific type of interaction.

Previously, diffusion coefficients of some refractory oxides were calculated via P-parameter of interacting atoms of diffusion and diffusion medium using initial spatial-energy notions [5].

Let us show that activation energy during the transition of atoms and molecules in different systems can be quite reliably and easily calculated with the help of the notion of P-parameter. For this we use the tabulated values of P_0-parameters calculated via the ionization energy by the equations (5-8), some of them are given in Table 1. In diffusion processes, where ionic structures prevail, the ionization energy of an atom (E_i) is used as the orbital energy, in contrast to purely covalence interactions, when the values of electron bond energy (W_i) are applied in calculations.

Let us define the averaged effective energy of paired interatomic interaction in the system M'- M" with the help of P-parameter. Summing up P_0 by valence electrons and dividing the value of $P_0/2r_i$ by the number of effective valence electrons of the atom (n), we obtain the atom affective energy falling on one valence electron: $Q = \dfrac{P_0}{2r_i n}$

Applying the previously established principle of adding inverse values of P_E-parameters – equation (10), we have the resultant value of the effective energy of paired interaction between atoms 1 and 2 in the process of diffusion and self-diffusion:

$$\frac{1}{E_a} = \frac{1}{Q_1} + \frac{1}{Q_2}; \text{ or: } \frac{1}{E_a} = 2\left[\left(\frac{r_i n}{P_0}\right)_1 + \left(\frac{r_i n}{P_0}\right)_2\right] \tag{12}$$

where E_a – activation energy.

Here and afterwards the introduction of the multiplier 2 for r_i usually characterizes the volume diffusion, and without it – the surface one.

The results of calculations E_a by the equation (12) for many binary systems are given [6].

To calculate the activation energy of diffusion processes in binary systems, first, the effective energy of interatomic interaction is found, for example, for the pair metal-oxygen (in oxides) or metal (1)-metal (2) in metal systems, based on the following equation:

$$\frac{1}{Q_C} = \frac{1}{Q_i^{'}} + \frac{1}{Q_i^{''}} \tag{13}$$

where $Q_i^{'} = \left(\frac{P_0}{r_i n}\right)_1$; $Q_i^{''} = \left(\frac{P_0}{r_i n}\right)_2$ (13a,13b)

Table 1. P-parameters of valence orbitals of neutral atoms in basic state (calculated via the energy of atom ionization)

Atom	Electrons of valence orbitals	E_i (eVB)	r_i (Å)	q_i^2 (eVÅ)	P_0 (eVÅ)	$\sum P_0$ (eVÅ)
1	2	3	4	5	6	7
H	$1S^1$	13.595	0.5295	14.394	4.7985	4.7985
Li	$2S^1$	5.390	1.586	5.890	3.487	3.487
C	$2P^1$ $2P^1$ $2S^1$ $2S^1$	11.260 24.383 47.86 64.48	0.596 0.596 0.620 0.620	35.395 35.395 37.243 37.243	5.641 10.302 16.515 19.281	51.739
O	$2P^1$ $2P^1$	13.618 35.118	0.414 0.414	71.380 71.380	5.225 12.079	17.304
Na	$3S^1$	5.138	1.713	10.058	4.694	4.694
Mg	$3S^1$ $3S^1$	7.469 15.035	1.279 1.279	17.501 17.501	6.274 9.162	15.436
P(III)	$3P^1$ $3P^1$ $3P^1$	10.487 19.73 30.16	0.919 0.916 0.916	38.199 38.199 38.199	7.696 12.268 16.038	35.996
K	$4S^1$	4.339	2.162	10.993	5.062	5.062
Ca	$4S^1$ $4S^1$	6.113 11.871	1.690 1.690	17.406 17.406	6.483 9.320	15.803
Cl	$3P^1$	12.268	0.728	59.842	8.125	8.125
N	$2P^1$ $2P^1$ $2P^1$	29.60 47.426 77.472	0.488 0.487 0.487	52.912 52.912 52.912	6.257 11.319 16.078	33.664
Mn(II)	$4S^1$ $4S^1$	7.435 154640	1.278 1.278	25.118 25.118	6.895 11.130	18.025
Mn(III)	$3d^1$	33.69	0.3885	177.33	12.200	30.225
Fe(II)	$4S^1$ $4S^1$	7.893 16.183	1.227 1.227	26.57 26.57	7.098 11.364	18.462
Fe(III)	$3d^1$	30.64	0.365	199.95	10.564	29.026
Fe(II)	$4S^1$ $3d^1$	7.893 16.183	1.227 0.365	26.57 199.95	7.098 5.7372	12.835
Fe(III)	$3d^1$	30.64	0.365	199.95	10.564	23.399
Tl	$6P^1$	6.108	1.319	60.054	7.037	7.037

here $Q_i^{'}$ and $Q_i^{''}$ - effective P-parameters of atoms 1 and 2, n_1 and n_2 – number of their valence electrons in chemical bond. Since the activation energy of diffusion in this model corresponds to the total effective energy of interatomic interaction in the system: diffusing atom – diffusion medium, we have:

$$\frac{1}{E_a} = \frac{1}{Q} = \frac{1}{Q_1} + \frac{1}{Q_C} \tag{14}$$

where Q_1 – value of effective P-parameter of the diffusing atom:

$$Q_1 = \frac{P_0}{r_i n} \tag{15}$$

ACTIVATION ENERGY OF DIFFUSION IN BIOSYSTEMS

Based on the theory by Volkenshtein M.V., the transition of ions through the membrane can be explained as the motion of special quasi-particles ("conformones") consisting of the carriers of electron charge or shift of electron density. "There is a reason to believe that conformation motion represents a limited continuous diffusion in extended potential pit" [11]. It means that the transport of atoms and molecules through biosystems is considered as the diffusion process defined by the changes in electron density during electron conformation interactions (ECI).

Spatial-energy parameter (P-parameter) is a direct characteristic of electron density of the atom at the given distance from the nucleus. Therefore, it is advisable to apply its methodology to evaluate the diffusion energy in biomembranes. The nature has chosen only several main elements participating in vital physical and chemical processes of biodiffusion. These are: sodium, potassium, chlorine, calcium, magnesium, hydrogen, phosphorus and their compounds. Transport can be passive, active and electrodiffusion.

Based on the equation (12) we calculated the energy parameters of paired bond (Q_C) in the systems M-O (where M-Fe, Mg, Ca), P-O and M′-M″. Orbital radii (r_i) for metal systems and ionic radii (R_I) – in the chemical bond P-O were used as dimensional characteristics. The calculation results are given in Table 2. In the last column of this Table in the denominator (for metal systems) we see experimental values of activation energy E_a [12]. The direction of diffusion process is marked with arrow.

The values of activation energy in some oxides of biosystems were calculated using the obtained values of Q_C in equations (14,15) Here (as before) for metals, when calculating Q_1, orbital radii (r_i) were used as a dimensional characteristic, and for oxygen ions – anion dimensions (1,36Å) – Table 3.

Table 2. Calculations of energy parameters of paired bond (Q_C)

Bond	First element					Second element					Q_i (eV)
	P_0 (eVÅ)	r_i (Å)	R_H (Å)	n	Q'_i (eV)	P_0 (eVÅ)	r_i (Å)	R_H (Å)	n	Q''_i (eV)	
Fe-O	29.026	1.227	($3d^1 4S^2$)	1	23.656	17.304	0.414		1	41.797	15.107
Mg-O	15.436	1.279	($3S^2$)	1	12.068	17.304	0.414		1	41.797	9.3646
Ca-O	15.803	1.690	($4S^2$)	1	9.3509	17.304	0.414		1	41.797	7.6415
Mg-O	15.436	1.279	($3S^2$)	2	6.0345	17.304	0.414		2	20.899	4.6825
P-O	7.696	($3P^1$)	2x1.86	5	0.4138	5.225	($2P^1$)	2x1.36	2	0.9605	0.2892
P-O	7.696	($3P^1$)	1.86	5	0.8275	5.225	($2P^1$)	1.36	2	1.9210	0.5784
P-O	7.796	($3P^1$)	1.86	3	1.3792	5.225	($2P^1$)	1.36	2	1.9210	0.8028
C→Ca	15.943	$2P^2$	2.60	2	3.066	15.803	($4S^2$)	1.04	2	7.5976	1.092/1.010
Fe→Ca	12.835	($4S^1 3d^1$)	2x0.80	2	4.011	15.803	($4S^2$)	1.04	2	7.5976	1.31/1.29
Mn→Fe	18.025	($4S^2$)	0.91	2	9.9038	29.026	($4S^2 3d^1$)	0.67	3	14.441	2.94/2.86
Mn→Fe	18.025	($4S^2$)	0.91	2	9.9038	18.462	($4S^2$)	0.80	2	11.539	2.67/2.71

For systems CaO and MgO experimental data [13,14] are given for comparison. The activation energy of diffusion of ions Ca^{2+} via phosphor-lipids of membranes were evaluated (in the first approximation) as structural interactions of calcium ions with the bond P-O. Experimental values of activation energy, energy of ATP and its free energy (ΔG) – at different valence-structural states (conformations) of initial systems are given for comparison. From Tables 2 and 3 we see that calculated values mainly correspond to the data in the limits of allowed errors of calculations and experiment.

It is known [15] that the transition of Ca^{2+} from one side of the membrane onto the other is connected with the consumption of energy 0.207 eV, and that the transition of two ions of calcium is connected with the hydrolysis of one ATP molecule. From the Table 3 we have, for two ions of Ca: $2E_a=2\cdot0.244=0.488$ eV, that somewhat exceeds the ATP hydrolysis energy that equals 0.414 eV. However, if we consider that the efficiency coefficient for actual processes during photosynthesis is 83%, for two ions of calcium we have: $2E_a=0.488\cdot0.83=0.405$ eV (and $E_a\approx0.203$ – for one ion of calcium), this is in good accordance with the reference data.

To evaluate the energy of diffusion transport, the values of P_E-parameters of diffusing atoms Na, K, Cl and Mg were calculated with the help of movable carriers (MC) (Table 4). The values of activation enthalpy (ΔH) of ionophores were taken from [11]. At the same time, it was considered that these values can be approximately (if the activation barrier equals the enthalpy of the process [16]) taken as equal to the activation energy (E_a):

Table 3. Activation energy of diffusion in oxides of biosystems

Diffusion element						Diffusion medium		E_a (eV)	
Atom	Orbitals	P_0 (eVÅ)	r_i (Å) $R_И$	n	Q_I (eV)	System	Q_C (eV)	Calculation	Experiment
Ca	$4S^2$	15.803	1.690	2	4.676	CaO	7.6415	2.90	2.79
O^{-2}	$2p^2$	17.304	$R_И$=1.36	2	6.3618	CaO	7.6415	3.47	3.50
Mg	$3S^2$	15.436	1.279	2	6.0345	MgO	9.3646	3.67	3.46
O^{-2}	$2p^2$	17.304	$R_И$=1.36	2	6.3618	MgO	9.3646	3.79	3.56
O^{-2}	$2p^1$	5.225	$R_И$=1.36	1	3.8419	MgO	9.3646	2.72	2.71
O^{-2}	$2p^1$	5.225	$R_И$=1.36	2	1.9210	MgO	4.6825	1.36	1.31
Mg^{2+}	$4S^2$	18.025	2x0.91	7	1.4149	MgO	9.3646	1.23	1.21
Ca^{2+}	$4S^1$	6.483	$2R_И$=2.08	2	1.5584	P-O	0.2892	0.244 (0.203)	0.207
Ca^{2+}	$4S^1$	6.483	$2R_И$=2.08	2	1.5584	P-O	0.5784	0.4218 (0.405)	0.414 ATP hydrolysis
Ca^{2+}	$4S^1$	6.483	$R_И$=1.04	2	3.1168	P-O	0.8028	0.638	0.59-0.67 ΔG ATP

$$E_a \approx \Delta H$$

According to the initial principles of this methodology or the nomogram (Fig. 1), the complete (ρ=100%) structural interaction takes place when α<6%, limited interaction – when 6%<α<(25-30)%. When $P''_E = \Delta H$, the coefficient α was calculated by the equation (11) and (by nomogram) – the degree of structural interaction of diffusing atoms (ρ) with ionophores valinomycine and monaktine. In Table 4 you can see the results of calculations and principal possibility of the process, its degree or absence.

K, Ca and Cl have total or almost total structural interaction with valinomycine, and Na – with monaktine. Na has limited conformation interaction with valinomycine, and Cl – with monaktine. The lack of the possibility of Mg to transit with these ionoforms shows, in particular, why the nature chose calcium but not magnesium as the main regulator of various intracellular processes. From Table 4 we see the complete absence of diffusion with the movable carrier of these elements with gramicidin (α>100%), but gramicidin forms a polar pore in biomembrane. The ion is carried not by one, but several molecules of ionophore. This is a forwarding mechanism where the ions of diffusing element, apparently, move by the principle of self-diffusion in energy field of gramicidin molecules.

Table 4. Diffusion via biomembrane with movable carriers (MC)

Diffusing atom	Orbital	P_0 (eVÅ)	r_i (Å)	n	$P'_E = \frac{P_0}{r_i n}$ (eV)	Carriers	$E_a \approx \Delta$ (eV)	α %	ρ (%)	Transport type
Na	$3S^1$	4.694	1.713	1	2.740	Valinomycine	2.393	13.5	15-18	MC
Na	$3S^1$	4.694	2x1.713	1	1.370	Monactine	1.414	3.16	100	MC
K	$4S^1$	5.062	2.162	1	2.341	Valinomycine	2.393	2.20	100	MC
Cl	$3p^1$	8.125	0.728	7	1.594	Monactine	1.414	12.0	17-22	MC
Cl	$3p^1$	8.125	0.728	5	2.232	Valinomycine	2.393	6.96	92-96	MC
Ca	$4S^2$	15.803	1.690	4	2.238	Valinomycine	2.393	6.69	93-97	MC
Mg	$3S^2$	15.436	1.279	4	3.017	Valinomycine	2.393	23.1	0-5	no MC
Na	$3S^1$	4.694	1.713	1	2.740	Gramicidin	0.404	149	0	no MC
K	$4S^1$	5.062	2.162	1	2.3414	Gramicidin	0.404	141	0	no MC
Cl	$3p^1$	8.125	0.728	7	1.594	Gramicidin	0.404	119	0	no MC

Table 5. Diffusion through the pores with gramicidin

Diffusing elements							Pores	α %	ρ %
Atom	Orbitals	P_0 (eVÅ)	r_i (Å)	n_1-n_2	Q_1 (eV) by eq. (12)	Process	ΔH (eV)		
Na	$3S^1$	4.694	1.713	1-2	0.457	Self-diffusion	0.404	12.3	17-22
K	$4S^1$	5.062	2.162	1-2	0.390	Self-diffusion	0.404	3.53	100
Cl	$3P^1$	8.125	0.728	7-7	0.399	Self-diffusion	0.404	1.25	100

In Table 5 you can see the calculations of activation energy (E_a=Q_1) of self-diffusion of sodium, potassium and chlorine ions – by the equation (12), where the possibility of generalizing two valence electrons of the medium atom for Na and K is taken into account (n_1=1 and n_2=2).

There you can also find the results of calculating the coefficient α and defining the degree of structural interaction (ρ) during the diffusion of ions Na, K and Cl through the pores with gramicidin. It is shown that ion Na^+ has a limited diffusion, and ions K^+ and Cl^- – total. This takes place under standard conditions. Conformation changes caused by external effects (for example, temperature) can produce certain shifts in optimal values of these criteria.

RELATIVE PENETRABILITY OF BIOMEMBRANE CHANNELS

In literature [3,4] much attention is paid to geometrical criteria of penetrability. Obviously, these criteria are necessary factors of penetrability, but, apparently, insufficient, and, perhaps, even not the main ones. Thus, we can see based on hydroxyl-ammonium and methyl-ammonium [4] that these structures, having practically the same dimensions, possess completely different penetrability via the biochannels of sodium and potassium.

Using the methodology of P-parameter, we can assume that the availability of direct functional dependence between the value of relative penetrability of membranes and degree of structural interaction (ρ):

$$\gamma \sim \rho \tag{16}$$

where ρ is determined by the nomogram based on the calculated values of relative difference of P-parameters of coefficient (α) of interacting structures – equation (11).

This can be easily demonstrated on the example of Na- and K- channels for 1-valence ions Li^+ and Tl^+ (Table 6,7).

Table 6. Structural P_S-parameters of hydroxyl-ammonium and methyl-ammonium

Molecule complex	1st component			2nd component			Complex	Calculation formulas
	P_0 (eVÅ)	R_H (Å)	P_E (eV)	P_0 (eVÅ)	R_H (Å)	P_E (eV)	n_c (eV)	
NH_3^+	33.664	1.48	22.746	3x4.7985	1.36	10.575	7.2190	$\frac{1}{P'_E}+\frac{1}{P''_E}=\frac{1}{P'_S}$
OH	5.225	1.36	3.8419	4.7985	1.36	3.525	1.8383	$\frac{1}{P'_E}+\frac{1}{P''_E}=\frac{1}{P''_S}$
NH_3^+-OH							9.057	$P_S = P'_S + P''_S$
CH_3	51.739	2.60	19.900	3x4.7985	1.36	10.575	6.9066	$\frac{1}{P'_E}+\frac{1}{P''_E}=\frac{1}{P''_S}$
NH_3^+-CH_3							14.125	$P_S - P'_S + P''_S$

Here, P-parameters are calculated by the equation (6a), in which the ion radius R_I is used as a dimensional characteristic, and the coefficient α is calculated by the equation (11). The value of ρ (which is not expressed in %) is defined by the nomogram (Fig.1).

For complicated complex structures it is necessary to determine their total structural P_S-parameter, using the above rules of adding P-parameters. In each separate fragment of the molecule P_S-parameter is calculated by the principle of adding the inverse values of initial parameters, and P-parameters of such fragments are added algebraically. In Table 6 you can see these formulas and results of calculating P-parameters for hydroxyl-ammonium and methyl-ammonium, and in Table 7 – the obtained values of α and ρ in these cases.

It should be noted that less accurate values of ρ are obtained by the nomogram for the values α>(25-30)%. In all other cases, the data obtained are rather close to the actual values of ρ and γ.

Table 7. Calculation of relative penetrability in the channels of biomembranes

Channels atoms	Main atom of the channel			Diffusion atom or complex			α %	ρ (not in %)	γ [14]
	P_0 (eVÅ)	R_H (Å)	P_E (eV)	P_0 (eVÅ)	R_H (Å)	P_E (eV)			
Na-Li	4.694	0.98	4.790	3.487	0.68	5.128	6.814	0.9-0.95	0.9
K-Li	5.062	1.33	3.806	3.487	0.68	5.128	29.60	below 0.01	0.02
Na-Tl	4.694	0.98	4.790	7.037	1.36	5.223	8.654	0.35-0.4	0.33
K-Tl	5.062	1.33	3.806	7.037	1.36	5.223	31.40	below 0.01	?
2Na-(H_3N-OH)	2x4.694	0.98	9.580			9.057	5.61	~1.0	~(1-0.9)
2Na-(H_3N-CH_3)	2x4.694	0.98	9.580			14.125	38.3	~0	0.007
2Na-(H_3N-CH_3)	2x5.062	1.33	7.612			14.125	59.9	~0	0.02

General Conclusions

1. The method for calculating the activation energy of diffusion processes in biomembranes is given based on the initial spatial-energy characteristics of free atoms.
2. Relative penetrability of biomembranes numerically equals the degree of structural interaction calculated by the nomogram via the relative difference of P-parameters of interacting structures.
3. The calculated values are mainly in accordance with experimental and reference data (in the limits of possible errors of experiments and calculations).

References

[1] Kometiani Z.P., Vekua M.G. Kinetics of membrane transport ferments – in col. Biochemistry of membranes – *M.: Book 5, Vysshaya shkola*, 1988, 111p.

[2] Remizov A.N. Medical and biological physics. *M.: Vysshaya shkola*, 1987, 638p.

[3] Rubin A.B. Biophysics. Book 1. Theoretical biophysics. *M.: Vysshaya shkola*,1987, 319p.

[4] Rubin A.B. Biophysics. Book 2. Biophysics of cellular processes. *M.: Vysshaya shkola*, 1987, 303p.

[5] Korablev G.A. Calculation of diffusion coefficient of refractory oxides of metals based on the use of spatial-energy parameters of neutral atoms // In col. Diffusion, sorption and phase transformations in the processes of metal reduction. *M.: Nauka*, 1981, p.4-6.

[6] Korablev G.A. Spatial – Energy Principles of Complex Structures Formation. Brill Academia Publishers and VSP, Leiden, *The Netherlands, 2005*,- 426p. (Monograph).
[7] Fischer C.F. Average-Energy of Configuration Hartree-Fock Results for the Atoms Helium to Radon.//*Atomic Data,-1972, -№ 4,* -p. 301-399.
[8] Alen K.U. Astrophysical values. *M.: Mir,* 1977. -446p.
[9] Waber J.T., Cromer D.T. Orbital Radii of Atoms and Ions//*J. Chem. Phys -1965*, -V 42, -№12, -p. 4116-4123.
[10] Clementi E., Raimondi D.L. Atomic Screening constants from S.C.F. Functions, 1.//*J.Chem. Phys.-1963*, -v.38, -№11, -p. 2686-2689.
[11] Volkenshtein M.V. Biophysics. *M.: Nauka*, 1988, 592p.
[12] Bokshtein B.S. Diffusion in metals. *M.: Metallurgia*, 1978. -248p.
[13] Physical and chemical properties of oxides. Reference-book, ed. by Samsonov G.V. *M.: Metallurgia*, 1978, 472p.
[14] Kofstad P. Deviation from stoichiometry, diffusion and electrical conductivity in simple oxides of metals. *M.: Mir*, 1975. -398p.
[15] Levitskiy D.O. Calcium and biological membranes – in col. Biochemistry of membranes. - *M.: Book 7, Vysshaya shkola*, 1990, 126p.
[16] Blyumenfeld L.A. Problems of biological physics. *M.: Nauka*, 1977, 336p.

In: Progress in Monomers, Oligomers, Polymers, Composites … ISBN 978-1-60456-878-3
Editors: R. A. Pethrick, G.E. Zaikov et al.

Chapter 2

SPATIAL-ENERGY PARAMETER AS A MATERIALIZED ANALOG OF WAVE FUNCTION

G.A. Korablev [1,*] ***and G.E. Zaikov*** [2,†]

[1]Basic Research-Educational Center of Chemical Physics and Mesoscopy, Udmurt Research Center, Ural Division, RAS, Izhevsk, Russian Federation
[2]Institute of Biochemical Physics after N.M. Emanuel, RAS, Moscow, Russian Federation

ABSTRACT

The notion of spatial-energy parameter (P-parameter), a complex characteristic of the most important atomic values, was introduced for relative motion of two interacting material points based on the modified Lagrangian equation. The physical sense of P-parameter as an averaged energy of valence electrons in atom responsible for inter-atom interactions and having a direct connection with electron density in atom is demonstrated. Wave properties of P-parameter are found, its wave equation formally analogous to the equation of Ψ-function is given.

Key words: Lagrangian equations, wave functions, spatial-energy parameter, electron density, Hamiltonian function.

1. SPATIAL-ENERGY PARAMETER

When summarizing, analyzing and introducing criteria of isomorphism, mutual solubility, phase-formation the quasi-classical notions referring to atom composition (or to more complex structures) are used, for instance, atom radius, electronegativity [1], electron density

* Correspondence to: E-mail: korablev@udm.net.

† Correspondence to: 2Institute of biochemical physics after N.M. Emanuel, RAS, Russia, Moscow 119991, 4 Kosygina St., E-mail: chembio@sky.chph.ras.ru.

or combined characteristics (atom radius - electronegativity). At the same time, the application of quantum-mechanical calculations for the same purposes is much less effective.

However the existing models often use only one parameter, though many properties can be interpreted only taking different atom characteristics into account. Therefore the use of the parameter of unilateral dependence does not appear to be quite universal, and criteria applied can succeed only for certain cases and classes of compounds.

A definite compensation of volume energy of interacting structures takes place during the interaction of differently charged heterogeneous systems leading to the decrease of resulting volume energy (for instance, during the hybridization of atom orbitals). But this is not a direct algebraic deduction of corresponding volume energies. The comparison of multiple regularities of physical and chemical processes allows assuming that in such and analogous cases a principle of adding inverse values of volume energies or kinetic parameters of interacting structures is fulfilled. Here are some examples:

1) During the ambipolar diffusion when mutual transition of differently charged particles takes place in the given medium (in plasma, in electrolyte), the diffusion coefficient is found using the following relation:

$$\frac{\eta}{Д} = \frac{1}{a_+} + \frac{1}{a_-}$$

where a_+ and a_- - mobilities of charges of both atoms, η – constant coefficient.

2) Total rate of topochemical reaction (υ) between a solid and gas is defined as follows:

$$\frac{1}{\upsilon} = \frac{1}{\upsilon_1} + \frac{1}{\upsilon_2}$$

where υ_1 – diffusion rate of a reagent, υ_2 – reaction rate between gaseous reagent and a solid.

3) Changes in the light velocity ($\Delta\upsilon$) during the transition from vacuum into the given medium is calculated based on the principle of algebraic deduction of inverse values of corresponding velocities:

$$\frac{1}{\Delta\upsilon} = \frac{1}{\upsilon} - \frac{1}{c},$$

where c – light velocity in vacuum.

4) Lagrangian equation for relative motion of the system of two interacting material points with masses m_1 and m_2 in coordinate x is a s follows:

$$m_r x'' = -\frac{\partial U}{\partial x} \quad \text{where} \quad \frac{1}{m_r} = \frac{1}{m_1} + \frac{1}{m_2} \qquad (1)\,(1a)$$

here U – mutual potential energy of material points; m_r – reduced mass. At the same time $x'' = a$ (characteristic of system acceleration).

For elementary parts of interactions Δx we can take:

$$\frac{\partial U}{\partial x} \approx \frac{\Delta U}{\Delta x}$$

Then:

$$m_r a\Delta x = -\Delta U; \quad \frac{1}{1/(a\Delta x)} \cdot \frac{1}{(1/m_1 + 1/m_2)} \approx -\Delta U \text{ or:}$$

$$\frac{1}{1/(m_1 a\Delta x) + 1/(m_2 a\Delta x)} \approx -\Delta U$$

Since in physical sense $m_i a\Delta x$ equals the potential energy of each material point ($-\Delta U_i$), then:

$$\frac{1}{\Delta U} \approx \frac{1}{\Delta U_1} + \frac{1}{\Delta U_2} \tag{2}$$

Thus the resulting energy characteristic of the interaction system of two material points is found when adding two inverse values of initial energies of interacting subsystems.

Therefore assuming that the energy of valence orbitals of an atom (responsible for inter-atom interactions) can be calculated by the principle of adding two inverse values of some initial energy components, we postulate the introduction of P-parameter as an averaged energy characteristic of valence orbitals according to the following equations:

$$\frac{1}{q^2/r_i} + \frac{1}{W_i n_i} = \frac{1}{P_E} \qquad \frac{1}{P_0} = \frac{1}{q^2} + \frac{1}{(Wrn)_i} \qquad P_E = P_0/r_i \tag{3),(4),(5}$$

where W_i – orbital energy of electrons; [2] r_i – orbital radius of i–orbital [3]; $q=Z^*/n^*$ [4,5], n_i – number of electrons of this orbital, Z^* and n^* - effective charge of the nucleus and effective main quantum number.

P_O will be called spatial-energy parameter (SEP), and P_E – effective P–parameter (effective SEP). Effective SEP has a physical sense of some averaged energy of valence electrons in atom and is measured in energy units, for instance, in electron-volts (eV).

The values of P_0-parameter are tabulated constants for electrons of the given orbital of the given atom.

For the dimensionality of SEP we can write down the following:

$$[P_0]=[q^2]=[E]\cdot[r]=[h]\cdot[\upsilon]=\frac{kgm^3}{c^2}=Jm$$

where [h] and [υ] – dimensions of Planck's constant and velocity.

The introduction of P-parameter should be discussed as further development of quasi-classical notions using quantum-mechanical data on atom composition for obtaining criteria of energy conditions of phase-formation. At the same time, for similarly charged homogeneous systems (for example, orbitals in the given atom) the principle of algebraic adding of such parameters is preserved:

$$\sum P_E=\sum(P_0/r_i)\quad \sum P_E=\frac{\sum P_0}{r} \tag{6),(7}$$

(where r – main dimension characteristic if atom), or:

$$\sum P_0=P_0'+P_0''+P_0'''+\ldots\quad r\sum P_E=\sum P_0 \tag{8),(9}$$

Here all P-parameters are summed up by all valence orbitals of atom.

To calculate P_E-parameter at the given distance from the nucleus, either atom radius (R) or ion radius (r_I) can be used depending upon the bond type.

Applying the equation (8) to hydrogen atom, we can write down the following:

$$K\left(\frac{e}{n}\right)_1^2=K\left(\frac{e}{n}\right)_2^2+mc^2\lambda \tag{10}$$

where: e – elementary charge, n_1 and n_2 – main quantum numbers, m – electron mass, c – electromagnetic wave velocity, λ – wave length, K – constant.

Using the known relations ν = C/λ and λ = h/mc (where h – Planck's constant, ν – wave frequency) from the formula (10), we can obtain the equation of spectral regularities in hydrogen atom, in which $2\pi^2e^2$ /hC = K.

2. Effective Energy of Valence Electrons in Atom Using SEP Method: Comparison with Statistic Model

Modified equation of Thomas-Fermi transformed into a simple form when introducing non-dimensional variables [6] looks as follows:

$$U=e(V_i-V_0+\tau_0^2) \tag{11}$$

where: V_o – count potential; e – elementary charge; τ_0 – exchange and correlation corrections; V_i – interatomic potential at the distance r_i from nucleus, U – total energy of valence electrons.

The given value U was compared with the values of P_E-parameter for 21 elements that are partly given in Table 1.

As it is shown in Table 1 the values of U and P_E-parameters are practically the same (with deviations in most cases below 1-2%) *without any transition coefficients*. Multiple corrections introduced into the statistic model are compensated with the application of simple rules of adding inverse values of energy parameters, and SEP rather accurately presents the known solutions of Thomas-Fermi equation for inter-atom potential of atoms at the distance r_i from the nucleus. Namely, the following equality takes place:

$$U = P_E = e\,(V_i - V_0 + \tau^2_0) \tag{12}$$

Using a known relation [6] between electron density (β_i) and inter-atom potential (V_i), we have:

$$\beta^{2/3}_i \approx (3e/5)\cdot(V_i - V_0);\ \beta^{2/3}_i \approx Ae\cdot(V_i - V_0 + \tau^2_0) = [Ae\cdot r_i\cdot(V_i - V_0 + \tau^2_0)]/r_i \tag{13}$$

where A – constant.

According to formulas (12 and 13) we have the following relation:

$$\beta_i^{2/3} = A\ P_0/r_i \tag{14}$$

that establishes the connection between P_0-parameter and electron density in the atom at the distance r_i from the nucleus.

Since the value $e(V_i - V_0 + \tau_0^2)$ in the model of Thomas-Fermi there is a function of charge density, P_0-parameter is an immediate characteristic of charge density in the atom.

3. Electron Density in Atom Calculated via P-Parameter

The correctness of the equality (14) was additionally checked using Clementi functions [7] to estimate electron density β_i at the distance r_i. The density was calculated for atoms from ^{3}Li to ^{35}Br for external s- and p- orbitals taking into consideration small contributions from electrons of inner shells of atom into β_i. For atoms of 3d-metals

β_i was calculated for $4S^2$-electrons taking into account the density of 3d-electrons, whose partial contribution into β_i is 30-40%. P-parameter was estimated either via the bond energy [8] or ionization energy [9]. The results of these calculations are given in Figure 1(a,b,c,d) with firm lines. With p- and d- orbitals electrons on them were distributed based on Hunda rule. The calculations for equations (12-14) are given in Fig. 1(a,b,c,d) with dots. As it is seen

Table 1. Comparison of the total energy of valence electrons in atom calculated in the statistic model of the atom Thomas-Fermi *(U)* and with the help of approximation.

Atom	Valence electrons	r_i(Å)	X	φ(X)	*U*(eV)	W(eV)	n_i	q^2(eV Å)	P_E(eV)
1	2	3	4	5	6	7	8	9	10
Ar	$3P^4$	0,659	3,548	0,09-0,084	35,36- 33,02	12	4	73,196	33,45
	$3S^2$	0,607	3,268	0,122-0,105	47,81- 44,81	34.41- 29,0	2 2	96,257 96,157	47,99 42,47
	$2P^4$	0,146	0,785	0,47	834,25	246	4	706,3	817,12
Kr	$4P^4$	0,795	5,501	0,0495-0,0459	32,28- 34,94	11	4	85,883	31,27
	$4S^2$	0,691	4,781	0,0629-0,0588	47,19- 37,50	27,5 24	2 2	115,22 115,22	41,36 37,26
	$3d^{10}$	0,2135	1,481	0.2976-0,296	722,2 718,5	89	10	680,68	695,6
	$3P^4$	0,244	1,688	0,267-0,265	567,2- 562,7	214	4	668,01	652,0
Xe	$5P^4$	0,986	7,8600	0,0311 0,0257	24,50 20,28	7	4	88,906	21,37
	$5P^2$	0,944	7,525	0,0214	17,62	11,1	2	88,906	17,97
	$5S^2$	0,837	6,672	0,0357 0,0402	33,21 37,28	18 23,4	2 2	116,43 116,43	28,60 35,01
V	$4S^2$	1,401 8,26	8,508* 8,26	0,0325 0,0345	7,680 8,151	7,5	2	22,33	7,730
Cr	$4S^2$	1,453	8,95 8,70	0,0295 0,0313	7,013 7,440	7	2	23,712	7,754
Mn	$4S^2$	1,278	7,76	0,0256м	10.89 7,214	6,6	2	25,12	7,895
Mg	3S2	1,279	5,90 5,90	0,055м 0,0183м	7,43 2,47	6 2	2 2	17,50 17,50	6,40 3,10
Sc	4S2	1,57	8,668	0,0315	6,07	6,6	2	19,34	6,37

Note: 1) Bond energy of electrons (W_i) are obtained: theoretically (by Hartri-Fock method), "+" – by ESCA method, the other – by the results of optical measurements; 2) * - energy of valence electrons (U) is calculated with Fermi-Amaldi correction.

there is a good correlation between the values of β_i, calculated via P_0 and obtained from atomic functions. Consequently it is possible to compare electron densities of orbitals of interacting atoms (at the given distance from the nucleus) via the values of P_0/r_i. The value, inverse to the electron density, characterizes the value of partial volume of the given spatial-energy parameter of the corresponding system, i.e. basically the energy intensity of the given structure ($1/P = V_E$). And the principle of inverse adding of P-parameters is a principle of algebraic addition of energy intensities of interacting subsystems:

$$\frac{1}{P_S} = \frac{1}{P_E^{'}} + \frac{1}{P_E^{''}} + \ldots \text{ or } V_E = V_E^{'} + V_E^{''} + \ldots \qquad (15),(16)$$

where P_C – resulting P-parameter of structure, V_E – its energy intensity, $P_E^{'}$ and $P_E^{''}$ - P_E-parameters of subsystems, $V_E^{'}$ and $V_E^{''}$ – their energy intensities.

Or taking into consideration the number of homogeneous atoms N_1 and N_2 in subsystems we have:

$$\frac{1}{P_S} = \frac{1}{N_1 P_E^{'}} + \frac{1}{N_1 P_E^{''}} + \ldots \qquad (17)$$

Such an equation (17) is further used to calculate the resulting P_S-parameter of complex structures.

4. Dependence of P-Parameter on the Module of Maximum Values of Radial Part of Ψ-Function

The modules of maximum values of radial part of Ψ-function Д (n,ℓ,r) with SEP values. Here Д (n,ℓ,r) – radial wave function of (n,ℓ,r). For initial data we used wave functions, bond energies and inner orbital radii [9], orbital radii of external orbitals [3], value q по [4,5].

Such calculations revealed a linear dependence between the relation Po/Д (n,ℓ,r) = N_X and serial number of an element (Z) as shown in Fig. 2. From the figure we can see that the relation N_X/Z is a characteristic value for all elements with the same orbital quantum number (ℓ) of valence orbitals. Therefore the inflections in the figure occur mainly in the points corresponding to the changes in valence electron structure of elements (by orbital quantum number), e.g.:

B($2P^1$), Ne($2P^6$), Sc($3S^1$),Ga($4P^1$), Cu($3d^{10}4S^1$).

When analyzing the figure, we see that $N_X = f(Z^*)$, where Z^* - effective charge of the orbital.

N_X was compared with the values of $(Wr)^{1/2}$ based on the following equation:

$$\beta_0 Po / Д(n,\ell,r) = [Wr\,(n+\ell+1)]^{1/2} \qquad (18)$$

Where β_0 – proportionality coefficient, and quantum numbers n and ℓ are introduced when considering the character of inflections of graphical dependence of N_X upon Z. The results of multiple calculations on initial data [2,7] show that coefficient β_0 for the most elements is a relatively constant value.

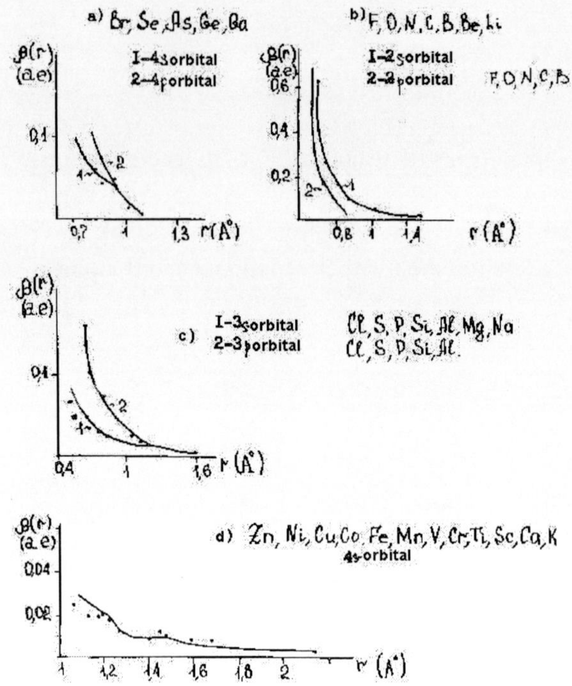

Figure 1. Electron density at the distance r_i calculated via Clementi functions (evan lines) and with the help of P-parameter (dots).

Thus for 64 elements coefficient β_0 is in the range of 0.59-0.66.

So we can consider that SEP has a direct connection with a maximum value of radial part of ψ-function and characterizes the distribution of electron charge density in the atom. But the redistribution of electron charge density has the main meaning when forming the chemical bond.

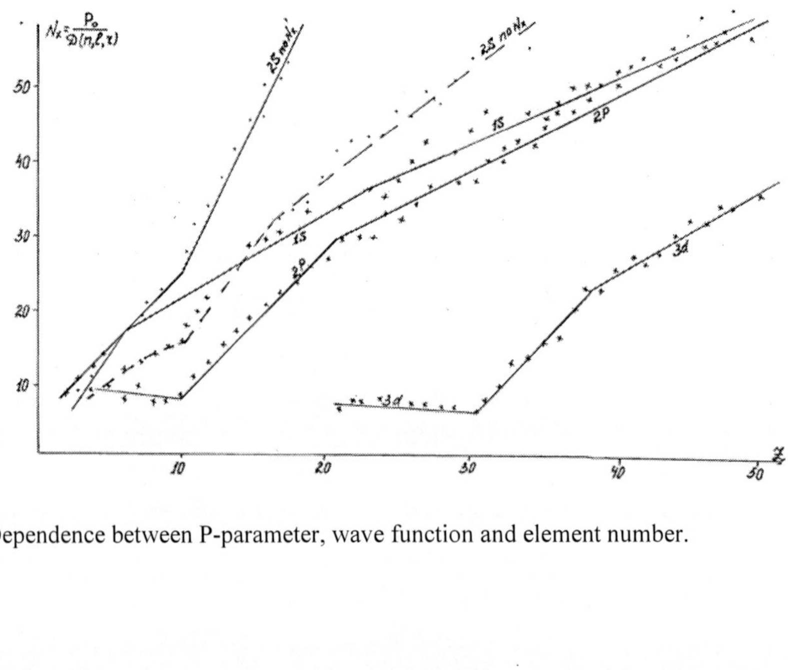

Figure 2. Dependence between P-parameter, wave function and element number.

5. Wave Equation of P-Parameter

To describe spatial-energy properties of an atom two types of P-parameters are introduced with simple connection between them:

$$P_E = \frac{P_0}{R}$$

where R – dimensional characteristic of an atom.

Taking into consideration additional quantum characteristics of atom sublevels, this equation in the coordinate x can be written down as follows:

$$\Delta P_E \approx \frac{\Delta P_0}{\Delta x} \qquad \text{or} \qquad \partial P_E = \frac{\partial P_0}{\partial x}$$

where ΔP equals the difference between P_0-parameter of i-orbital and P_I–parameter of indication (parameter of main state at the given set of quantum numbers).

According to the established rule of adding P-parameters [10] of similarly charged or homogeneous systems for two orbitals in the given atom with different quantum characteristics and in accordance with the law of energy conservation we have:

$$\Delta P_E'' - \Delta P_E' = P_{E,\lambda}$$

where $P_{E,\lambda}$ – spatial-energy parameter of quantum transition.

Taking $\Delta\lambda=\Delta x$ as the dimensional characteristic of the interaction, we have:

$$\frac{\Delta P_0''}{\Delta\lambda} - \frac{\Delta P_0'}{\Delta\lambda} = \frac{P_0}{\Delta\lambda} \text{ or: } \frac{\Delta P_0'}{\Delta\lambda} - \frac{\Delta P_0''}{\Delta\lambda} = -\frac{P_0\lambda}{\Delta\lambda}$$

We will divide it termwise by $\Delta\lambda$:

$$\left(\frac{\Delta P_0'}{\Delta\lambda} - \frac{\Delta P_0''}{\Delta\lambda}\right)\Big/\Delta\lambda = -\frac{P_0}{\Delta\lambda^2}, \qquad \text{where:}$$

$$\left(\frac{\Delta P_0'}{\Delta\lambda} - \frac{\Delta P_0''}{\Delta\lambda}\right)\Big/\Delta\lambda \sim \frac{d^2 P_0}{d\lambda^2}, \qquad \text{then:} \qquad \frac{d^2 P_0}{d\lambda^2} + \frac{P_0}{\Delta\lambda^2} \approx 0$$

Taking into consideration only those interactions when $2\pi\Delta x=\Delta\lambda$ (closed oscillator), we have the following equation:

$$\frac{d^2 P_0}{dx^2} + 4\pi^2 \frac{P_0}{\Delta\lambda^2} \approx 0$$

Since $\Delta\lambda = \frac{h}{mv}$, then:

$$\frac{d^2 P_0}{dx^2} + 4\pi^2 \frac{P_0}{h^2} m^2 v^2 \approx 0$$

Or

$$\frac{d^2 P_0}{dx^2} + \frac{8\pi^2 m}{h^2} P_0 E_k = 0 \qquad (19)$$

where $E_k = \frac{mV^2}{2}$ - kinetic energy of an electron.

Schredinger equation for the stationary state in coordinate x:

$$\frac{d^2 \psi}{dx^2} + \frac{8\pi^2 m}{h^2} \psi E_k = 0 \qquad (20)$$

When comparing the equations (19 and 20) we see that P_0-parameter is numerically approximately equal to the value of Ψ-function, $P_0 \approx \Psi$

and in general is proportional to it: $P_0 \sim \Psi$. Taking into consideration the practical capabilities of applying the methodology of P-parameter, we can consider that this parameters is a materialized analog of Ψ-function.

Since P_0-parameters as well as Ψ-function have wave properties, the principles of superposition have to be executed for them, this determines a linear character of equations of adding and changing P-parameters.

6. Spectral Characteristics of Some Atoms and Their Connection with Spatial-Energy Parameter

We use some properties of wave function as applied to P-parameter. In particular, by the analogy with wave function we can assume that the square of P_E-parameter multiplied by the volume element (Δr^3) has to characterize the electron density in this element of the volume – (e^2).

$$\text{or: } P_E^2(\Delta r)^3 = \frac{P_0^2}{\Delta r^2}\Delta r^3 = P_0^2\Delta r \sim e^2 = \frac{\hbar c}{137} \tag{21}$$

where c – velocity of speed; $\hbar = \frac{h}{2\pi}$. Here h – Planck's constant.

When estimating quantum transitions between energy levels w_2 and w_1, the substitution of Δr for the wavelength of radiation (λ) is possible: $\Delta r = \lambda$

The calculations made for simple atomic quantum transitions ns^1-nP^1 in elements C, N, O, Al and Si [9] showed that the following relation is fulfilled:

$$(\Sigma P_0)^2\,\lambda = 48\,\frac{c\hbar^3}{2m_e} \text{ or: } (\Sigma P_0)^2\,\lambda = 24\frac{c\hbar^3}{m_e}; \tag{22}$$

$\frac{2m_e}{\hbar^2}$ - coefficient of Schredinger equation, where: $\Sigma P_0 = P_0'(ns^1) + P_0''(np'')$

here n – main quantum number.

The average value of $(\Sigma P_0)^2\,\lambda$ equals 364118 $eV^2Å^3 = 9.3184 \cdot 10^{-63}\ J^2m^3$ with deviations in the range of ± 6% from calculated values (Table 2).

For complex multiplet transitions the equation (22) was modified:

$$(\Sigma P_0)^2\,\lambda\,\frac{N_H}{N_K} = 24\frac{c\hbar^3}{m_e} \tag{22a}$$

where $N_н$ and $N_к$ – number of registered electrons on nS and nP sublevels at the beginning and end of quantum transition.

The analysis of all these calculations by the equations (22,22a) confirms that P-parameter has a direct connection with the wave function Ψ and can characterize energy values of atom spectral transitions, in particular – for the simplest transitions between sublevels with the same main quantum numbers (n).

7. Wave Properties of P-Parameter and Principles of Adding P-Parameters

Since P-parameter has wave properties (as analogous to Ψ'-function), in the process of structural interactions mainly the regularities for the interference of corresponding waves have to be carried out.

The minimum interference, oscillation weakening (in antiphase) takes place if the difference of wave motion (Δ) equals the odd number of semi-waves:

Table 2. P-parameters of some atoms and their spectral characteristics of S-P transitions

Atom	Valence electrons	W(eV)	r_i (Å)	q_i^2 (eVÅ)	P_0 (eVÅ)	ΣP_0 (eVÅ)	Transition	λ (Å)	$\Sigma P_0 \lambda$ ($eV^2Å^3$)
C	$2P^1$	11.792	0.596	35.395	5.8680	14.889	$2S^1$-$2P^1$	1549	343386
	$2P^2$	11.792	0.596	35.395	10.061				
	$2S^1$	19.201	0.620	37.240	9.0209				
	$2S^2$	19.201	0.620	37.240	14.524				
N	$2P^1$	15.445	0.4875	52.912	6.5916	17.301	$2S^1$-$2P^1$	1238	370564
	$2P^2$	15.445	0.4875	52.912	11.710				
	$2P^3$	15.445	0.4875	52.912	15.830				
	$2S^1$	25.724	0.521	52.283	10.709				
	$2S^2$	25.724	0.521	52.883	17.833				
O	$2P^1$	17.195	0.4135	71.383	6.4663	$2S^1$+$2P^1$	$2S^1$-$2P^1$	1031	374557
	$2P^2$	17.195	0.4135	71.383	11.858				
	$2P^3$	17.195	0.4135	71.383	16.423				
	$2P^4$	17.195	0.4135	71.383	20.338				
	$2S^1$	33.859	0.450	72.620	12.594				
	$2S^2$	33.859	0.450	72.620	21.466				
Al	$3P^1$	5.7130	1.312	26.443	5.840	13.755	$3S^1$-$3P^1$	1854	350777
	$3S^1$	10.706	1.044	27.119	7.9149				
	$3S^2$	10.706	1.044	27.119	12.253				
Si	$3P^1$	8.0848	1.068	29.377	6.6732	16.5458	$3S^1$-$3P^1$	1393	381306
	$3P^2$	8.0848	1.068	29.377	10.876				
	$3S^1$	14.690	0.904	38.462	9.8716				
	$3S^2$	14.690	0.904	38.462	15.711				

Average value: 364118 ($eV^2Å^3$)

$$\Delta = (2n + 1)\frac{\lambda}{2} = \lambda(n + \frac{1}{2}), \text{ where } n = 0, 1, 2, 3, \ldots \tag{23}$$

As applied to P-parameter, this rule means that the minimum of interaction occurs if P-parameters of interacting structures are also in "anti-phase" – there is an interaction of either differently charged systems or heterogeneous atoms (for example, during the formation of valence-active radicals CH, CH_2, CH_3, NO_2 …, etc).

The summation of P-parameters in this case follows the principle of adding inverse values of P-parameters – equations (3,4,17).

The difference in wave motion (Δ) for P-parameters can be estimated via their relative value ($\gamma = \frac{P_2}{P_1}$) or via the relative difference of P-parameters (coefficient α), which at minimum interactions give an odd number:

$$\gamma = \frac{P_2}{P_1} = (n + \frac{1}{2}) = \frac{3}{2}; \frac{5}{2} \ \ldots \text{ At } n=0 \text{ (main state) } \frac{P_2}{P_1} = \frac{1}{2} \tag{24}$$

Let us point out that for stationary levels of one-dimensional harmonic oscillator the energy of these levels $\varepsilon = h\nu(n + \frac{1}{2})$, therefore in quantum oscillator, in contrast to the classical one, the least possible energy value does not equal zero.

In this model the minimum of interaction does not produce energy equal to zero corresponding to the principle of adding inverse values of P-parameters – equations (3,4,17).

Maximum of interference, increase of oscillations (in phase) takes place if the difference of wave motion equals the even number of semi-waves: $\Delta = 2n\frac{\lambda}{2} = \lambda n$ or $\Delta = \lambda(n+1)$.

As appropriate to P-parameters the maximum intensification of interaction in the phase corresponds to the interactions of similarly charged systems or systems homogeneous by their properties and functions (for example, between fragments or blocks of complex organic structures, such as CH_2 and NNO_2 in octogen).

Then:

$$\gamma = \frac{P_2}{P_1} = (n+1) \tag{25}$$

Analogously for "confluent" systems (with the same values of functions) of 2D harmonic oscillator the energy of stationary states is as follows: $\varepsilon = h\nu(n+1)$

By this model the maximum of interactions corresponds to the principle of algebraic adding of P-parameters – equations (6-8). When n=0 (basic state) we have $P_2 = P_1$, or:

maximum of interactions of structures occurs if the have the same P-parameters. This concept was used [10] as the main condition of isomorphic replacements.

8. Functions of Lagrange-Hamilton and Analogous Comparisons

The Lagrange function is the difference between kinetic (*T*) and potential (*U*) energies of the system:

$$L = T - U \tag{26}$$

Hamilton function can be considered as the total of potential and kinetic energies, i.e. as the total mechanical energy of the system:

$$H = T + U \tag{27}$$

From equations (1) and (2) and in accordance with the law of energy preservation we have:

$$H + L = 2T \quad H - L = 2U \qquad (28)\,(29)$$

We try to evaluate the motion of isolated system of free atom as a relative motion of its two subsystems: nucleus and orbital.

An atom structure is formed from differently charged masses of nucleus and electrons. In this system the orbital energy of electrons (W_i) and effective energy of atom nucleus, that takes into consideration screening effects, form an atom structure.

In a free atom its electrons move in coulomb field of nuclear charge. The effective charge nucleus, characterizing the potential energy of such subsystem, with screening effects equals:

$$q^2/r_i \quad \text{where} \quad q = Z^*/n^* \qquad (30)\,(31)$$

We can assume that orbital energy of electrons during their motion in coulomb field of nucleus charge is mainly determined by the value of kinetic motion of such motion.

Thus it is assumed that:

$$\text{T} \sim \text{W and U} \sim q^2/r_i \qquad (32,33)$$

In such approach the total of values of W and q^2/r_i is analogous to Hamilton function (H):

$$W + q^2/r_i \sim \mathrm{H} \tag{34}$$

Since the value of P_E-parameter has to be proportional to L, then

$$P_E \sim L \tag{35}$$

Equation (28) is: $\left(W + \dfrac{q^2}{r_i}\right) + P_E \approx 2W$ (36)

P_E-parameters of free atoms [10] were calculated based on the equations (3-5) using the values of bond energies of electrons as their orbital energies [2]. When calculating the values of effective P_E-parameter as dimensional characteristics of the atom (R), mainly the values of their radii by Belov-Bokiy or covalence radii (for non-metals) were applied.

Mutual overlapping of orbitals of valence electrons in the atom can be considered, in the first approximation, via the averaging of their total energy dividing it by the number of valence electrons recorded (N):

$$\left(\frac{q^2}{r_i} + W\right)\frac{1}{N} + P_E \approx 2W \tag{37}$$

Thus, $\left(\dfrac{q^2}{ri} + W\right)\dfrac{1}{N}$ characterizes the averaged value of the total energy of *i*-orbital as converted to one valence electron and in this approach it is the analog of Hamilton function – H.

In free atoms of 1a and 2a groups of periodic System S-orbital is the only valence orbital the peculiarities of deformation of electron cloud of which were considered through the introduction of coefficient $K = n/n^*$, where n – main quantum number, n^* – effective main quantum number by the following equation:

$$\left(\frac{q^2}{r_i}+W\right)\frac{1}{KN}+P_E\approx 2W \tag{38}$$

This means that for subgroups 1a and 2a in small periods K=1 and then K=$\frac{4}{3,7};\frac{5}{4};\frac{6}{4,2}$ - for periods 4,5 and 6 of the System of these subgroups. For the rest of the cases K=1.

And one more peculiarity as applied only to group 1a of periodic System: $2r_i$ (orbital diameter of *i*-orbital) was used as dimensional characteristic in the first term of equation (38).

Taking into consideration the aforesaid observations by initial equations (37) and (38) for valence orbitals of practically all the elements, the calculations and comparisons of the values $\left(\frac{q^2}{r_i}+W\right)\frac{1}{KN}+P_E$ and $2W$ were carried out, some results are given in Table 3. Besides, the equation (38) can be transformed as follows:

$$\left(\frac{q^2}{r_i}+W\right)\frac{1}{KN}-P_E\approx 2(W-P_E) \tag{39}$$

And we can consider the relation obtained in the same way as the equation (39). Actually, following the given analogy, W ~ T and P_E ~ L. Thus (W-P_E) ~ (T-L), and in the equation (26) T-L=*U*. Therefore (W-P_E) ~ *U*.

The analysis of all the data shows that the nearness of values of the investigated magnitudes compared is mostly within 5% and only sometimes – over 5% but never exceeds 10%.

All this proves that there is a certain analogy of the equations (28) and (38), as well as (29) and (39), therefore in the first approximation P_E-parameter can be the analog of Lagrange function and $\left(\frac{q^2}{ri}+W\right)\frac{1}{KN}$ - the analog of Hamilton function [11].

Table 3. Comparison of main energy atomic characteristics

Element	Valence electrons	W (eV)	r_i (Å)	q^2 (eVÅ)	P_0 (eVÅ)	R (Å)	$P_E=P_0/R$ (eV)	N	$\left(\frac{q^2}{r_i}+W\right)\frac{1}{N}+P_E$	2W (eV)
1	2	3	4	5	6	7	8	9	10	11
Li	$2S^1$	5.3416	1.586·2	5.8892	3.475	1.55	2.2419	1	9.440	10.683
Be	$2S^2$	8.4157	1.040	13.159	7.512	1.13	6.6478	2	17.182	16.831
B	$2P^1$	8.4315	0.776	21.105	4.9945	0.91	5.4885	3	17.365	16.863
	$2S^2$	13.462	0.769	23.890	11.092	0.91	12.189	3	27.032	26.924
C	$2P^2$	11.792	0.596	35.395	10.061	0.86	11.699/2	4	23.645	23.584
	$2S^2$	19.201	0.620	37.240	14.524	0.77	18.862	4	38.824	38.402
N	$2P^3$	15.445	0.4875	52.912	15.830	0.71	22.296/3	5	32.228	30.890
	$2S^2$	25.724	0.521	53.283	17.833	0.71	19.788	5	50.775	51.448
O	$2P^1$	17.195	0.4135	71.383	6.4663	0.66	9.7979/4	6	34.087	34.390
	$2S^2$	33.859	0.450	72.620	21.466	0.66	32.524	6	65.064	67.718
F	$2P^1$	19.864	0.3595	93.625	6.6350	0.64	10.367/5	7	42.115	39.728
	$2S^2$	42.792	0.396	94.641	24.961	0.64	39.002	7	79.257	85.584
Na	$3S^1$	4.9552	1.713·2	10.058	4.6034	1.89	2.4357	1	10.327	9.9104
Mg	$3S^1$	6.8859	1.279	17.501	5.8588	1.60	3.6618	2	13.946	13.772
Al	$3P^1$	5.713	1.312	26.443	5.840	1.43	4.084	3	12.707	11.426
	$3S^2$	10.706	1.044	27.119	12.253	1.43	8.5685	3	20.796	21.412
Si	$3P^1$	8.0848	1.068	29.377	6.6732	1.17	5.7036	4	14.600	16.170
	$3S^2$	14.690	0.904	38.462	15.711	1.17	13.428	4	27.737	29.380
P	$3P^3$	10.659	0.9175	38.199	16.594	1.30	12.765/3	3	21.686	21.318
	$3S^1$	18.951	0.803	50.922	11.716	1.10	10.651	3	38.106	37.902

Conclusions

1. The notion of spatial-energy parameter (P-parameter) that numerically equals energy expression in Thomas-Fermi equation is introduced based on the simultaneous recording of the most important atomic characteristics.
2. It is shown that P-parameter is a direct characteristic of electron density in atom. The dependence of P-parameter on maximum value of radial part of ψ-function is obtained.
3. Wave equation of P-parameter, which is formally analogous to the equation of ψ-function, is given.
4. Since P-parameter comprises main constants characterizing the state of valence electrons inside the atom, it helps to correlate and predict physical and chemical properties of compounds and processes of phase formation in simple and complex systems, by bond and hybridization energy of atom orbitals, as well [12-14].

References

[1] Batsanov S.S., Zvyagina R.A. Overlapping integrals and problem of effective charges. – Novosibirsk: *Nauka, 1966*, - 386 p.

[2] Fischer C.F. Average-Energy of Configuration Hartree-Fock Results for the Atoms Helium to Radon.// *Atomic Data,*-1972,- № 4,- p. 301-399.

[3] Waber J.T., Cromer D.T. Orbital Radii of Atoms and Ions //*J.Chem. Phys* —1965,- V 42,-№ 12,-p. 4116- 4123.

[4] Clementi E., Raimondi D.L. Atomic Screening constants from S.C.F. Functions, 1.// *J.Chem. Phys*.-1963,-v.38,-№11,-p.2686-2689.

[5] Clementi E., Raimondi D.L. Atomic Screening Constants from S.C.F. Functions, II.//*J. Chem. Phys.*-1967,-V.47, № 4,-p. 1300-1307.

[6] Gombash P. Statistic theory of atom and its applications. – *M.: I.L.,1951*, - 398 p.

[7] Clementi E. Tables of atomic functions // *J.B.M. S. Res. Develop. Suppl.,* 1965, V. 9, № 2, p.76.

[8] Zigban K., Nordling K., Falman A., et al. Electron spectroscopy. - *M.: Mir, 1971*,- 493 p.

[9] Allen K.U. Astrophysical magnitudes. - *M.: Mir*, 1977, 446 p.

[10] Korablev G.A. *Spatial Energy Principles of Complex Structures Formation.* Netherlands, Leiden, Brill Academic Publishers and VSP. 2005. 426 p. (Monograph).

[11] Korablev G.A., Kodolov V.I., Lipanov A.M. Analogous comparison of Lagrange and Hamilton functions with spatial-energy parameter. – Chemical physics and mesoscopy, *URC RAS, 2004*, №1, v.6, p. 5-18.

[12] Korablev G.A., Zaikov G.E. Energy of chemical bond and spatial-energy principles of hybridization of atom orbitals.//*J. of Applied Polymer Science. USA*, 2006, V.101, n.3, p.2101-2107.

[13] Korablev G.A., Zaikov G.E. Formation of Carbon Nanostructures and Spatial-Energy Stabilition Criterion//Polymers and Composites (*Polymer Year book Series, V.22*), Nova Science Publishers, Inc., New York, 2007, p.31-42.

[14] Korablev G.A., Zaikov G.E. P-Parameter as and Objective Characteristics of Electronegativity//Reactions and Properties of Monomers Polymers, Nova Science Publishers, Inc., New York, 2007, p.203-213.

In: Progress in Monomers, Oligomers, Polymers, Composites ... ISBN 978-1-60456-878-3
Editors: R. A. Pethrick, G.E. Zaikov et al.

Chapter 3

RESEARCH OF THE MECHANISM OF THE MOTION TO LIGUIDS IN NEW DEVICE FOR CLEANINGS INDUSTRIAL EFFLUENT

R.R. Usmanova[1,*], G.E. Zaikov[2,†] and I.G. Ibragimov[3,‡]
[1]Ufa State Technical University of Aviation, Ufa, Bashkortostan, Russian Federation
[2]N.M. Emanuel Institute of Biochemical Physics,
Russian Academy of Sciences, Moscow, Russian Federation
[3]Ufa State Technical University of Oil, Ufa, Bashkortostan, Russian Federation

INTRODUCTION

Now in connection with intensive development chemical, oil refining and other industries all rises a problem of preservation of the environment more sharply.

Millions tons of harmful gaseous substances are thrown out annually in an atmosphere. Creation of high-power plants results in necessity of application of clearing constructions of high efficiency [1].

One of perspective directions of gas purification are separation the devices using effect of action of a field of centrifugal forces, allowing most full to realize advantages new power - intensive technologies.

Centrifugal devices are characterized by high efficiency, simplicity of a design and low metal-intensive. Application of such equipment allows also essentially to intensify process mass transfer due to increase of speed of movement of phases [2].

* Correspondence to: Ufa State Technical University of Aviation, 12 Karl Marx str., Ufa 450000 Bashkortostan, Russia, usmanovarr@mail.ru

† Correspondence to: N.M. Emanuel Institute of Biochemical Physics, Russian Academy of Sciences, 4 Kosygin str., Moscow 119334 Russia, chembio@sky.chph.ras.ru

‡ Correspondence to: Ufa State Technical University of Oil, 1 Cosmonaut str., Ufa 450062 Bashkortostan, Russia, ibragimov@rusoil.net

As efficiency of separation is defined by requirements of concrete process, the big value gets correct calculation of speeds of movement of phases at which the high degree of clearing without occurrence of secondary ablation is achieved.

Without the decision of these, and also of some other questions, the further optimization processes of gas purification becomes difficult a problem.

1. Development of a New Design Dynamic Gas Washer

The new design highly effective dynamic gas washer, directed on decrease secondary ablation and a radial drain of particles [3] is developed.

The circuit dynamic gas washer is submitted in *figure 1.*

The gas stream containing mechanical or gaseous impurity, acts in the device on a tangential branch pipe 3. The liquid acts by means of an axial branch pipe 10. The bottom basis swirler 8 is placed at a level top cut an axial branch pipe 10, forming with it a backlash.

Due to this the liquid acts in the device as flat radial jets which, being reflected from the case 1, form a liquid veil that causes high-intensity contact of phases. Swirler contains blank off bottom 8 and the cylindrical top 6 bases connected with each other by means of unidirectional blades 9. Swirler it is fixed on a rotor 12 which is resulted in rotation by means of a belt drive 13. The centrifugal forces arising at rotation of a rotor 12, provide intensive crushing water gas a stream that results in updating a surface of contact of phases and an intensification of processes of an interphase exchange.

The basic advantage of the offered device is that rotation in a field of centrifugal forces, and also the offered organization of liquid jets, allow essentially to change a principle of work of the device and on the basis of it to improve the basic technical and economic parameters (a degree of clearing of gas, productivity, metal consumption).

2. An Experimental Research of Aerodynamics of a Stream

Research of aerodynamic laws of a biphase stream was carried out(spent) on the device in diameter of 0,25 m, height 0,6 m, on system air - water. During researches speed changed within the limits of 15-20 м/с. The charge of a liquid 0-3 m^3 / hour, the charge of gas 200-600 m^3 / hour.

Measurement of a field of speeds and static pressure was carried out with the help of a focused single-channel cylindrical probe in diameter of 4 mm. Techniques calibration a probe and realization of measurements were standard [4]. The chosen measurement technique does not allow to determine radial making speeds of the gas, however known experimental data specify that radial making speeds is scornfully small, in comparison with axial and tangential [5].

Results of measurements were represented in relative sizes.

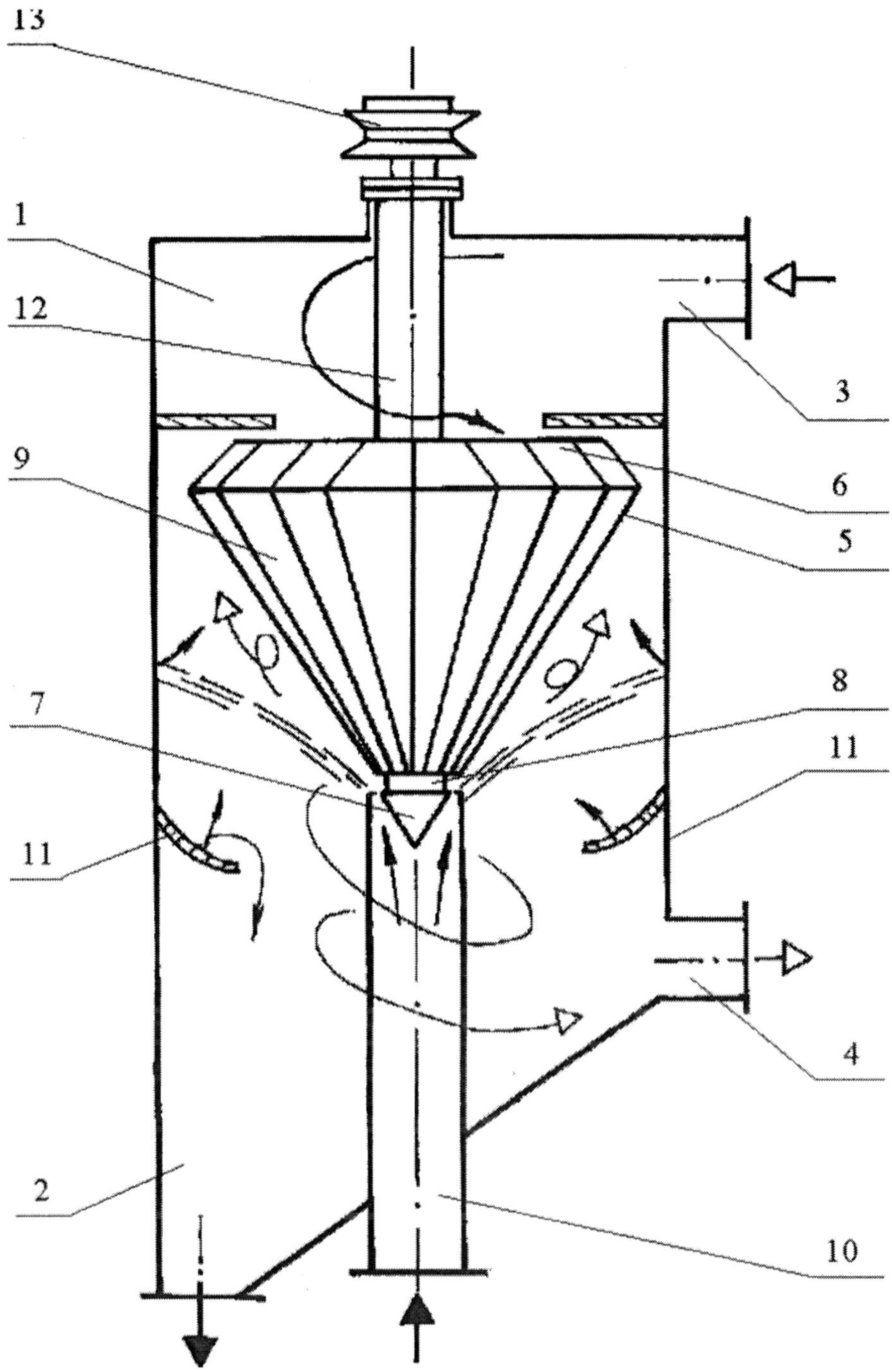

The 1-case; 2-tap(removal) slime; 3,4-branch pipes of an input / conclusion of a gas stream; 5-conic swirler; 6-cylindrical swirler; 7-dissectior; 8-basis swirler; 9-blades; a 10-axial branch pipe 11-entrainment separator; a 13-pulley.

Figure 1. Dynamic gas washer.

$$\overline{\vartheta_t} = \frac{\vartheta_t}{\vartheta},$$
$$\overline{\vartheta}_Z = \frac{\vartheta_Z}{\vartheta}, \quad (1)$$
$$\overline{\rho} = \frac{2\rho}{\rho \cdot \vartheta^2},$$

Where ϑ_z, ϑ_t - axial and tangential making speeds of gas, m/s;

- p - Static pressure, Pa.

The analysis of experimental data has shown, that distinction in structure of the stream, caused by design features of the device, is shown only at a level swirler. For the device repeatability is characteristic Character of structures of speeds at return on a new coil of rotation, breakdown axiall symmetric, and also insignificant decrease of size tangential speed of a stream in process of his promotion.

3. Mathematical Description of the Motion Water Gas Flow

At mode of the usages dynamic gas washer, not providing spraying to liquids with formation liquid cover, hydrodynamics of the condition of the functioning the device are broken. Occurs the sewer to liquids in slime tank and breach of the reasonable conditions of the work device.

For stable work dynamic gas washer necessary determined correlation between axial velocity of the gas and density of the irrigation.

Zapishem equation Bernoulli, characterizing balance to energy required for moving the liquids on axial branch pipe [6]

$$\frac{P_1}{\rho_l} + g \cdot z_1 + \frac{\omega_{z_1}^2}{2} = \frac{P_2}{\rho_l} + g \cdot z_2 + \frac{\omega_{z_2}^2}{2} + \Delta H \quad (2)$$

The Loss of the pressure on length axial branch pipe neglect that is to say $\Delta H = 0$.

The Velocity to liquids is not changed on length of the considered area $\omega_{z_1} = \omega_{z_2}$.

Then loss of the pressure on length axial branch pipe write with provision for (2) as

$$\Delta P = P_1 - P_2 = \rho_l \cdot g \cdot (z_1 - z_2) = \rho_l \cdot g \cdot h \quad (3)$$

In zone of contact phases of the loss of the pressure to liquids is compensated by speed pressure of the gas.

For finding relationship between velocity of the gas and liquids, shall analyse motion to liquids in coordinate system - z, r.

Motion consider formed and axially symmetric.

For determination axial forming velocities use the equation continuity flow.

$$\frac{d\upsilon_r}{dr}+\frac{\upsilon_r}{r}+\frac{d\upsilon_z}{dz}=0 \tag{4}$$

Considering that thickness of the film to liquids very small in contrast with radius of the device, consider that transverse forming velocities of the gas possible to neglect. Then equation continuity will be of the form of $\frac{d\upsilon_z}{dz}=0.$.:

Supposing that direction of power to gravity g complies with direction of the axis z, record the equation of the motion Navie - Stoksa:

$$\rho_l \cdot \frac{d\upsilon_z}{dz}=\rho_l \cdot g-\frac{dP}{dz}+\mu \cdot \nabla^2 \cdot \omega_z, \tag{5}$$

Where ∇^2 - an operator Laplasa, presents itself amount second derived:

$$\nabla^2=\frac{d^2}{dr^2}+\frac{1}{r}\cdot\frac{d}{dr}+\frac{d^2}{dz^2} \tag{6}$$

Solving together equations (4) and (5), shall get

$$\frac{d^2\upsilon_z}{dr^2}+\frac{1}{r}\cdot\frac{d\upsilon_z}{dr}-\frac{g}{v_l}=0 \tag{7}$$

For decision of the equation (7) record the border conditions:

beside walls , consequently, r = R;

on surfaces of the liquids $r=R-\sigma$, the gradient to velocities is absent

$$\frac{d\upsilon_z}{dz}=0.$$

With provision for taken border conditions, integration equality (7) brings about correlation:

$$\upsilon_z = \frac{g\cdot(R-\sigma)^2}{2\cdot\nu_l}\cdot In\frac{R}{r} - \frac{g}{4\cdot\nu_l}\cdot\left(R^2 - r^2\right) \tag{8}$$

The Formula (8) allows to analyse distribution axial forming velocities of the gas $\overline{\upsilon_z}$ on transverse coordinate r.

The Determination average on thickness of the film of importance of the axial velocity to liquids $\overline{\upsilon_z}$ is conducted on area of a certain section of the flow S.

$$\overline{\upsilon_z} = \frac{1}{S}\cdot\int_S \upsilon_z\cdot dS \tag{9}$$

The Correlation (9) contains the unknown value S.

However, it can be used for determination parameter, characterizing stable work dynamic gas washer.

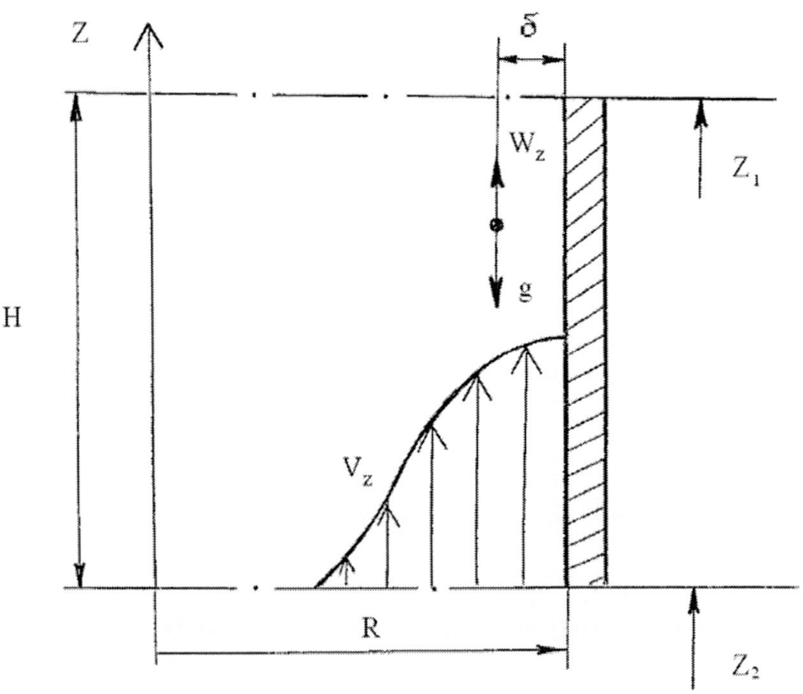

Figure 2. Distribution speeds at biphase film current

CONCLUSIONS

1. The new design highly effective dynamic gas washer, directed on improvement of the basic technical and economic parameters (a degree of clearing of gas, productivity, metal consumption) is developed.
2. On the grounds of studies aerohydrodynamics flow is created mathematical model of the motion to liquids in contact zone of the device.
3. The Got dependencies allow to analyse distribution to velocities to liquids on volume of the device and check the stability of his work.

REFERENCES

[1] Kuznetsov I.E., Troizkaia T.M. Protection of air pool from pollution in harmful substance. - *Moscow: Chemistry, 1979*, 344p.

[2] Uzhov V.N., Valdberg A.J. Clearing of gases wet filter.-*Moscow: Chemistry, 1972*, 245p.

[3] Usmanova R.R. Dynamic gas washer. *The application for the invention №2007* 120001 from 29.05.07.

[4] Petunin A.N. Technique and engineering of measurements of parameters of gas stream.-*Moscow: Mechanical engineering, 1974*, 250p.

[5] Stim A.N. Aerodynamics vortex chamber. - Vladivostoc, 1984, 200p.

[6] Prostodiyakonov I. O., Ryzhkov A. E. Hydrodynamics and mass transfer in dispersion system. Leningrad: *Chemistry,* 1987, 336 p.

In: Progress in Monomers, Oligomers, Polymers, Composites ... ISBN 978-1-60456-878-3
Editors: R. A. Pethrick, G.E. Zaikov et al.

Chapter 4

RELAXATION AND STRUCTURAL PROPERTIES OF THE ANISOTROPIC POLYMER COMPOSITES

*G.M. Magomedov**
Daghestan State Pedagogical University,
Makhachkala-city, Dagestan, Russian Federation

ABSTRACT

The results of carbo-, organo-, glass-, hybrid- and stratified plastic composites are generalized in the paper. A number of new phenomena and effects conditioned by the interaction and influence upon each other of polymer matrix and armiring fibre in a composite are detected and described.

New methods to evaluate the structural characteristics of interphase strata and polymer matrix are suggested. Scientific bases of creating polymer composites possessing high relaxation ability on the basis of separate coating technology of components are worked up.

Keywords: composite, armiring, fibre, polymer matrix, relaxation process, anisotropy, interphase strata, structure, heterogeneity.

From the scientific and practical point of view the problem of anisotropic polymer composites (APC) creation with high relaxation capacity of conserving their high elastic, solid and heat-valued features is the most actual [7-3]. The properties of these materials depend upon many factors: nature, volume, form and content of dispersoid (armiring filler) and continuous polymer matrix (PM), and also upon interphase correlation, temperature, baro-temporal and humidity conditions of material forming, technology and others [4-6]. All these factors influence on the polymer matrix structure and property changing in composite in comparison with the basic and new forming gradients both by structure and by properties.

---

* Correspondence to: Daghestan State Pedagogical University, 57 Yaragskogo str., 367003 Makhachkala-city, Dagestan, Russian Federation, DGPU@mail.ru

Therefore the theoretical calculation and purpose foreseeing of required level of APC properties on the base of properties of components is much troubled.

The information about the influence character of armiring filler on the PM structure has a contradictory character. There are such opinions that the presence of armiring filler causes the lowering of molecular activity of PM, the increasing of glassing temperature and shifting of relax spectrum in the direction of higher movement temperature [4]. These changes are explained as the result of adsorbing interaction between armiring filler and PM in forming interphase levels, conformation limit of polymer chains close to hard surface of armiring filler, in forming regulated structures of PM like clusters under the influence of fiber.

But in some works they state the process of relax spectrum shifting of composites in the direction of lower temperature and falling of the glassing temperature in comparison with the pure PM. The contradictory character of data of the armiring filler influence on the structure and relaxation properties of PM may be caused by several factors sometimes not taken into account, e.g. by the absence of control over the completeness of hardening process, while armiring filler depending on its nature, type of surface processing, and also the quantity of sorbent water may speed up or inhibit the process of hardening [7]. Its role may play the selective sorption of armiring filler of one polymer matrix components [9]. Less attention is paid to the direction of measuring and character of inner power acting [2, 10].

Incompleteness and contradictoriness of research results is explained also by absence of fundamental and structural approach in studying this process. Today there are not worked out conceptions of establishing interrelation between the PM structure and macroscopic physiomechanical properties of composites and their anisotropy. Therefore the investigation of influence of polymer matrix structure, interphase correlation, technology on the structure and properties of APC, providing their high relaxation capacity are actual and fundamentally important.

Relaxation properties define the capacity of composites to disperse the energy under loading exter-mechanical fields and it is the base of such exploitation properties as vibro-crack stability, static and dynamic fatigue and others [8]. These important properties of modern polymer composites (carbo-, organo-, glass-, hybrid and stratified plastic) equally with high property of specific hardness determine the progress (development) of modern machinery and airo-cosmic engineering.

Presently such technologies of separate component loading are used to form (produce) composites gradient polymer matrix.

The existing opinion that armiring of polymer with fiber fillers may be taken as a specific case of their filling with disperse filler, is not well-grounded, because connecting plastic have specific macrostructure of filler, high anisotropy degree of properties, great variety of properties of armiring fiber filler in comparison with disperse filler even for one and the same material [2]. There is a kind of difference between the character of structure forming PM and polymer properties of powder filler. The technologies of forming these two types of materials are different too.

To study the anisotropy of solid and relaxation properties of hardened materials it is efficient to use methods of winding resonance waves and dilatometery together with the other methods of relaxation spectrometery.

In the process of its studying new factors and effects were found out: multiplicity of the α-relaxation process, caused by interphase stratum close to the surface of armiring fiber (the gradient of its structure and properties in comparison with the base polymer) [2,4], anomalous dependence of dynamic module of composites hardness on the degree of connecting of polymer matrix; anomalous dependence of hardness under winding epoxy fiber polymer on the degree of connecting; forming additionally to α_1 and α_2 – processes two new processes α'_1 and α'_2 – relaxation in organoplastic, caused by two interphase stratum in the polymer matrix and organic fiber and other many factors.

New technology of separate loading of components in carbo-plastic realizes micro heterogeneous gradient structure of mosaic type assisting the more effective relaxation of remain inertensions, intensified with deep hardness in all measure directions[7-9].

Methods of estimating of polymer matrix structural characteristics and interphase stratums of composites are worked out on the base of ductile-elastic properties of anisotropy, and also such methods of estimating the interphase correlation in the system of "polymer-fiber" by dynamic method and method of thermostimulated depolarization are suggested [1,2,11].

Composites with gradient polymer matrix by their main characteristics (elasticity, hardness, dissipation) are numerically surpass the composites on the base of traditional prepreg blends [8]. It is conditioned by the peculiar structure of PM of former, which provides the high relaxation level of remain inner tensions, especially in longitudinal direction of fiber.

On the base of investigation of organoplastic methods of torsinal and winding waves, and also of currents of thermostimulating depolarization it is suggested to estimate the following:

- the solidity of adhesion of polymer matrix and armiring fiber by the value of dynamic shift module (hardness in transversal direction) in its high elastic condition;
- the effect of penetration of polymer matrix components into organic fiber by volume

$$\Delta G2 = \frac{G_{VEL}}{G_{573}}$$

- the electric index of hardness of adhesion by intensivity of inverse maximum of currents of thermostimulating depolarization.

In the course of experiments optimal selecting of components, technological conditions (pressure, temperature, time), physico-chemical modifications of AB, polymer matrix, forming composites with definite properties are suggested [10,11].

Temperature intervals of exploitation of anisotropic polymer composites, which are different even for one and the same material depending on the angle between the directions of connecting and laying the loading are defined.

REFERENCES

[1] Magomedov G.M., Ulyanenko S.N., Lebedev L.B., Mashinskaya G.P., Zelenev Yu.V. Mechanical Relaxation Properties of Organoplastics // *Mechanics of Composite Materials*. – 1984 – №5 – P.52-61.

[2] Magomedov G.M., Zadorina E.N. Anisotropy of Relaxation Properties of Fiber Polymer Composites// *Papers of the USSR AS*. – 1986. – vol.286, №3. – P.630-633.

[3] Zadorina E.N., Magomedov G.M., Kestelman V.N. Enflup der Grenzschichten aus die Ausbilding Viscoelasticher und Thervicher Eigen Shaften Verstakter Epoxihj Jymer// *Plaste und Kautschuk*. – 1989. - №5- P.154-156.

[4] Magomedov G.M., Radjabov Z.R., Mashinskaya G.P., Aivazov A.B. Anisotropy of Tough Elastic Propertiesof Flaky Metals with Organoplastics// *Mechanics of Composite Materials*. – 1989.-№5.- P.771-775.

[5] Magomedov G.M., Gamaeva N.M., Abakarov S.A., Smirnov Yu.N. Manifestations of the Anisotropic Properties of Carbon Fiber Epoxi Composites in the Region of a-Relaxation // *Polymer Science. Ser. B. – Vol. 40*. – Nos. 1-2. – 1998. – P.49-52.

[6] Smirnov Yu.N., Magomedov G.M., Shatzkaya T.E., Amiraliyev A.D., Djamaeva N.M., Natrusov V.I. Modificator's Influence on the Static and Dynamic Mechanical. Properties of Glass Plastic on the Basis of Epoxiamin Cohesive//*Plastic Masses*. – 1998.- №6. – P.8 – 10.

[7] Smirnov Yu.N., Magomedov G.M., Djamaeva N.M. Investigation of Relaxation Properties of Epoxi Fenol Cohesive and Carbon Plastic on its Base in the Course of Hardening Process. // *Plastic Masses. – 1999*. – №7. P. 36-42.

[8] Smirnov Yu.N., Magomedov G.M., Djamaeva N.M. On the Connection of Elastic and Dissipative Properties of Epoxi Fenol Carbon plastics on the Base of Mixed and Binary Prepregs. // *Plastic Masses*. – 1999. – №9. – P.21-25.

[9] Smirnov Yu.N., Magomedov G.M., Djamaeva N.M. Investigation of Epoxi Fenol Carbon Plastics Relaxation Properties on the Basic of Binary Prepregs.// *Plastic Masses*. – 1999. - №9. – P. 43-49.

[10] Smirnov Yu.N., Magomedov G.M., Tarasov V.P. Complex Investigation of rexation Properties of Glass Plastics on the Basis of Various Prepreg Technologies. // *Plastic Masses. – 2000*. - №12. – P.7-15.

[11] Magomedov G.M. Relaxation Phenomena in Anisotropy Polymer Composites: Papers of VIII Scientific session MAI. *Makhachkala, 2002*. P.71-76.

In: Progress in Monomers, Oligomers, Polymers, Composites ... ISBN 978-1-60456-878-3
Editors: R. A. Pethrick, G.E. Zaikov et al.

Chapter 5

THE INFLUENCE OF POLYAZOMETHINES ON BASIC OF 4,4'-DIAMINOTRIPHENYLMETHANE ON PHYSICOMECHANICAL PROPERTIES OF POLYBUTYLENETEREPHTHALATE.

B.S. Mashukova [*] ***and T.A. Borukaev*** [†]
Kabardino-Balkarian State University, Nalchic city, Russian Federation

ABSTRACT

The modifying influence of polyazomethines on main physicomechanical properties of polybutyleneterephthalate has been investigated. The improvement of operation characteristics has been found out. It has been shown that the main reason of changing properties of polymeric mould is plastifying influence of polymeric add.

Keywords: polybutyleneterephthalate, polyazomethine, modification, properties.

Like with mineral dispersed and fibred fillers for modification of polybutyleneterephthalate (PBT) else are use large polymeric adds [1, 2]. Use such adds have certain advantages. The materials which modified polymeric adds on basic of PBT are characterize on higher impact strength lesser density, heightened water resistance and etc. [3, 4]. Besides that thermal expansion coefficients of polymeric adds and polymeric matrixes are very like, that create additional solid effect of compounds in consequence of thermal strains lowering [5]. However receive of high-quality compound materials on basic of different polymers (for example PBT) and different polymeric adds connect with decision of problem elaborate optimum correlation of components.

---

[*] Correspondence to: Kabardino-Balkarian State university, 360004,Russia, Nalchic city, Chernyshevsky st. 173; Mahsuk_bela@mail.ru

[†] Correspondence to: Kabardino-Balkarian State university, 360004,Russia, Nalchic city, Chernyshevsky st. 173; . boruk-chemical@mail.ru

The results of physicomechanical investigations of PBT modified by polymeric adds have been examined in this work. The synthesized polyazomethine (PAM) has been used as polymeric add. It has the following chemical structure [6]:

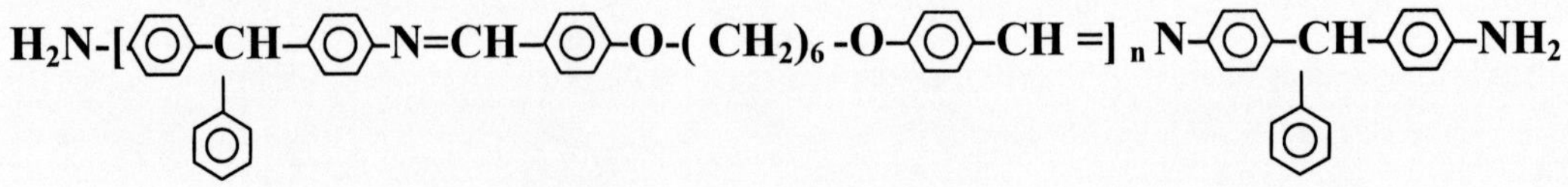

The polymeric add has been introduced in PBT according to method this has been shown in the work [7]. The content of modifier is C_{PAM} = 0,1-1,0 mass %. It should be noted that the choice of such PAM concentration is not accidental. During this process research results have been used which have been shown in work [7].

The introduction of PAM add into the PBT brings to considerable improvement of mechanical properties of initial polymer. Impact tests, which have been conducted on the method proved it [8]. During this process it has been discovered that composites PBT+PAM (where C_{PAM} = 0,1-0,3 mass %) without incision have not been destroyed that is their impact elasticity may be evaluated by quantity $A_P > 140$ kJ/m^2 (where A_P – impact strength). In its turn initial PBT is destroyed without incision. The PBT+PAM composites have been tested with incision for evaluating of strength deformation properties. The results have been showed in table 1.

Table 1. Deformation-strength properties of PBT, which is modified by PAM determined during impact tests according Sharpy's method.

Samples	A_p, кJ/m^2	σ_p, MPa	E, GPa	ε_P, %
$PBT_{industrial}$	18	25	1,1	2,4
PBT + 0,1% PAM	22	35	1,3	3,0
PBT + 0,3% PAM	29	46	0,9	7,0
PBT + 1,0% PAM	21	38	1,2	3,5

Note: PBT samples of type B-110 with incision 0,2 mm have been tested.

The analysis of table 1 shows that PBT + 0,3% PAM composite has highest significance of A_P. It can be explained by plasticity increasing of composite of this compound that is introduction in PBT more elastic phase of PAM leads to plastic destruction of material. The destruction surfaces of initial polymer and PBT + 0,3% PAM composite at the high-speed load has been studied to find out in details the rising mechanism of PBT impact strength, modified by PAM.

The photographs have been taken with the help of optical microscopy at increasing × 100.

The analysis of destruction surfaces of initial PBT and PBT+0,3%PAM composite on figure 1 shows that destruction nature of initial polymer considerably differs from the composite[9]. As you can see from figure 1,a the destruction of initial PBT starts on all border of surface layer then we can observe the zone of low rough, which is replaced by zone of the relief breaks. Taking in consideration such surface of destruction we can come to the conclusion that initial PBT has fragile destruction. Unlike of initial sample the destruction surface of PBT + 0,3% PAM composite has different character. So, from figure 1,b we can see that destruction process origin starts in artificial defect (incision). However the destruction has began not along the whole incision length, but from the definite single defect. The typical elements of destruction surface are mirror zone and lines (roughs), which fan radiated from mirror zone. Such character of destruction surface of PBT + 0,3% PAM composite apparently have been defined by origin of displacement deformations because of PAM macromolecules. Consequently, displacement deformation brings essential contribution into rise of A_P of PBT + 0,3 % PAM composite rising during impact tests owing to dissipation impact energy and owing to cracks of growth prevention.

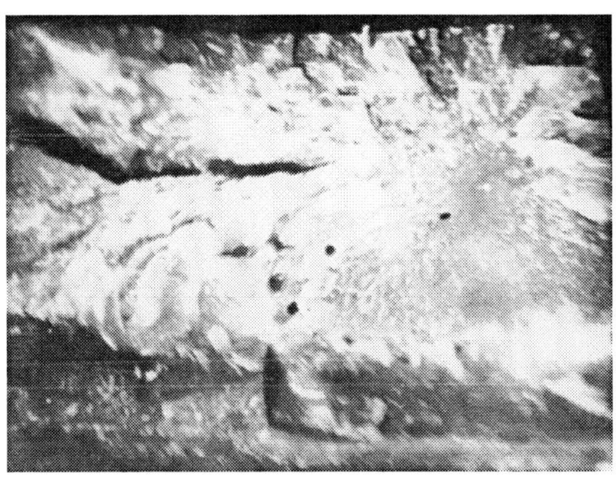

a)

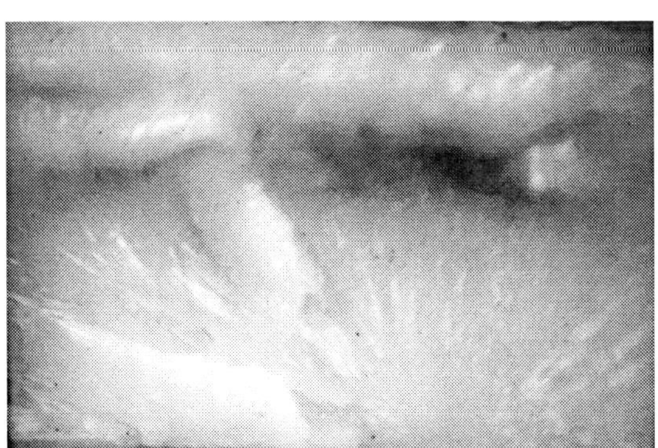

b)

Figure 1. Destruction surfaces of samples on basic of initial PBT (a) and PBT + 0,3 mass % PAM composite (b).

PAM influence on character of PBT destruction, apparently, is connected with structure of polymeric modifier. So investigations of PAM structure have shown the availability of definite regulation structure of macromolecules. In particular in thermogram of differently-scanning calorimetry (DSC) endothermic effect ($\Delta H \geq 0,3$ J/g) has been discovered , which may be attributed to so-called transition of second kind, that is to nematic phase (figure 2). The availability of such phase has been discovered also with help of optical microscope, where higher of T_{fusion} of this PAM have been shown its mesomorphic state having nematic structure. Accounting to this fact we can tell that PAM influence armouring action on polymeric matrix of PBT.

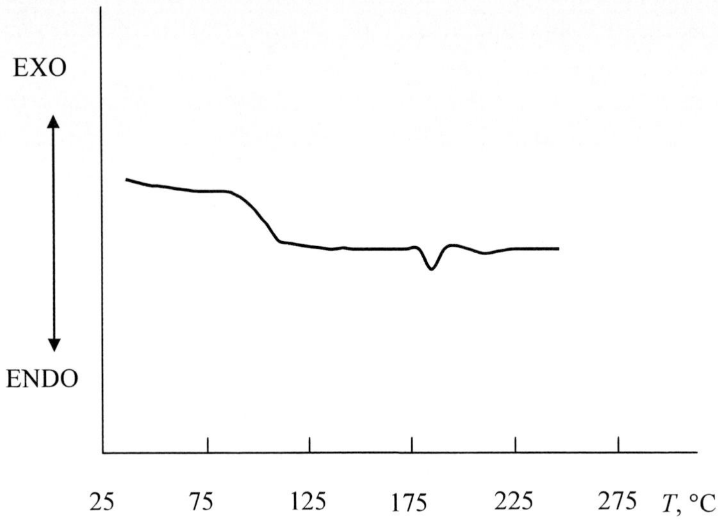

Figure 2. DSC(differently-scanning calorimetry) thermogram of PAM on basic of 4,4$^{/}$-diaminotriphenylmethane and 4,4$^{/}$-diformyldiphenoxyhexane.

It should be mentioned that in practice nematic polymers are often used for improvement of physicochemical properties of PBT. Here nematic polymers act as armouring agents [10]. Consequently displacement deformations owing to PAM particles arise thanks to their armouring influence on PBT. The temperatures on which die casting of samples have been conducted and existence of nematic state of PAM are in one interval (220-250 C). This permits to assume that anisotropy of PAM structure with axises of orientation which are parallel to flow melt of polymeric matrix direction raise compound strength.

In favour of considerable orientation of PAM macromolecules at its processing from melt in temperature interval 220-240 C and results of x-ray structure analysis of extrudates on basic of PAM are shown (figure 3). From figure we can see that diffractogram there are some reflexes which shows considerable orientation of PAM macromolecules.

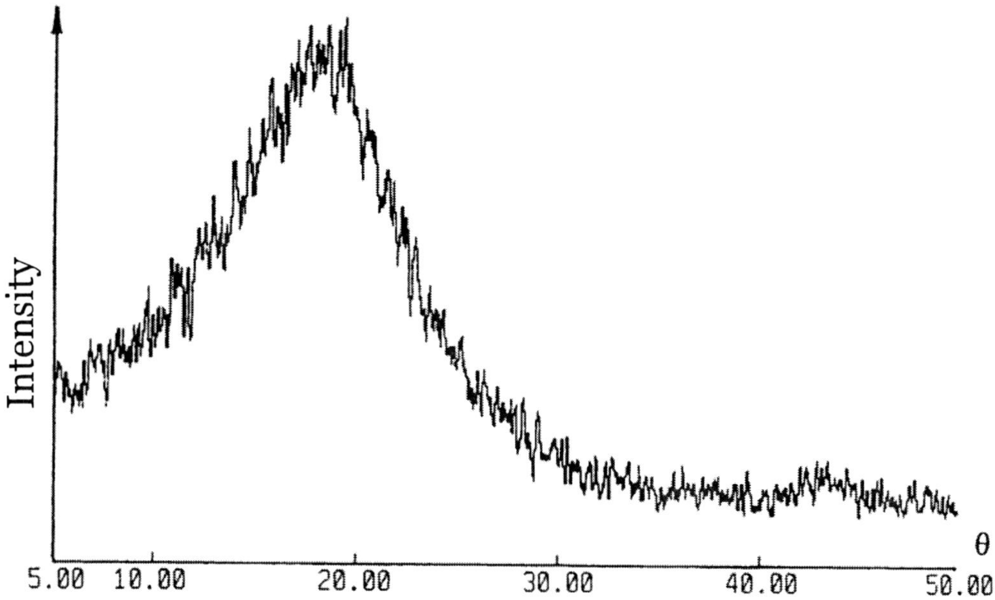

Figure 3. Diffractogram of PAM on basic of 4,4/-diaminotriphenylmethane and 4,4/-diformyldiphenoxyhexane with temperature 220°C.

In its turn these results confirm decrease of modulus elasticity (E) significance at growth accordingly of deformation (ε_P) (table 1).Besides introduction PAM into PBT brings to increasing of macromolecules mobility relatively each other which is expressed in falling of melt viscosity (increasing of values of index melt flow, determining in standard conditions), in melting temperature and in density of composites (table 2). Consequently add of PAM into PBT furthers plastifying that leads of mechanical properties of initial polymer changing.

Table 2. Physical properties of PBT and of PBT+PAM compounds

Samples	*FMI_5^{230}, g/10 мин	$T_{softening}$, °C	ρ, g/см^3
$PBT_{industrial}$	36	214	1,245
PBT + 0,1% PAM	45	210	1,240
PBT + 0,3% PAM	65	197	1,230
PBT + 1,0% PAM	48	198	1,234

*F M I = flow melt Index

Together with increasing of A_P and ε_P of PBT when introducing PAM in it in quantity of 0,3 mass % as shown in table 1, value and destructive stress (σ_P) is increasing. So value σ_P of PBT + 0,3% PAM composite in 2 and more times is higher than the initial polymer.

Such growth may be explained by the fact that introducing PAM into PBT leads to getting of more homogeneous (less defective) structure (figure 4). This structure can disperse impact energy better. This, in its turn, means some raise of composite molecular mobility level. It is the confirmation of above mentioned reasoning about plastifying role of PAM.

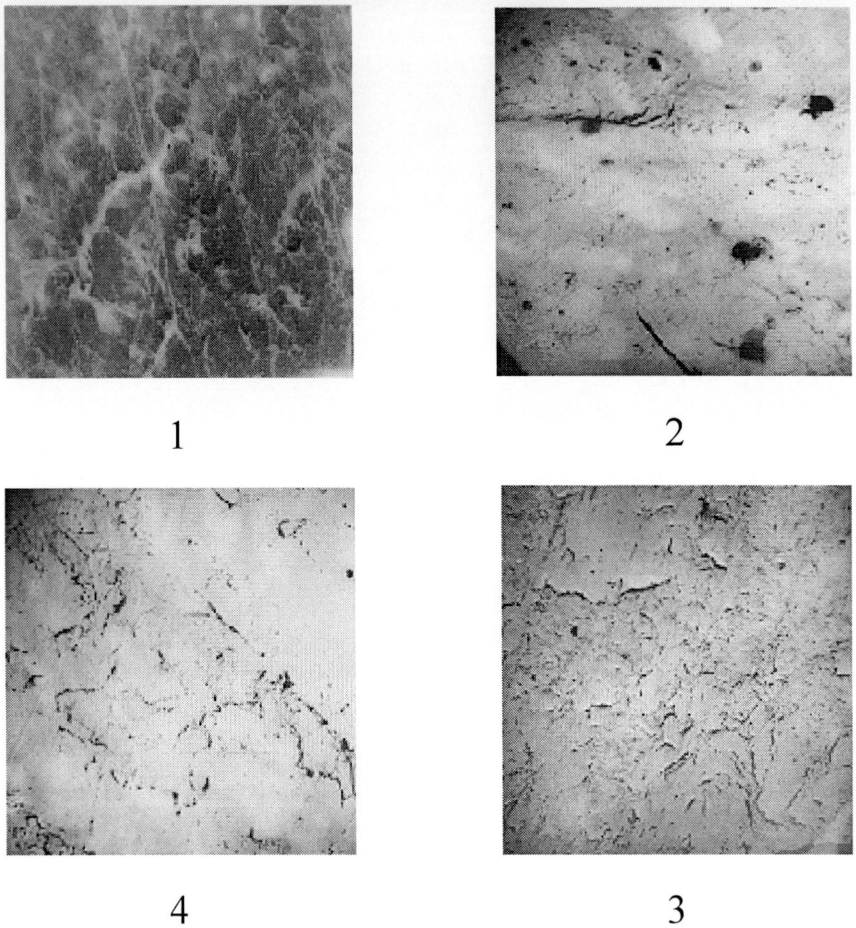

Figure 4. Electronic microphotography of film surface: 1 – initial PBT; 2 – PBT + 0,1% mass PAM; 3 - PBT + 0,3% mass PAM; 4 - PBT + 1,0 % mass PAM. (accuracy of survey is 25 micron).

In its favour results of dielectric investigations of PBT + PAM composites confirm (figure 5). Dielectric investigations have been conducted according to method [8]. From figure 5 it can be seen that molecular mobility of PBT+PAM composite is somewhat higher than in initial PBT in temperature limits which are lower than second order transition temperature. Besides value of temperature through conduction beginning of PBT + 0,3 % mass PAM composite is higher than the temperature of initial polymer.

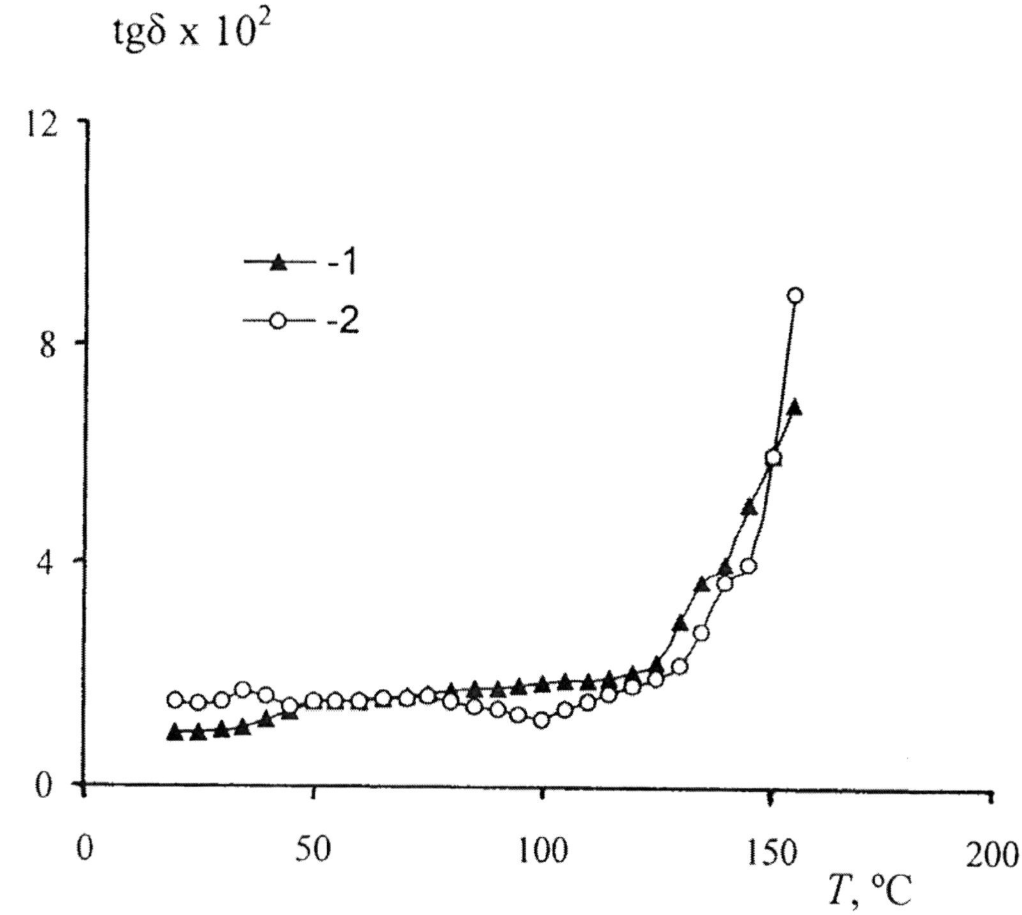

Figure 5. Dependence angle tangent of dielectric losses tg δ from temperature for initial PBT (1) and PBT + 0,3% mass PAM (2).

Thus obtained results permit to make the following conclusions:

- Introduction of a few PAM adds (C_{PAM} = 0,1-0,3 mass %) into PBT renders modifying influence on main physicomechanical;
- It has been discovered the change of main physicochemical properties of PBT stipulated with plastifying influence of PAM;
- The obtained results permit to create materials with the complex of improved deformation–strength characteristics, which can be applied in difference branches of industry.

References

[1] Pan L., Liang B., Jang J. // *J. China Text. Univ*. 1994. V. 20. №4. P. 9-14.

[2] Zhang H.X., Hourston D.J. // *J. Appl. Polym. Sci.* 1994. V. 71. № 12. P. 2049-2057.

[3] Joshi M., Misia A., Maiti S. N. // *J. Appl. Polym. Sci.* 1991. V. 43. № 2. P. 311-328.

[4] Chin S., Gastins T., Durning C. J. // *J. Polym. Sci.* B. 1996. V. 34. № 16. P 2689 – 2699.

[5] Manson J.,Sparling L. Polymeric compounds and composites. *M.Chemistry*, 1979. 434 p.

[6] Borukaev T. A., Tlenkopatchev M. A., Bondarenko G. N., Filatov M. P., Dzubina M. A., Parenago O.P., Mikitaev A. K., Bekanov M. H. // *J. Plastic masses 1996*. № 2. P. 17-19.

[7] Borukaev T. A., Shkalenko L.V., Mashukov N. I., Mikitaev A. K. // *J. Plastic masses 2002*. №7. P. 16-17.

[8] Borukaev T.A., Mashukova B.S., Mashukov N.I., Tlenkopatchev M.A., Mikitaev A. K // *J Plastic masses 2004*. № 4. P. 18-19.

[9] Borukaev T. A., Mashukov N. I., Mikitaev A. K. // *J. Material-authority . 2000*. № 10. C. 42- 45.

[10] Maksimov R., Steshinsky T., Garbarchik I // *J. Mech. Composite mater*. 1996. T. 32 . № 5. P. 676-689.

In: Progress in Monomers, Oligomers, Polymers, Composites … ISBN 978-1-60456-878-3
Editors: R. A. Pethrick, G.E. Zaikov et al.

Chapter 6

REGULARITIES OF ORGANIC SOLVENTS PENETRATION INTO TETRAFLUOROETHYLENE-PROPYLENE COPOLYMER

I.Yu. Yevchuk[1,*], G.G. Midyana[1], R.G. Makitra[2,†], G.E. Zaikov[3,‡], G.I. Khovanets'[1], O.Ya. Palchykova[2]

[1]Department of Physical Chemistry of Combustible Minerals L. M. Lytvynenko Institute of Physico-organic Chemistry and Coal Chemistry NAS of Ukraine Lviv, Ukraine
[2]Institute of Geology and Geochemistry of Fossil Fuels NAS of Ukraine, Lviv , Ukraine
[3]N. M. Emanuel Institute of Biochemical Physics RAS, Moscow, Russian Federation

ABSTRACT

The processes of swelling and diffusion of solvents into the structure of tetraflurethylene-propylene coolymer can not be described quantatively using only one characteristic of a solvent. In all cases a molar volume of liquid has the determining effect, hampering its penetration into the structure of polymer. However, the effects of other factors – both solvation and density of cohesion energy of solvents – are significant. Adequate quantitative generalization of abovementioned processes can be obtained on the basis of free energies linearity concept by using of linear multiparameter equations taking into account the effects of different factors.

Key-words: regularities, solvends, duffusion, copolymers, swelling, transport, membranes, liquids.

* Correspondence to: Department of Physical Chemistry of Combustible Minerals L. M. Lytvynenko Institute of Physico-organic Chemistry and Coal Chemistry NAS of Ukraine 79060 Lviv Naukova St., 3a, e-mail: vfh@org.lviv.net

† Institute of Geology and Geochemistry of Fossil Fuels NAS of Ukraine, 79060 Lviv Naukova St., 3a, e-mail: vfh@org.lviv.net

‡ N. M. Emanuel Institute of Biochemical Physics RAS, 19991 Moscow, Kosygin St., 4, e-mail: chembio@sky.chph.ras.ru

Growing interest to investigation of liquids transport through fluoropolymer membranes processes is due to their high strength and stability to external effects. Therefore studing the behaviour of such membranes in different liquids, including aggressive ones, becomes especially actual.

Parameters of polymers swelling in organic solvents and diffusion of the last ones into polymer structure can not be described unambiguously by means of simple dependences on physico-chemical properties of liquids. As in the most widespread model of these processes, based on the regular solvents theory, the solvation effects are not taken into account, the complicated bell-shaped relations between the equilibrium degree of swelling S_M and the Hildebrand solubility parameter δ are revealed even in the case of such simple carbon-chain polymers as poly(ethylene) or poly(propylene) [1, 2] accompanied by the numerous deviations. Introduction of correction coefficients does not lead to essential improvement of generalizing relations [3]. The attempts to describe the swelling processes by means of another characteristics of solvents, such as their molar volume V_M [4] or the Reichardt's parameter of specific electrophilic solvation E_T [5], give positive results only for the series of similar in properties solvents, mainly homologues.

We have shown since the processes of swelling of polymers and diffusion of the solvents into their structure are accompanied by solvation processes they should depend on summary effect of some factors in complicated way and acceptable relation between characteristics of these processes and properties of liquids may be obtained only by means of multiparameter equations [6, 7]. It was found as effective the use of the Koppel-Palm "solvation" equation [8], which takes into account the influence of medium on kinetics of chemical reactions , supplemented with some parameters which characterize the structure of the solvent:

$$\lg S_M = a_0 + a_1\frac{n^2-1}{n^2+2} + a_2\frac{\varepsilon-1}{2\varepsilon+1} + a_3B + a_4E_T + a_5\delta^2 + a_6V_M\,. \qquad (1).$$

In equation (1) S_M is the degree of swelling (expressed in moles of solvent, adsorbed by 100 g of polymer); n is refractive index and ε is permittivity of liquids, which characterize their polarizability and polarity, determining their ability to nonspecific solvation; B is the Koppel-Palm basicity and E_T is the Reichardt's electrophilicity, that characterize the ability to specific (acid-base) interactions; square root of solubility parameter δ^2 is proportional to cohesion energy of liquids and reflects inputs of energy on disturbance of their structure; V_M is the molar volume of liquids. Here it is necessary to accentuate the advisability of representation of the values of swelling in moles of adsorbed liquid, because if calculations are accomplished with the values of S expressed in weight or volume parts, usually cited in literature, the results of their generalizations are essentially worse.

The use of the equation (1) gives the possibility of the satisfactory generalization of the data on swelling in organic solvents of series of polymers – poly(ethylene) [6], poly(*cis*-isoprene) [9], poly(butadiene) [10], as well as of correlation of them with the properties of the solvents. So, in accordance with elementary logic, it was found that in all three cases the coefficients at V_M have negative value, that is, the larger is molar volume of the solvent, the more difficult is its penetration into polymer structure. In the case of poly(ethylene) the factors f(n) and f(ε) (ability of solvents to nonspecific solvation of polymer) promote its swelling. The signs "+" at solvation members point out on this fact. At the same time parameter E_T (ability of solvents to electrophilic solvatation) counteracts this process, what is no wonder, taking into account hydrophobicity and nonpolarity of poly(ethylene), and high values of E_T are typical for alcohols, protoncontaining amines etc. Similar generalizations were revealed for swelling relatively low-polar poly(*cis*-isoprene) and poly(butadiene).

In [11-15] the experimental data on swelling of membranes made from tetrafluoroethylene-propylene copolymer Aflas FA 150P containing carbon black (30 parts per 100 parts of polymer), d =1.55g/cm^3 in 22 solvents at 30, 45 and 60°C are given. Although in original works high termal and chemical stabilities of this polymer are mentioned, nevertheless it can absorb up to 100% and even 200% of solvent from polymer weight. At the same time in the cited works reliable relationship between the properties of solvents and the swelling degree of polymer was not established. That is why the attempt of generalization of these data by means of equation (1) appears to be advisible.

In Table 1 corresponding data on swelling abovementioned copolymer are represented. Since in [11-15] these data were given in g of solvent per 100 g of polymer, they were recalculated in moles of solvent absorbed by 100 g of polymer according to our previous works [9, 10]. Solvents characteristics were taken from reviews [16, 17], calculation procedure was realized according [18].

Generalization of available data obtained at 30°C for all 22 solvents leads to expression with unsatisfactory low values of multiple correlation coefficient R = 0.891. However, the exclusion (according [18]) of the most deviating data for only three solvents – cyclohexane, dimethylacetamide and dichloroethane – makes it possible to obtain expression (2) with acceptable value of R:

$$\lg (S_M \cdot 10^2) = 0.527 + (11.506 \pm 2.047) f(n^2) + (1.020 \pm 1.188) f(\varepsilon) + (1.484 \pm 0.431) \cdot 10^{-3} B + (0.062 \pm 0.042) E_T - (7.305 \pm 0.935)\, 10^{-3} \delta^2 - (0.014 \pm 0.002)\, V_M \qquad (2)$$

N = 19, R = 0.964, s = ± 0.127.

Low values of pair correlation coefficients of S_M values with separate parameters of equation r (the order of magnitude is 0.2-0.6) do not allow to define their significance. It may be determined (according to [18]) by alternating exception of individual members of the regression equation with calculating of the value of R of the equations with lesser amount or members in every case. If R of such equations falls down unsufficiently the exceptioned member is considered as insignificant. In this way insignificance of the factors of polarity f(ε) and electrophilic solvation E_T was found and so the degree of swelling of polymer can be characterized adequately by fourparameter equation (3):

$$\lg (S_M \cdot 10^2) = 2.736 + (9.155 \pm 1.515) f(n^2) + (1.366 \pm 0.411)\, 10^{-3} B - (5.997 \pm 0.624)\, 10^{-3} \delta^2 - (0.016 \pm 0.001)\, V_M \qquad (3)$$

R = 0.957, s = ± 0.139.

Table 1. Logarithms of swelling degree of copolymer Aflas FA 150P at 30, 45 and 60°C according to [11-15] and values, calculated by means of equations (3), (5) and (7)

N	Solvent	$lg(S_M \times 10^2)$								
		30°C			45°C			60°C		
		exp.	calc.	$\Delta lg(S_M \times 10^2)$	exp.	calc.	$\Delta lg(S_M \times 10^2)$	exp.	calc.	$\Delta lg(S_M \times 10^2)$
1	Hexane	1.1526	1.4252	0.2726	1.2497	1.4261	0.1765	1.3339	1.4899	0.1560
2	Heptane	0.9638	1.1805	0.2167	1.0645	1.2458	0.1813	1.1511	1.3118	0.1607
3	Octane	0.8698	0.9175	0.0477	0.9703	1.0435	0.0731	1.0558	1.1124	0.0566
4	Isooctane	1.2370	1.1116	-0.1254	1.3164	1.2234	-0.0930	1.4028	1.2888	-0.1139
5	Nonane	0.7709	0.6827	-0.0882	0.8739	0.8658	-0.0081	0.9518	0.9371	-0.0147
6	Decane	0.6042	0.4177	-0.1865	0.8482	0.6599	-0.1883	0.8609	0.7340	-0.1269
7	1,2-Dichlorethane	1.2993*	1.5750	0.2757	1.3981	1.6455	0.2474	1.5120	1.7191	0.2071
8	Chloroform	1.9666	1.7866	-0.1800	2.0342	1.8046	-0.2296	2.0644	1.8727	-0.1918
9	Trichlorethylene	1.8878	1.7867	-0.1011	1.9669	1.8052	-0.1617	2.0057	1.8724	-0.1333
10	Tetrachlormethane	1.8017	1.8717	0.0701	1.8842	1.8475	-0.0367	1.9473	1.9095	-0.0378
11	1,1,1-Trichlorethane	1.9258	1.8312	-0.0946	1.9794	1.8816	-0.0978	2.0072	1.9440	-0.0632
12	Tetrachlorethylene	1.5821	1.6623	0.0803	1.6736	1.6771	0.0035	1.7517	1.7446	-0.0071
13	Tetrachlorethane	1.1316	1.3528	0.2212	1.3122	1.4782	0.1660	1.4077	1.5580	0.1503
14	Dichlormethane	1.8132	1.6273	-0.1859	-	-	-	-	-	-
15	Acetone	1.5563	1.4630	-0.0933	1.6104	1.5117	-0.0987	1.6570	1.5743	-0.0827
16	Methylethylketone	1.5038	1.5426	0.0388	1.6397	1.6461	0.0064	1.7088	1.7017	-0.0071
17	Acetonitrile	0.4800	0.5453	0.0653	0.6212	0.5837	-0.0375	0.7016	0.6704	-0.0312
18	Tetrahydrofuran	1.9671	1.9379	-0.0293	2.0054	2.0080	0.0026	2.0242	2.0447	0.0205
19	1,4-Dioxane	1.5561	1.5855	0.0294	1.6307	1.6621	0.0313	1.6627	1.6924	0.0297
20	Dimethyl acetamide	-	-	-	0.9263*	1.5199	0.5935	1.0599*	1.5584	0.4985
21	Cyclohexanone	1.2543	1.2964	0.0421	1.4244	1.4876	0.0632	1.5159	1.5444	0.0286
22	Cyclohexane	1.0777*	1.6764	0.5986	1.1870*	1.6378	0.4509		1.7031	0.3961

* Data excepted from calculation

The most significant and decreasing the swelling degree is here the influence of molar volume V_M, the value R between lg $(S_M \cdot 10^2)$ and V_M is equal to 0.569, and exception of this member from equation (3) destroys the correlation at all: R of threeparameter equation obtained gets infeasible low value of 0.572. Obviously, the larger is the size of solvent molecule, the more difficult it will be penetrate into polymer structure. The other members of equation have only certain correcting effect, at that the members with $f(n^2)$ and B, which characterize nonspecific and specific solvation of polymer molecules containing electronegative fluorine atoms, promote an adsorption of solvent by polymer. At the same time the increasing of solvent cohesion diminishes adsorption degree, probably, due to necessity of input of energy for tearing off a solvent molecule from the structure of liquid. Adequacy of equations (2), (3) and of all consequent ones is confirmed by coordination of the values of Fisher criterium, obtained for them, with table values for corresponding number of points at the degree of reliability $\alpha = 0.95$.

In Table 1 the values of lg $(S_M \cdot 10^2)$, calculated by means of equation (3) and also their deviations from experimental values - Δ lg $(S_M \cdot 10^2)$ are shown. As one can see the majority of deviations (with the exception of the results for three solvents, excepted from calculation) are within the limits of mean-root-square error $s = \pm 0.139$ or are only slightly larger than it, not exceeding the value of 2s.

Similar results were obtained at generalization of the data on swelling of copolymer Aflas FA 150 P at 45°C and 60°C (Table 1). The values of R of the equations, summarizing the data for all solvents, solvents which must be excluded from calculations for achievement of acceptable values of R, six- and fourparameter equations, adequately summarizing dependences of lg $(S_M \cdot 10^2)$ on properties of organic liquids, are shown below.

At 45°C R of sixparameter equation for 21 solvents is equal 0.894. After the exclusion of the data for cyclohexane and dymethylacetamide adequate equations (4) and (5) were obtained:

$$\lg (S_M \cdot 10^2) = 1.433 + (10.456 \pm 1.983) f(n^2) - (0.188 \pm 1.164) f(\varepsilon) + (1.761 \pm 0.479) \cdot 10^{-3} B + (0.019 \pm 0.038) E_T - (5.918 \pm 0.892) \cdot 10^{-3} \delta^2 - (0.013 \pm 0.002) V_M \quad (4)$$

N = 19, R = 0.953, s = ± 0.131.

$$\lg (S_M \cdot 10^2) = 2.146 + (9.728 \pm 1.476) f(n^2) - (1.732 \pm 0.413) \cdot 10^{-3} B - (5.441 \pm 0.600)\ 10^{-3} \delta^2 - (0.013 \pm 0.001) V_M \quad (5)$$

R = 0.951, s = ± 0.134.

Accordingly, at 60°C for 21 solvents R=0.918, and after the exclusion of the data for cyclohexane and dymethylacetamide we have:

$$\lg (S_M \cdot 10^2) = 1.567 + (10.187 \pm 1.668) f(n^2) - (0.028 \pm 0.979) f(\varepsilon) + (1.585 \pm 0.403) \cdot 10^{-3} B + (0.016 \pm 0.032) E_T - (5.762 \pm 0.750) \cdot 10^{-3} \delta^2 - (0.012 \pm 0.001) V_M \quad (6)$$

N = 19, R = 0.965, s = ± 0.100.

$$\lg (S_M \cdot 10^2) = 2.202 + (9.550 \pm 1.260) f(n^2) + (1.594 \pm 0.352) \cdot 10^{-3} B - (5.294 \pm 0.513) \cdot 10^{-3} \delta^2 - (0.013 \pm 0.001) V_M \quad (7)$$

$N = 19$, $R = 0.962$, $s = \pm 0.114$.

As one can see, in both cases signs and significances of individual members of equations are similar with those in the equations revealed at 25°C.

The maximal degree of swelling was determined in works [11-15] by constructing of the curves S vs time τ before achievement of equilibrium, what allowed to calculate the proper coefficients of diffusion D from the slopes of the linear parts of adsorption curves, and also, as the process of adsorption is formally described by the first order equation of kinetics, to define the rate constants of penetration of solvents into polymer structure k.

Diffusion of substances in homogeneous medium was interrogated and described by the known Fick, Einstein-Smolukhovsky and others' formulas, however the relations between the characteristics of both membrane-making polymer and penetrating substance and coefficient of diffusion (i.e. its rate), were not obtained with the reliable quantitative decision until now [19, 20, 21]. So, in the case of diffusion of esters into ethylene-propylene membranes we observe $n = 0.5$ in expression $Q_\tau/Q_\infty = k\tau^n$ (where Q_τ and Q_∞ are the amount of adsorbed solvent during the time τ and at saturation; k and n are constants, relatively), i.e. submission to the Fick law. However, for films from other materials (natural and synthetic rubbers, polyurethane) $n \approx 0.6$, i.e. we deal with partial transition into the region of so-called anomalous transport conditioned by different influences of separate areas of polymer chain on adsorption processes [21]. Such deviations at diffusion of aromatic hydrocarbons or dymethylformamide into polyurethane membrane are yet more noticeable [19, 22]. It is due to the fact that between diffusing solvent and polymer structure the solvation interactions take place influencing on the energetics and the rate of the process of diffusion. As well as in the case of swelling only semiempirical account of them is possible on the basis of the concept of free energies linearity (LFE). The similar approach is used at generalization of the data on the effects of solvents on the rate of chemical reactions, solubility of gases, distribution of substances between two phases and so on. With this respect the Koppel-Palm equation, taking into account the effects of polarity and polarizability of solvents, as well as their capacity for acid-base interaction seems to be the most widespread. In the case of swelling of polymers and diffusion of liquids into their structure the additional account of geometric factor is necessary since the larger is a molecule of a solvent, the more difficult will be its penetration into a polymer structure, while solvatation of some areas of polymer structure by a solvent will facilitate diffusion. This supposition follows logically from the Einstein-Smolukhovsky formula:

$$D = RT/N \cdot 6\pi\eta r, \quad (8)$$

in accordance to which the diffusion coefficient D in homogeneous media is inversely proportional to the radius r of diffusing particle. Judgements similar to foregoing ones were expressed earlier as well [23, 24].

Table 2. Logarithms of coefficients of diffusion of solvents into copolymer Aflas FA150P at 30, 45 and 60^0 according to [11-15] and values, calculated by means of equations (10), (12) and (14)

N	Solvent	lg (D×10^7)								
		30°C			45°C			60°C		
		exp.	calc.	Δlg (D×10^7)	exp.	calc.	Δlg (D×10^7)	exp.	calc.	Δlg (D×10^7)
1	Hexane	0.2455	0.3255	0.0799	0.4900	0.5532	0.0632	0.5502	0.6163	0.0661
2	Heptane	-0.1024	-0.0722	0.0301	0.3263	0.2576	-0.0687	0.3820	0.4261	0.0441
3	Octane	-0.5528	-0.4698	0.0831	0.0828	-0.0533	-0.1361	0.2455	0.2180	-0.0275
4	Isooctane	-0.3188	-0.2492	0.0695	-0.0605	0.1270	0.1875	0.3032	0.3642	0.0610
5	Nonane	-0.9586	-0.8287	0.1299	-0.2518	-0.3327	-0.0809	0.0374	0.0355	-0.0020
6	Decane	-0.9208	-1.2168	-0.2960	-0.6576	-0.6242	0.0334	-0.0506	-0.1489	-0.0983
7	1,2-Dichlorethane	0.2355	0.3875	0.1520	0.6405	0.7657	0.1253	0.8235	0.0877	0.0543
8	Chloroform	0.7016	0.6183	-0.0832	0.7839	0.9427	0.1588	0.9513	0.9603	0.0090
9	Trichlorethylene	0.8363	0.5031	-0.3332	0.8927	0.5230	-0.3697	0.9415*	0.4269	-0.5146
10	Tetrachlormethane	0.3598	0.6186	0.2588	0.4014	0.5107	0.1093	0.5729	0.3297	-0.2432
11	1,1,1-Trichlorethane	0.4502	0.6468	0.1965	0.4669	0.4690	0.0021	0.4969	0.3911	-0.1059
12	Tetrachlorethylene	0.4757	0.2767	-0.1990	0.7267*	-0.0251	-0.7519	0.8169*	0.4269	-0.5146
13	Tetrachlorethane	-0..1938	-0.0364	0.1574	0.0086	-0.0092	-0.0178	0.0334	0.0549	0.0215
14	Diclormethane	0.9745	0.7272	-0.2960	-	-	-	-	-	-
15	Acetone	0.6866	0.5836	-0.1030	0.7612	0.6325	-0.1287	0.7973	0.7018	-0.0955
16	Methylethylketone	0.5527	0.4597	-0.0929	0.6170	0.5542	-0.0628	0.6355	0.6615	0.0261
17	Acetonitrile	-0.3372	-0.1320	0.2053	-0.1135	-0.0786	0.0349	0.0607	0.0794	0.0187
18	Tetrahydrofuran	0.7482	0.7681	0.0199	0.8451*	0.2255	-0.6196	0.8531*	0.0335	-0.8196
19	1,4-Dioxane	0.1492	0.1480	-0.0012	0.2041	0.2482	0.0441	0.2227	0.2082	-0.0145
20	Dimethyl acetamide	-0.2924	-0.3190	-0.0266	-0.0809	-0.1557	-0.0748	0.0492	0.0498	0.0006
21	Cyclohexanone	-1.3010*	-0.1181	1.1830	-0.5528	-0.3719	0.1809	-0.3979	-0.3165	0.0814
22	Cyclohexane	-0.7696*	0.5588	1.3284	-0.0605*	0.3998	0.4603	0.0453	0.2495	0.2042

* Data excepted from calculation

In Table 2 the values of diffusion coefficients of solvents into polymer Aflas FA 150 P at 30, 45, 60°C $D\cdot 10^7 (cm^2/s^{-1})$ are given (from [11-15]). They may be generalized by means of equation of the type (1) with acceptable exactness as well. For 22 solvents at 30°C the value of R is equal 0.825, but after the exclusion of the data for cyclohexane and cyclohexanone we obtain the adequate equation (9):

$$\lg (D\cdot 10^7) = 5.015 + (1.815\pm 2.183)f(n^2) + (1.751\pm 1.445)f(\varepsilon) + (0.074\pm 0.479)\cdot 10^{-3}B - (0.045\pm 0.046) E_T -(5.548\pm 1.154)\cdot 10^{-3}\delta^2 -(0.022\pm 0.002)V_M \quad (9)$$

N = 20, R = 0.956, s = ± 0.168.

And after exception of parameters of little significance:

$$\lg (D\cdot 10^7) = 5.957 + (2.235\pm 1.311)f(\varepsilon) - (0.065\pm 0.039) E_T - (5.157\pm 1.032)\cdot 10^{-3}\delta^2 - (0.022\pm 0.002) V_M \quad (10)$$

R = 0.954, s = ± 0.171.

In contrast to generalization of the data on S_M parameters B and $f(n^2)$ appear to be insignificant here, although positive signs at these parameters show, that possible formation of interaction between electronegative atoms of fluorine and electrodonor solvents may hamper their penetration into a polymer. The effect of the factor of electrophilic solvation is also little significant – the value of R falls down to 0.947 at the exception of parameter E_T. As well as in the previous case, the increase of solvents cohesion and of their molecule sizes reduces their capacity for penetration into polymer structure. The single factor promoting the process of diffusion into polar copolymer structure is polarity of solvents.

Similar to the case of S_M, the values of $\lg (D\cdot 10^7)$, calculated by equation (10), and their deviations with the experiment results are given (Table 2). In this case most of deviations do not exceed the region of s ± 0.171 as well. The values of $\lg (D\cdot 10^7)$ for temperatures 45 and 60°C may also be generalized by means of the equation of type (1). However, for these temperatures is insignificant the effect of only one parameter – polarizability $f(n^2)$. The effect of the factor of basicity is negative, possibly, because the process of penetration of the basic solvent into the polymer structure is slowed as a result of bonding of basic solvent by electronegative atoms of fluorine, however, in general this factor is relatively insignificant. At the same time, electrophilicity of solvents promotes their diffusion. In both cases the density of the cohesion energy and molar volume of a solvent have the most essential effect on the process.

So, R of sixparameter equation for 21 solvents at 45°C is equal 0.848, and after exclusion of the data for cyclohexane, trichlorethylene and tetrahydrofuran we get the adequate equation (11):

$$\lg (D\cdot 10^7) = 0.047 + (1.204\pm 1.839)f(n^2) - (3.184\pm 1.260)f(\varepsilon) - (0.707\pm 0.459)\cdot 10^{-3}B+(0.143\pm 0.044)E_T -(7.567\pm 1.017)\cdot 10^{-3}\delta^2 -(0.015\pm 0.002)V_M \quad (11)$$

N = 18, R = 0.953, s = ± 0.138.

$$\lg (D\cdot10^7) = 0.705 - (2.780\pm1.113)f(\varepsilon) - (0.784\pm0.450)\cdot10^{-3}B + (0.128\pm0.039) E_T - (7.305\pm0.946)\cdot10^{-3}\delta^2 - (0.015\pm0.002) V_M \quad (12)$$

$R = 0.952$, $s = \pm 0.139$.

The subsequent exception of the parameter of basicity leads to equation with $R = 0.943$.

For 21 solvents at 60°C $R = 0.786$ and after exception of the data for trichlorethylene, tetrahydrofuran and tetrachlorethylene we obtain the adequate equation (13):

$$\lg(D\cdot10^7) = -2.774 + (0.728\pm1.280)f(n^2) - (4.604\pm0.828)f(\varepsilon) - (0.887\pm0.317)\cdot10^{-3}B + (0.218\pm0.028) E_T - (7.648\pm0.672)\cdot10^{-3}\delta^2 - (0.008\pm0.001) V_M \quad (13)$$

$R = 0.965$, $s = \pm 0.094$.

$$\lg (D\cdot10^7) = -2.425 - (4.387\pm0.742)f(\varepsilon) - (0.923\pm0.314)\cdot10^{-3}B + (0.211\pm0.025) E_T - (7.517\pm0.638)\cdot10^{-3}\delta^2 - (0.008\pm0.001) V_M \quad (14)$$

$R = 0.964$, $s = \pm 0.095$.

As well as in the previous case, the exclusion of basicity results in decrease of the value of R to 0.946.

On the basis of the data on lg S_M and lg D the authors [11-15] calculate the values of the effects of enthalpy of absorption of solvents at swelling ΔH (kJ/mol) and activation energies of the process of diffusion E_D (kJ/mol) (Table 3). These data can not be consider as fully reliable, because they were defined using the values only for three temperatures. Indeed, at generalization of 21 values of ΔH for the receipt of adequate equation with $R > 0.95$ it is necessary to exclude from consideration the data for 5 solvents, that is 25% of their common number, and all parameters in resulting equation appears to be significant at the simultaneously extremely low values of pair correlation coefficients of separate terms with ΔH – about 0.15-0.30. The best results were got at generalization of the data on activation energies of the process of diffusion E_D. In spite of the fact that R is equal only 0.825 at generalization of the data for 21 solvents, exception of the data for only three solvents (cyclohexane, dichlorethane and chloroform) allows to get the adequate eqution (15):

$$E_D = 64.26 - (300.59\pm74.92)f(n^2) + (66.80\pm46.83)f(\varepsilon) - (2.85\pm1.60)\cdot10^{-2}B - (3.81\pm1.73) E_T + (0.25\pm0.04)\delta^2 + (0.61\pm0.05) V_M \quad (15)$$

$N = 18$, $R = 0.965$, $s = \pm 4.98$.

The factor of molar volume ($r = 0.82$) has the defining effect here too, that can be explained logically – the larger is the size of molecule, the more is the expense of energy for its penetration into polymer structure, i. e. the activation energy of the process. The equation can be simplified to fourparameter one:

Table 3. Values of enthalpy, activation energy and logarithms of rate constants of the process of diffusion of solvents into copolymer Aflas FA 150P at 30°C according to [11-15] and calculated by means of equations (16) and (17)

N	Solvent	Δ H	exp. E_D	calc. E_D	Δ E_D	exp. lg (k·10³)	calc. lg (k·10³)	Δlg (k·10³)
1	Hexane	11.68	19.7700	24.0639	4.2939	0.4314	0.6147	0.6675
2	Heptane	12.06	31.6600	34.0768	2.4168	0.0682	0.4330	0.6542
3	Octane	11.98	51.9100	44.7810	-7.1290	-.0.3098	0.1673	0.4456
4	Isooctane	10.67	39.9400	39.5000	-0.4400	-0.2147	0.2330	0.3598
5	Nonane	11.70	63.5600	54.6109	-8.9491	-0.7447	0.0212	0.2856
6	Decane	16.81	56.8300	65.2269	8.3969	-0.6990	-0.5229	0.1367
7	1,2-Dichlorethane	13.69	38.0300*	-1.1306	-39.1606	0.2405	0.5933	0.7093
8	Chloroform	7.54	16.0000*	-4.1227	-20.1227	0.7672	0.5999	0.8825
9	Trichlorthylene	7.63	6.8000	4.3870	-2.4130	0.8248	0.6637*	0.7513
10	Tetrachlorethane	9.40	14.2800	7.9458	-6.3342	0.3636*	0.3096	0.4183
11	1,1,1-Trichlorethane	5.26	2.9500	4.9618	2.0118	0.3243	0.3284	0.3979
12	Tetrachlorethylene	10.93	22.1200	18.5788	-3.5412	0.4249	0.5185*	0.6730*
13	Tetrachlorethane	17.86	13.6800	16.5198	2.8398	-0.2218	-0.0757	-0.0177
14	Dichlormethane	-	-	-	-	0.8645	-	-
15	Acetone	11.18	7.6600	4.7200	-2.9400	-0.1549	0.8306	0.6722
16	Methylethylketone	13.20	4.4400	6.5789	2.1389	0.6721*	0.5539	0.7067
17	Acetonitrile	14.34	31.4700	26.9686	-4.5014	-0.1549	-0.0177	0.3598
18	Tetrahydrofuran	3.70	6.8300	6.1880	-0.6420	-	-	-
19	1,4-Dioxane	6.91	4.7400	19.0154	14.2754	-	-	-
20	Dimethyl acetamide	25.04	21.8600	20.9678	-0.8922	-	-	-
21	Cyclohexnone	16.92	22.8700	24.2784	1.4084	-0.4685	-0.1675	-0.0655
22	Cyclohexane	14.77	52.5500*	16.6218	-35.5500	-0.5850*	0.0969	0.4232

* Data excepted from calculation

$$E_D = 11.29 - (225.69 \pm 65.61) f(n^2) - (2.02 \pm 0.73)\, E_T + (0.21 \pm 0.04)\,)\, \delta^2 + +(0.62 \pm 0.06) V_M \quad (16)$$

$R = 0.955$, $s = \pm 5.62$.

Along with a molar volume the factor of cohesion energy of liquid also influences on the value of E_D substantially, promoting it. At the same time the solvation factors reduce the value of E_D. However, they are relatively insignificant. In Table 3 along with the experimental values of E_D, the values calculated by the equation (16) and their deviations from experimental data are shown.

As it was mentioned above, the relative rate constants of the process of penetration of liquid into polymer structure $k \cdot 10^3$ (min^{-1}) were determined in [11-15]. These values are also represented in Table 3 and may be generalized adequately by the equations of type (1) after the exclusion of the data for 1-3 solvents. In all cases the factor determining and reducing the penetration rate of liquid into polymer structure is the molar volume of a solvent (r about 0.50-0.60). Polarity and density of cohesion energy of liquids reduce the penetration rate. At the same time their capacity for electrophilic solvation increases the values lg k; the factors of polarizability and capacity to nucleophilic solvation are insignificant. The simplified fourparameter equations, generalized data on lg k at three temperatures, the values of R and s and also the solvents excluded from calculation are represented below.

At 30°C after exception of the data for cyclohexane, methylethylketone and tetrachlormethane we obtain equation (17):

$$\lg (k\cdot 10^3) = 5.16 - (2.72\pm1.66)f(\varepsilon) + (1.66\pm4.65)10^{-2} E_T - (3.35\pm1.15)10^{-3}\delta^2 - (2.19\pm0.23)10^{-2} V_M \quad (17)$$

R = 0.933, s = ± 0.186.

At 45°C after exception of the data for tetra- and trichlorethylene we get (18):

$$\lg (k\cdot 10^3) = 1.052 - (3.36\pm1.00)f(\varepsilon) + (0.16\pm0.03)10^{-2} E_T - (7.067\pm0.745)10^{-3}\delta^2 - (0.908\pm0.128)10^{-2} V_M \quad (18)$$

R = 0.949, s = ± 0.112.

At 60°C after exception of the data for tetrachlorethylene we obtain (19):

$$\lg (k\cdot 10^3) = -2.213 - (5.59\pm0.84)f(\varepsilon) + (0.194\pm0.025)10^{-2} E_T - (5.509\pm0.020)10^{-3}\delta^2 - (0.626\pm0.109)10^{-2} V_M \quad (19)$$

R = 0.944, s = ± 0.094.

Thus, consideration of the data from [11-15] show, that the processes of swelling and diffusion of solvents into the structure of tetraflurethylene-propylene coolymer can not be described quantatively using only one characteristic of a solvent. In all cases a molar volume of liquid has the determining effect, hampering its penetration into the structure of polymer. However, the effects of other factors – both solvation and density of cohesion energy of solvents – are significant. Adequate quantitative generalization of abovementioned processes can be obtained on the basis of free energies linearity concept by using of linear multiparameter equations taking into account the effects of different factors.

References

[1] Richards R. B. // *Trans. Faraday Soc.* 1946. V. 42. N 1. P. 20-28.

[2] Briston G. V., Watson W. F. // *Trans. Faraday Soc.* 1958. V. 54. N 11. P. 1731-1741.

[3] *Mc* Kenna G.B., Flynn K. M., Chen Gihong. // *Polymer. 1990*. V. 31. N 10. P. 1937-1945.

[4] Aminabhavi T. M., Harogoppad S. B., Khinnavar R. S. et al. // *IMS – Rev. Macromol. Chem. Phys.* 1991. V. C31. N. 4. P. 433-448.

[5] Jonquieres A., Roizard D., Luchon P. J. // *J. Appl. Polymer Sci.* 1974. V. 54. N 11. P. 1673-1682.

[6] Makitra R., Pyrig I., Zagladko E. et al. // *J. Appl. Polymer Sci. 2001*. V. 81. P. 3133-3140.

[7] Makitra R. G., Zagladko E. A., Midyana G. G., Protsaylo L. V. // *Zhurn. fiz. khim.* 2001. V. 75. N 12. P. 2283-2287.

[8] Koppel I. A., Palm V. A. // Advances in Linear Free Energ y Relationships / Ed. Chapman N.B., Shorter J. London; New York: Plenum Press, 1973. P. 203-280.

[9] Makitra R. G., Zagladko E. A., // *Zhurn. fiz. khim*. 2002. V. 76. N 10. P. 1797.

[10] Makitra R. G., Zagladko E. A., Turovsky A. A., Zaikov G. E. // *Zhurn. prikl. khim.* 2004. V. 77. N 2. P. 324.

[11] Aminabhavi T. M., Harlapur S. F., Balundgi R.H. et al. // *Polymer. 1998*. V. 39. N 5. P. 1067-1074.

[12] Aminabhavi T. M., Harlapur S. F., Balundgi R.H. et al. // *J. Appl. Polym. Sci.* 1996. V. 59. P.1857-1870.

[13] Aminabhavi T. M., Harlapur S. F., Balundgi R.H. // *J. Polym. Eng*. 1996/1997. V. 16. N 3. P. 181-202.

[14] Aminabhavi T. M., Harlapur S. F. // *Chem. Eng. Proc*. 1997. V. 36. P. 363-370.

[15] Aminabhavi T. M., Phayde H. T. // *J. Chem. Eng. Data*. 1996. V. 41. P. 813-818.

[16] Makitra R. G., Pyrig Y. M., Kivelyuk R. B. // *Major characteristics of solvents*, using in LFE equations. M. 1986. 34 P. – Dep. in VINITI 26.03.86. N 628-B86.

[17] Abboud J. L. M., Notario R. // *Pure Appl. Chem*. 1999. V. 71. N 4 P. 645.

[18] Recommendations for Reporting the Results of Correlation Analysis in Chemistry // *Quant. Struct.-Act. Relat*. 1985. V. 4. N 1. P. 29.

[19] Aithal U. S., Aminabhavi T. M. // *Indian J. Technology*. 1990. V. 28. N 10. P. 592.

[20] Charlesworth J. M., Riddel S. Z., Mathews R. J. // *J. Appl. Polymer Sci.* 1993. V. 47. P. 653.

[21] Khinnavar R. S., Aminabhavi T. M. // *J. Appl. Polymer Sci.* 1992. V. 46. P. 909.

[22] Aithal U. S., Aminabhavi T. M. // *Polymer. 1990*. V. 31. N 9. P. 1757.

[23] Kotyukov I. M. // Fizicheskaya khimiya. Tomsk: *Sibirskaya nauchnaya mysl. 1993*. V. 2. P. 713–725.

[24] Gerasimov Ya. I., Dreving V. P., Eryomin E. I. at al. // Kurs fizicheskoy khimiyi. *M.: Khimiya*. 1979. P. 239-246.

In: Progress in Monomers, Oligomers, Polymers, Composites … ISBN 978-1-60456-878-3
Editors: R. A. Pethrick, G.E. Zaikov et al.

Chapter 7

CRYSTALLOHYDRATES OF CARBOXY- AND SULFO-SUBSTITUTED COBALT(II)PHTHALOCYANINES

***Nataliya Sh. Lebedeva*[1,*], *Nataliya A. Pavlycheva*[1], *Anatoliy I. Vyugin*[1], *Elena V. Parfenyuk*[1], *Vladimir P. Kulinich*[2,†] and *Gennagiy P. Shaposhnikov*[2]**

[1]Institute of Solution Chemistry, Russian Academy of Sciences, Ivanovo, Russian Federation,
[2]Ivanovo State University of Chemistry and Technology, Ivanovo, Russian Federation,

ABSTRACT

Physical chemical characteristics (composition, energetic stability and temperature of destruction) of crystallohydrates of cobalt(II)phthalocyanines were obtained by thermogravimetric analysis method.

Keywords: Metallophthalocyanines · Crystallohydrates · Molecular complexes · Thermogravimetry

INTRODUCTION

Applications of water soluble macroheterocyclic compounds such as porphyrins and phthalocyanines are many and diverse, as catalysts, photo- and chemosensors, materials for photoelectric transformers, etc. However, according to literature data, properties of these compounds are affected significantly by content of water in crystal samples [1]. Thus,

* Correspondence to: Institute of Solution Chemistry, Russian Academy of Sciences, Akademicheska st. 1, Ivanovo 153045, Russia, tel.+7(4932)336272, e-mail: nsl@isc-ras.ru

† Correspondence to: Ivanovo State University of Chemistry and Technology, Fr. Engels st., 7, Ivanovo 153460, Russia, tel. +7(4932)307346, e-mail: tocc@isuct.ru

quantitative estimation of metallophthalocyanine-water interactions is required [1,2]. At the present time information about the interactions are still controversial. Specific interactions between metallophthalocyanine and water molecules were studied by IR spectroscopy [3]. The crystal samples of α- and β-polymorphs of phthalocyanine and its metallocomplexes (M = Mg^{2+}, Zn^{2+}, Cu^{2+}, Fe^{3+}) were studied. Only the Mg(II) and Zn(II)complexes interact specifically with water while phthalocyanine itself and Cu(II) and Fe(III)complexes are not hydrated. It is known that water molecules form complexes with metallophthalocyanines in which the bond between metal ion and macrocycle is preferentially electrostatic. In the case of metalloporphyrins, their molecular complexes with water were isolated and characterized by X-ray diffraction method [4]. Water molecule in such complexes is axial ligand (water molecule is an electron donor, Mg^{2+} is an acceptor with weak neutralized charge) [5]. On the contrary, according to data [3], Mg(II)phthalocyanine is hydrated through nitrogen *meso*-atoms of phthalocyanine macrocycle. This conclusion was made on the basis of the shift of 20 cm^{-1} of valent ν(O-H) vibrations of water molecule to low-frequency region. The conclusion is unconvincing because the band of valent vibrations of O-H bond is intensive, wide and its position differs significantly according to different literature data [3,6]. Generally, having a single meaning conclusion about binding with one or another fragment of phthalocyanine molecule on the IR data are doubtful because the IR spectrum of phthalocyanine is very complicated. The characteristic bands assigned to vibrations of bonds between atoms in these molecules are absent. Therefore it is difficult to make some conclusions about specific interactions with ligands on the basis of shifts of these bands [7]. In the number of works [3,6,8] the bands sensitive to vibrations of some atomic bonds in phthalocyanine molecule were registered including the bands sensitive to state of *meso*-atoms [6,8]. Nevertheless changes of intensity and position of the indicated bands are not discussed [3]. Moreover the obtained IR spectra do not allow estimation any changes associated with hydration of the macrocycle. From the point of view of these authors, β-polymorph of Mg(II)phthalocyanine does not interact specifically with water. On the contrary, the results obtained in work [9] testify about formation of thermally stable 1 : 1 complex of β-polymorph of this metallophthalocyanine with water. The data confirmed hydration of the Zn(II) complex of phthalocyanine are not presented in this work but blocking of α→β polymorphous transformation is indirect evidence of the hydration.

One more evidence of the interaction of water molecules with metallophthalocyanines is presented in [10]. In this work association of cobalt(II)tetrasulfophthalocyanine (CoTSPc) in aqueous media was studied. "Unique" role of water was found in dimerisation process of CoTSPc. Since the dimerisation was not observed in N,N-dimethylformamide and mixed solvent water-urea, water molecules are not only promoters but participants of this process:

$$2\,\text{CoTSPc} + 12\,H_2O \xleftrightarrow{K_{dim}} (\text{CoTSPc})_2\cdot(H_2O)_{12}.$$

The dimerisation was studied also in aqueous alcohols (methanol, ethanol) but influence of the alcohols on the mentioned equilibrium was not taken into account although it is well known that alcohol molecules can form hydrogen bonds with oxygen containing peripheral substitutes of macroheterocyclic compounds [11,12]. Besides, the authors exclude a possible hydration of monomer species of metallophthalocyanine. They present the following evidences of specific hydration of the dimer species: (i) no linear correlation of $\log(K_{dim})$ with the activity of water (activity of water was considered as partial pressure of water vapor in mixed solvent water-alcohol)) was found; (ii) no linear correlation of log(K_{dim}) with $1/\varepsilon$ where ε is the medium permittivity. In a number of the studies on metallophthalocyanine molecules with O-containing substitutes [13-16] hydrogen bonding of the peripheral substitutes with water is demonstrated. However no quantitative characteristics of the interactions of metallophthalocyanines with water were found in literature. Therefore, the aim of this work is to study the physical chemical characteristics (composition, energetic stability and temperature of destruction) of crystallohydrates of cobalt(II)phthalocyanines (CoPc).

EXPERIMENTAL

The oxygen containing cobalt(II)phthalocyanines: Co(4-$C_6H_4SO_3H$)$_4$Pc (I), Co(4-Br)$_4$(5-SO_3H)$_4$Pc (II), Co(4-Cl)$_4$(5-SO_3H)$_4$Pc (III), Co(4-COOH)$_4$(5-SO_3H)$_4$Pc (IV), Co(4-COO^-)$_4$(5-SO_3^-)$_4$(Ba^{2+})$_4$Pc (V) were synthesized and purified according to [17-19].

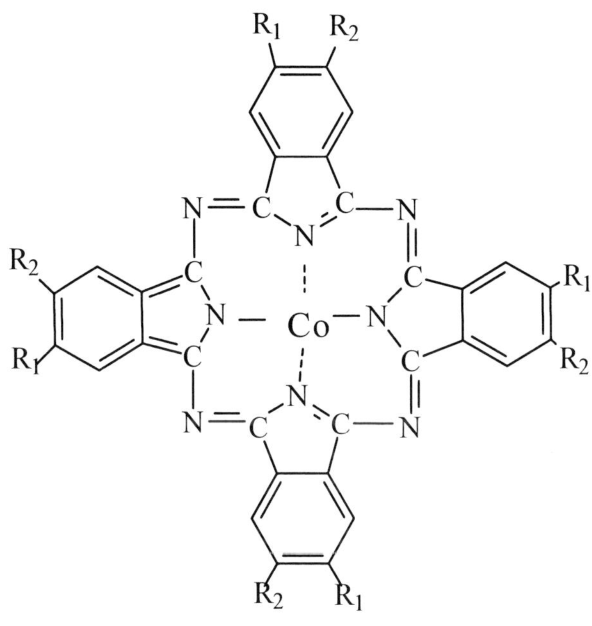

I $R_1 = H$, $R_2 = C_6H_4SO_3H$
II $R_1 = Br$, $R_2 = SO_3H$
III $R_1 = Cl$, $R_2 = SO_3H$
IV $R_1 = COOH$, $R_2 = SO_3H$
V R_1 = COO-, $R_2 = SO_3^-$, Ba^{2+}

PREPARATION OF THE SAMPLES

The crystallohydrates of CoPc were prepared by slow crystallization from concentrated solutions at 298-303 K as for DSC and X-ray diffraction analysis [20]. In the case of Co(4-$COO^-)_4(5-SO_3^-)_4(Ba^{2+})_4$Pc, the crystal samples were pounded and treated by water vapor in Co(4-$COO^-)_4(5-SO_3^-)_4$Pc apparatus during 12 hours.

METHODS

Thermogravimetric Analysis

Thermogravimetric measurements were made with thermoanalytical set [21] including a software-hardware unit for measuring and recording signals from a derivatograph (model 1000 D, MOM, Hungry) and computer-assisted processing of thermogravimetric data.

The approach that had been successfully applied to the anlyisis of crystallosolvates of metallophthalocyanines [22] and metalloporphyrins [23] was used. The functional dependences of the evaporation rate and the pressure of the saturated vapor of a solvent, which are sensitive to differences in states of molecules vaporized, served as a theoretical basis for thermogravimetric analysis. The measurements were carried out in air atmosphere and constant pressure. Samples of the crystallosolvates were placed in the Knudsen cell. At minimal heating rate (1.25°C·min^{-1}) assumption of quasi-equilibrium vaporization process is valid. Therefore, the Knudsen (1) and Clapeyron-Clausis (2) equations can be applied to the process under study:

$$\frac{dW}{d\tau} = \alpha_k \cdot S \cdot \frac{N_A}{\sqrt{2\pi \cdot R \cdot T \cdot M}} \cdot P \,, \tag{1}$$

$$\ln P = -\frac{\Delta_{ucn} H}{R \cdot T} + B \tag{2}$$

Simultaneous solution of eqs (1) and (2) gives the following relation:

$$\ln\left(\frac{dW}{d\tau} T^{1/2} \right) = -\frac{\Delta_{ucn} H}{R} \cdot \frac{10^3}{T} + C \,, \tag{3}$$

where N_A is the Avogadro number, M is the molecular mass of evaporated substance, α_k is the accommodation coefficient, R is the universal gas constant, S is the vaporation area, P is the pressure of saturated vapor at the sample temperature T, $dW/d\tau$ is the mass loss rate which is proportional to the deviation of DTG curve from the basis line, $\Delta_{evp}H$ is the change of enthalpy of evaporation, B and C are the constants of integration. The error of indirect measurements was calculated by adding up its components according to the addition rules for random errors; in this study, their distribution laws were considered via entropy coefficients as recommended in [24]. All measurements were run five times. To check reliability of the obtained data, enthalpies of evaporation of water and some organic solvents as well as vaporization of water from $BaCl_2 \cdot 2H_2O$ crystallohydrate were measured. The obtained data are in agreement within uncertainty with literature data [25].

RESULTS AND DISCUSSION

Correct identification of the registered changes on thermogravimetric curves to destruction process of the crystallohydrates of CoPc requires information about thermal stability of CoPc under study. Their thermal oxidative destruction starts at temperatures above 483 K [24]. These temperatures are greatly higher than those for removal of crystallization water from the samples studied. It is interesting to note that percentage of the absorbed water in the crystal CoPc samples increases in the following order: $Co(4\text{-}Br)_4(5\text{-}SO_3H)_4Pc \leq Co(4\text{-}C_6H_4SO_3H)_4Pc \leq Co(4\text{-}Cl)_4(5\text{-}SO_3H)_4Pc \leq Co(4\text{-}COOH)_4(5\text{-}SO_3H)_4Pc \leq Co(4\text{-}COO^-)_4(5\text{-}SO_3^-)_4(Ba^{2+})_4Pc$ and is from 8 to 18% (mass %). As example, the TG, DTG, DTA curves of the crystallohydrates of $Co(4\text{-}C_6H_4SO_3H)_4Pc$ and $Co(4\text{-}COOH)_4(5\text{-}SO_3H)_4Pc$ in equilibrium with their saturated solutions are presented in Fig.1 and Fig.2, respectively.

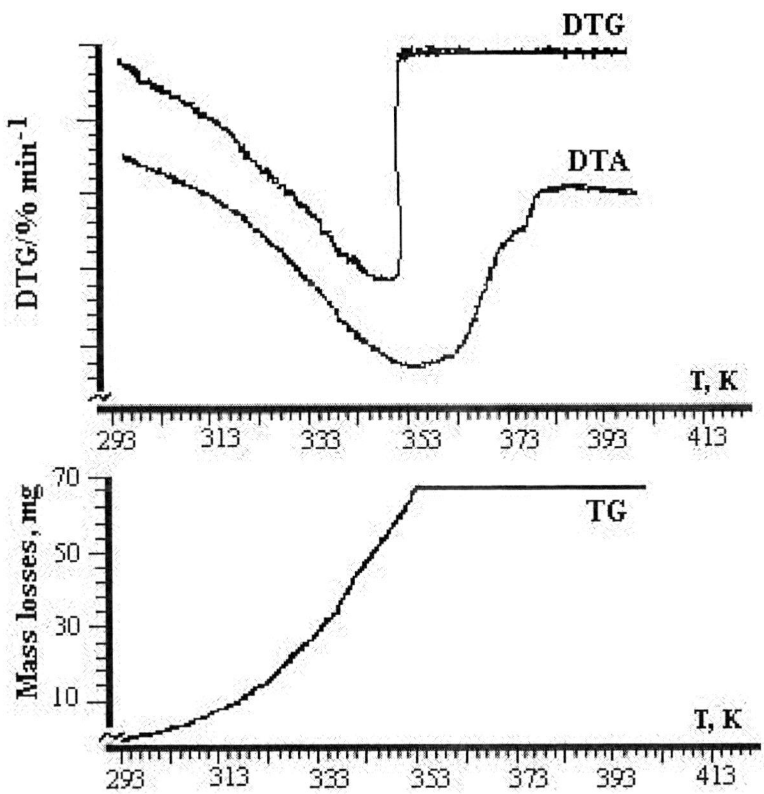

Figure 1. Thermogram of the crystallohydrate of $Co(4\text{-}C_6H_4SO_3H)_4Pc \cdot n\ H_2O$.

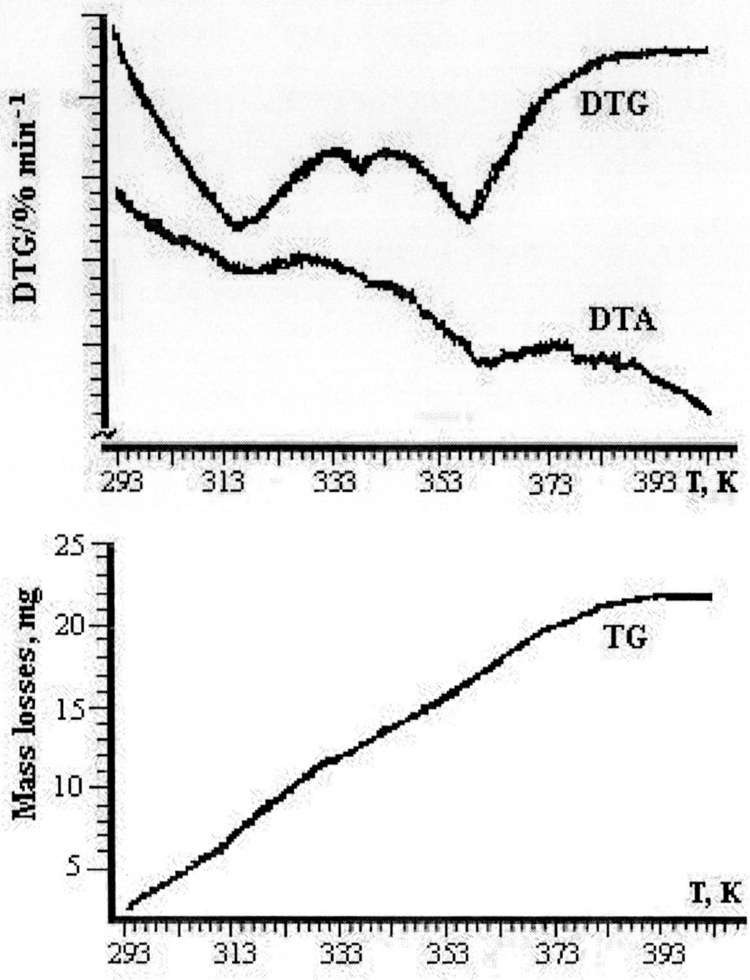

Figure 2. Thermogram of the crystallohydrate of Co(4-COOH)$_4$(5-SO$_3$H)$_4$Pc · n H_2O.

The process of removal of water from the crystallohydrates has several steps. For all systems studied except Co(4-CI)$_4$(5-SO$_3$H)$_4$Pc shapes of the DTA and DTG curves are the same. Hence, the endothermic process of destruction of the crystallohydrates is connected with evaporation of the solvent only and is not accompanied by conformational changes or any phase changes. Reasons of some differences in the DTA and DTG curves for the crystallohydrates of Co(4-CI)$_4$(5-SO$_3$H)$_4$Pc will be discussed below.

The $\Delta_{evp}H$ values allow to check the dehydration process. The enthalpy of evaporation values contain two contributions: the energy of intermolecular interaction in liquid or liquid-crystal phase and work of expansion at the phase transition of liquid to the gaseous state [26]. As the latter contribution is small (1-3 kJ·mol^{-1}), in the first approximation, the $\Delta_{evp}H$ values can be considered as characteristics of energy of water-water interaction (evaporation of water molecules unbounded with crystallosolvate) or energy of CoPc-water interactions (evaporation of water from crystallosolvate) [23,26]. According to eq.(3), the temperature dependence of the mass loss rate increase with increasing the $\Delta_{evp}H$ values from one step of the destruction process to another at non-isothermal conditions (Fig.1,2). On the thermoanalytical curves such changes are reflected as peaks or sharp changes of slope of the mass loss curve. The treatment of the thermogravimetric data according to eq.(3) allows to divide the dehydration process into individual steps. The molar ratios CoPc:H_2O at the boarder of the steps are stoichiometric composition of the crystallohydrates.

Analysis of the thermogravimetric data for Co(4-$C_6H_4SO_3H)_4$Pc – water system show that the Co(4-$C_6H_4SO_3H)_4$Pc · 62 H_2O crystallohydrate is formed. In the first step the $\Delta_{evp}H$ values are very close to that for pure water. This may be explained by removal of the solvent molecules unbound to the crystallohydrate. At higher temperatures up to 327 K thirty energetically equivalent molecules of water ($\Delta_{evp}H$ = 30.93 kJ·mol^{-1}) are removed from the hydrate. Relatively low value of $\Delta_{evp}H$ indicates that the hydrate is formed preferentially due to Van-der-Waals interactions of water with Co(4-$C_6H_4SO_3H)_4$Pc and/or weak hydrogen bonding. At the 341-350 K range twenty four water molecules are removed. Low energy of water - Co(4-$C_6H_4SO_3H)_4$Pc interaction in this step may be explained by steric locking of water molecules. It is well known that phenyl groups are efficient breakers of dense crystal packing of metallophthalocyanines (α- and β-polymorphs). On the other hand, SO_3H-groups of CoPc can form intermolecular hydrogen bonds. It is likely the mentioned above factors lead to formation of cavities between metallophthalocyanine layers in their crystals. Analogous conclusion was made in work [1]. It is interesting to note that at higher temperature of 350 K, as individual step, residuary eight water molecules are removed. The $\Delta_{evp}H$ value on this step was not calculated in consequence of high rate of the process. This step is characterized by a high evaporation rate and absence of its dependence on temperature. According to the data [27], this may be connected with diffusion braking of these eight water molecules at their removal from the Co(4-$C_6H_4SO_3H)_4$Pc ·8 H_2O crystallohydrate and, hence, with their steric blocking.

Analysis of the thermogravimetric data for the crystallohydrates of Co(4-Br)$_4$(5-$SO_3H)_4$Pc and Co(4-CI)$_4$(5-$SO_3H)_4$Pc (Table) shows that up to a temperature of 352 K and 319 K, respectively, the interactions between the macrocycles and water are universal. At higher temperatures the crystallohydrates of clear stoichiometry are destructed. As in the case of the Co(4-$C_6H_4SO_3H)_4$Pc·32 H_2O and Co(4-$C_6H_4SO_3H)_4$Pc·8 H_2O crystallohydrates, the $\Delta_{evp}H$ values were not calculated at the 352-360, 360-373, 373-393 K for the hydrates of Co(4-Br)$_4$(5-$SO_3H)_4$Pc and 319-341 and 341-395 K for the hydrates of Co(4-CI)$_4$(5-$SO_3H)_4$Pc in consequence of low evaporation rates of these processes. It may be concluded that in these steps steric blocked in the crystallohydrates water molecules are removed in these steps. Clear repeatability of the thermoanalytical curves for halogen derivatives of metallocomplexes of sulfo-phthalocyanines testifies about a high structuredness of the samples. Comparison of compositions of the crystallohydrates and temperature ranges (Table) indicates on a significant effect of CI^- (Br^-) substitutes on the density of packing of the metallophthalocyanine molecules in their crystals. It is known that introduction of any substitute to phthalocyanine molecule results in redistribution of electron density and different intermolecular interaction in crystal state. It should be noted some differences in the DTA and DTG curves. This fact as well as an extremum at 331 K on the DTA curve for Co(4-CI)$_4$(5-$SO_3H)_4$Pc (Fig.3) testifies about conformational transitions of the metallophthalocyanine.

Table. Physical chemical characteristics of the crystallohydrates of cobalt(II)phthalocyanines

Cobalt(II)-phthalocyanine	Composition nCoPc : $n H_2O$	Temperature range of decomposition, K	$\Delta_{evp}H$, kJ·mol^{-1}	$\Delta_{evp}H°$, kJ·mol^{-1}
	> 1: 62	294 - 327	44.61 ± 1.40	44.19 - 42.69
$Co(4\text{-}C_6H_4SO_3H)_4Pc$	1:62 - 1:32	327 - 341	30.93 ± 4.13	42.69 - 42.05
	1:32 - 1:8	341 - 350	-	42.05 - 41.64
	1:8 - 1:0	350 - 353	-	41.64 - 40.50
	> 1: 38	294 - 348	45.85 ± 1.23	44.19 - 41.55
$Co(4\text{-}Br)_4(5\text{-}SO_3H)_4Pc$	1:38 - 1:16	352 - 360	-	41.55 - 41.19
	1:16 - 1:4	360 - 373	-	41.19 - 40.59
	1:4 - 1:0	373 - 393	-	40.59 - 39.68
	> 1: 62	292 - 306	43.09 ± 0.74	44.28 - 43.64
$Co(4\text{-}Cl)_4(5\text{-}SO_3H)_4Pc$	1:62 - 1:49	306 - 319	49.30 ± 8.75	43.64 - 43.05
	1:49 - 1:18	319 - 341	-	43.05 - 42.05
	1:18 - 1:0	341 - 395	-	42.05 - 39.59
	> 1: 24	293 - 323	-	44.23 - 43.32
$Co(4\text{-}COOH)_4(5\text{-}SO_3H)_4Pc$	1:24 - 1:16	323 - 327	45.56 ± 1.12	43.32 - 42.69
	1:16 - 1:8	327 - 349	-	42.69 - 41.69
	1:8 - 1:0	349 - 382	65.21 ± 1.11	41.69 - 40.18
	> 1: 61	293 - 317	-	44.23 - 43.14
$Co(4\text{-}COO^-)_4(5\text{-}SO_3^-)_4(Ba^{2+})_4Pc$	1:62 - 1:28	317 - 339	40.57 ± 0.69	43.14 - 42.14
	1:28 - 1:14	339 - 349	-	42.14 - 41.69
	1:14 - 1:8	349 - 362	-	41.69 - 41.09
	1:8 - 1:4	362 - 368	-	41.09 - 40.82
	1:4 - 1:0	368 - 381	55.74 ± 10.69	40.82 - 40.23

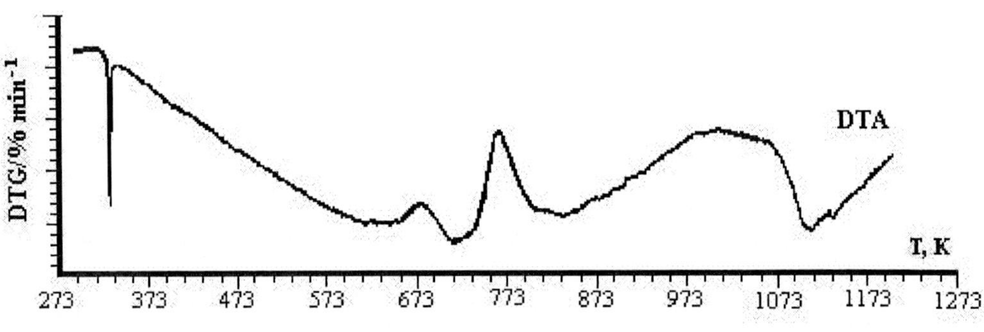

Figure 3. DTA curve of thermal oxidative destruction of $Co(4\text{-}Cl)_4(5\text{-}SO_3H)_4Pc$.

The thermoanalytical curves for $Co(4\text{-}COOH)_4(5\text{-}SO_3H)_4Pc$ (Fig.2) differ from others. The process of removal of water has several steps. The latter (349-382 K) is accompanied by an individual peak on the DTG curve. The $\Delta_{evp}H$ value for the $Co(4\text{-}COOH)_4(5\text{-}SO_3H)_4Pc \cdot n H_2O$ crystallohydrates at molar ratios higher than 1:24 show that water molecules bonded with the metallophthalocyanine due to universal interactions are removed. At the 313-327 K range eight water molecules ($\Delta_{evp}H$ = 45.56 kJ·mol^{-1}) are removed. This enthalpy of evaporation is slightly higher than that for pure water (Table). It is likely these molecules interact with the macrocycle due to electrostatic forces and/or hydrogen bonding. Using the equation proposed in work [28], the enthalpy change of specific solvation ($\Delta_{solv}H$) of Co(4-COOH)$_4$(5-SO$_3$H)$_4$Pc by water molecules can be calculated:

$$\Delta_{сол}H = - n \cdot (\Delta_{исп}H - \Delta_{исп}H^{\circ}), \tag{4}$$

where n is the number of water molecules interacting specifically with CoPc, $\Delta_{evp}H$ is the change of enthalpy of evaporation of water from crystallohydrate, $\Delta_{evp}H^{\circ}$ is the change of enthalpy of evaporation of pure solvent calculated as $\Delta_{evp}H^{\circ} = - 4.55\cdot10^{-2} \cdot T + 45.14$ (R^2 = 0.998) [29].

The heat of solvation of Co(4-COOH)$_4$(5-SO$_3$H)$_4$Pc by water molecules calculated according to eq.(4) is 20.4 kJ·mol^{-1}. As can be seen from Fig.2, the step at the 327-349 K range on the TG curve reflects destruction of Co(4-COOH)$_4$(5-SO$_3$H)$_4$Pc · 16 H_2O hydrate up to molar ratio 1:8. The TG and DTG curves show that at the 327-349 K range sterically blocked water molecules are removed. At higher temperatures (368-381 K) (the DTG curve) the destruction of Co(4-COOH)$_4$(5-SO$_3$H)$_4$Pc · 8 H_2O molecular complex occurs. The energy of interaction between the metallophthalocyanine and water in this specific complex is 65.21 kJ·mol^{-1}. The enthalpy change of the specific hydration of Co(4-COOH)$_4$(5-SO$_3$H)$_4$Pc calculated according to eq.(4) is 194.24 kJ·mol^{-1}. The higher ability of Co(4-COOH)$_4$(5-SO$_3$H)$_4$Pc to complex formation with water in comparison with another CoPc under study may be explained by presence of carboxyl groups in the 4 position of the macrocycle. Analogous conclusion was made on the basis of the studies of solvation interactions of water soluble metallophthalocyanines with proton donor solvents [13-16].

The thermoanalytical curves for Co(4-COO$^-$)$_4$(5-SO$_3^-$)$_4$(Ba^{2+})$_4$Pc are similar to those for Co(4-C$_6$H$_4$SO$_3$H)$_4$Pc (Fig.1).Co(4-COO$^-$)$_4$(5-SO$_3^-$)$_4$(Ba^{2+})$_4$Pc forms with water 1:62 crystallohydrate. The first step is characterized by the $\Delta_{evp}H$ value that is close to the enthalpy change at evaporation of pure water, i.e. destruction of the structures formed due to universal interactions occurs. Then at the 317-339, 339-349, 349-362, 362-368 K ranges, the sterically blocked water molecules from the 1:62, 1:28, 1:14, 1:8 crystallohydrates, respectively, are removed. At 368-381 K range the Co(4-COO$^-$)$_4$(5-SO$_3^-$)$_4$(Ba^{2+})$_4$Pc · 4 H_2O crystallohydrate formed due to specific molecular interactions between peripheral *O*-containing substitutes and water molecules is destructed.

The higher ability of Co(4-COOH)$_4$(5-SO$_3$H)$_4$Pc to complex formation with water as compared to its baric salt results from both electron and steric effects of COOH- and COO$^-$-groups. It is well known that energy of hydrogen bond consists of energy of electrostatic attraction and energy of exchange repulsion. Besides, polarization of interacting particles and dispersion interactions contribute significantly to complex stability [30]. According to Hammet constants [31], COOH-group is a stronger electron acceptor (σ = +0.37(*meta*), +0.45 (*para*)) in comparison with COO-group (σ = +0.1 (*meta*) +0.00 (*para*)). Therefore, partial negative charge will located on the oxygen atoms of COOH-groups and their affinity to proton donor molecules will be higher.

Conclusions

It can be concluded on the basis of the obtained data that the studied cobalt(II)phthalocyanines in crystal state form ordered structures in which water molecules interact due to universal forces with sulfo-substituted CoPc or due to hydrogen bonding with carboxy-substituted CoPc. The obtained results may be useful for interpretation of the data on photo- and electroconductivity of phthalocyanine samples.

Acknowledgements

This work was supported by Russian Academy of Sciences (the program of fundamental studies of Presidium of RAS "Directional synthesis of compounds with prescribed properties and creation of functional materials on their basis" № 8P) and Russian science Support Foundation.

References

[1] Kerp, H.R., Van Faassen, E.E.: *Solar Energy Materials & Solar Cells*. 63, 15 (2000)

[2] Oda, K., Ogura, Sh.-I., Okura, I.: *J. Photochem. Photobiol. B: Biology*. 59, 20 (2000)

[3] Stymne, B., Sauvaget, F.X., Wettermark, G.: *Spectrocopy Acta.* 35A, 1195 (1979)

[4] Storuse, C.E.: Proc. *Nat. Acad. Sci. USA*. 2, 325 (1973)

[5] Askarov, K.A., Berezin, B.D., Evstigneeva, R.P.: Porfiriny: struktura, svoystva, sintezh [Porphyrins: Structure, Properties, Synthesis]. *Nauka, Moscow* (1985) 333 pp (in Russian)

[6] Sidorov, A.N., Kotlyar, I.P.: Optika I spektroskopia. 11, 175 (1961) (in Russian); *Russ. J. Opt. Spectrosc.* 11, 165 (1961) (Engl. Transl.)

[7] Solov'ev, K.N., Gladkov, L.L., Starukhin, A.S., Skirman, S.F.: Spektroskopia porfirinov: kolebatelinye sostoyaniya [Spectroscopy of Porphyrins: Vibration States]. *Science and Technics, Minsk* (1985) 414 pp (in Russian)

[8] Starke, M., Wagner, H.: *J. Chem*. 9, 193 (1963)

[9] Moser, F.H., Thomas, A.L.: Phthalocyanine Compounds. Reinhold Publ. Corp., New York, London (1963) 365 pp

[10] Abel, E.W., Pratt, J.M., Whelan. R.: *J. C. S. Dalton* (1976) 509

[11] Sheidt, R., Geiger, D.K., Lee, Y.J.: *J. Am. Chem. Soc. 31*, 2660 (1992)

[12] Lebedeva, N.Sh., Petrova, O.V., Vyugin, A.I., Maizlish, V.E., Shaposhnikov, G.P.: Russ. Zh. Fiz. Khim. 77, 1594 (2003); Russ. *J. Phys. Chem.* 77, 1577 (2003) (Engl. Transl.)

[13] Galaup, J.-P., Fraigne, S., Landraud, N.: *J. Luminescence 94-95,* 719 (2001)

[14] Reinot, T., Kim, W.-H., Hayes, J.M., Small, G.J.: *J. Chem. Phys*. 107, 457 (1997)

[15] Reinot, T., Zazubovitch, V., Hayes, J.M., Small, G.J.: *J. Phys. Chem*. 105, 5083 (2001)

[16] Hayes, J.M., Reinot, T., Small, G.J.: *J. Chem. Phys. Let.* 312, 362 (1999)

[17] Lucenko, O.G., Kulinich, V.P., Shaposhnikov, G.P., L'ubimcev, A.V.: Russ. Zh. Obsh. Khim. 74, 491 (2004); *Russ. J. General Chem*. 74, 485 (2004) (Engl. Transl.)

[18] Lucenko, O.G., Kulinich, V.P., Shaposhnikov, G.P.: Russ. Zh. Obsh. Khim. 73, 1548 (2003); Russ. *J. General Chem.* 73, 1538 (2003) (Engl. Transl.)
[19] Lucenko, O.G., Kulinich, V.P., Shaposhnikov, G.P.: Russ. Zh. Obsh. Khim. 74, 319 (2004); *Russ. J. General Chem. 74*, 308 (2004) (Engl. Transl.)
[20] Mink, L.M., Polam, J.R., Christensen, K.A.: *J. Am. Chem. Soc.* 117, 9329 (1995)
[21] Lebedeva, N.Sh., Yakubov, S.P., Kinchin, A.N., Vyugin, A.I.: Russ. Zh. Fiz. Khim. 79, 955 (2005); Russ. *J. Phys. Chem.* 79, 949 (2005) (Engl. Transl.)
[22] Lebedeva, N.Sh., Trofimova, E.V., Pavlycheva, N.A., Vyugin, A.I.: Russ. Zh. Org. Khim. 38, 1246 (2002); *Russ. J. Org. Chem. 38, 1289* (2002) (Engl. Transl.)
[23] Antina, E.V., Barannikov, V.P., Vyugin, A.I.: *Russ. Zh. Neorg. Khim.* 35, 400 (1990); Russ. J. Inorg. Chem. 35, 395 (1990) (Engl. Transl.)
[24] Novitskii, P.V., Zhograf, I.A.: Otsenka Pogreshnostei Rezhultatov Izhmerenii [Error Eestimates of Results Measurements]. *Energoatomizdat, Leningrad* (1985) (in Russian)
[25] Wendlandt, W.W.: Thermal Methods of Analysis. Wiley, New York (1974); *Mir, Moscow* (1978), 2rd edn., p.526 (in Russian)
[26] Khimicheskaya enciklopedia [Chemical Encyclopedia]. *Sovetskaiya enciklopedia,* Moscow (1988), 2, p.540 (in Russian)
[27] Panchenkov, G.M., Lebedev, V.P.: Khimicheskaya kinetika I kataliz [Chemical Kinetics and Catalysis]. *Khimiya, Moscow* (1985) 292 pp (in Russian)
[28] Krestov, G.A.: Termodinamika ionnykh processov v rastvorakh [Thermodynamics of ion process in solution]. *Khimiya, Lenindrad* (1973) 304 pp (in Russian)
[29] Eizenberg, D., Kautsman, W.: *Struktura I svoystva vody* [Structure and properties of water]. Leningrad (1975)
[30] Khimicheskaya enciklopedia [Chemical Encyclopedia]. *Sovetskaya enciklopedia, Moscow* (1988), 1, p.404 (in Russian)
[31] Johnson, C.D.: The Hammett Equation. Cambridge University Press. 1973.

Inscription to drawings to article
"Crystallohydrates of carboxy- and sulfo-substituted cobalt(II)phthalocyanines"
Nataliya Sh. Lebedeva, Nataliya A. Pavlycheva, Anatoliy I. Vyugin,
Elena V. Parfenyuk, Vladimir P. Kulinich, Gennagiy P. Shaposhnikov

In: Progress in Monomers, Oligomers, Polymers, Composites … ISBN 978-1-60456-878-3
Editors: R. A. Pethrick, G.E. Zaikov et al.

Chapter 8

MODIFICATION OF CELLULOSE BY THE BIOCIDAL POLYELECTROLITE

S.Yu. Khashirova[1,*], A.M. Esmurziev[2,†] and Y.A. Malkanduev[1]

[1]Kabardino-Balkarian University, Faculty of Chemistry. Nalchik, Russian Federation
[2]Sør-Trøndelag University College, Faculty of Technology, Trondheim, Norway

ABSTRACT

The opportunity of the modification of the cellulose using biocidal polyelectrolite on the basis of metacrylate guanidine was investigated. Optimum contitions of modification, such as concentration and duration of modifying are found, dependence of the bactericidal properties on the maintenance of guanidine groups is investigated.

Keywords: guanidine containing polymers; modification of cellulose; bactericidal activity; implanted polymers.

INTRODUCTION

Because of the variety of properties (such as biological activity, formation of chelates, solubility in water, ability to a biologigal degradation and floculation) of guanidine containing polymers, they are widely used of late as biocides and floculants in the industry, agriculture, medicine and pharmacy [1-6].

The conciderable scientific and practical interest represents the research of an opportunity of modification of cellulose by water-soluble polymers, in particular, by the guanidine containing polymers. Such compounds can be used for creation of the cellulose membranes that are processing simultaneously bactericidal properties.

---

* Correspondence to: [1]Kabardino-Balkarian University, Faculty of Chemistry. Russia, Nalchik, Chernyshevskogo street, 173, e-mail.: sveta_daova@ mail.ru

† Correspondence to: [2]Sør-Trøndelag University College, Faculty of Technology E. C. Dahls gt. 2, 7004 Trondheim, aslan.esmurziev@hist.no

Since such guanidinecontaining macromolecules compound contains the ionogenic groups it can be expected that the modified cellulose can be used as materiales having the sorption activity , materials of the medical functions with the prolonged action. Especially a valuable quality of compounds on the basis of cellulose and guanidine containing polymers is that they do not accumulate in an environment since are capable to biobegradation.

The aim of the given work was the research of an opportunity of a modification of the cellulose with metacrylate guanidine. For the modifying have been taken cellulose cotton and the unsterile cotton wool used for packing of the laboratory glassware.

CHEMISTRY AND EXPERIMENTAL PART

Reactions of the interactions of cellulose with various oxidizing agents (such as NaClO, $KMnO_4$, NO_2, $HclO_4$, H_2O_2) lead to formation of the derivatives containing of th carbonyl and carboxyl groups. Thus the greatest interest represents an action of the compounds that provid a primary oxidation of specified hydroxyl groups of an elementary part. To such oxidizers applies $NaIO_4$, at interaction of cellulose with wich leads to an oxidation of secondary hydroxyl groups of an elementary part to aldehyde, accompanied by the break of the link between the second and third atoms of the carbon [7] (scheme 1).

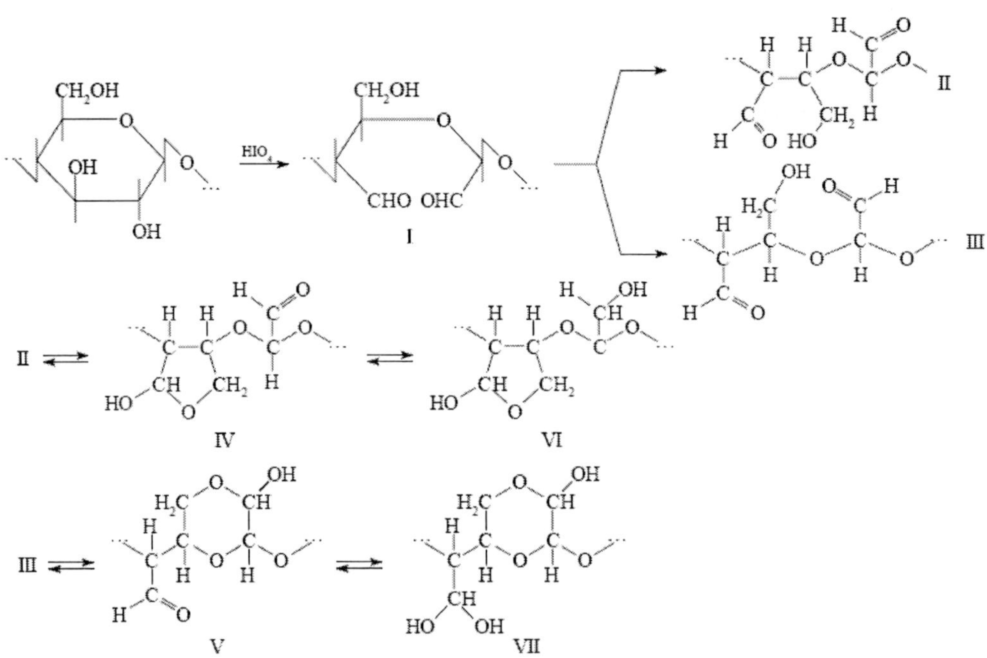

Scheme 1.

The formed dialdehyde part I can be in configurations II and III, where one of the aldehyde gropus reacts with the nonoxidized primary hydroxyl group of this part with formation of the thermodynamically stable penta- or hexamerous cyclic acetal IV and V, while the second aldehyde group forms structures VI and VII as a result of hydratation [8]. The activation of cellulose and cotton carried out by oxidation with HIO_4 and H_2O_2.

Modifying of samples spent to three stages:

1) Oxidation in a 0.1M solution of HIO_4 (H_2O_2) during 1 hour at temperature 60^0C;
2) Treathment of the cellulose by a solution of a monomer containing 1% of $(NH_4)_2S_2O_8$ at an ambient temperature;
3) Radical polymerization in ampoules at temperature 60^0C

The reaction of dialdehyde cellulose with methacrylate guanidine is shown on the scheme 2.

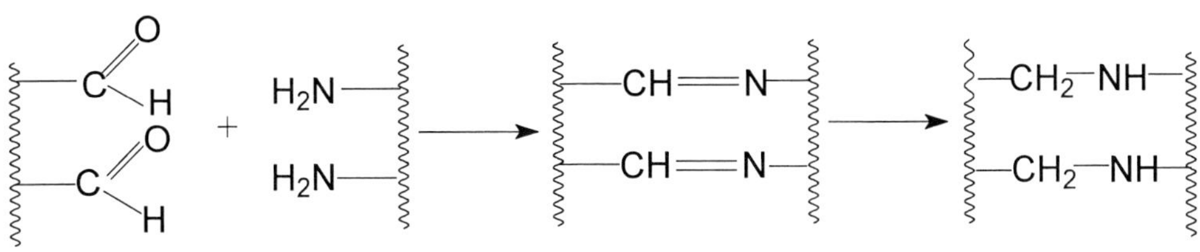

Scheme 2.

The time of modifying of cellulose was changed from 1 up to 20 h and the concentration of a monomeric solution within the limits of 10 up to 50% (weights). After modifying, the cellulose was washed in the distilled water during 4 hours for the removal of unreacted monomeric salt and also formed homopolymer. The quatity of metacrylate guanidine implanted to the cellulose defined according to the element analyse – under the maintenance of nitrogen.

The investigation of a bactericidal properties of the modified samples spent with use of the bacterial culture collections of the microorganisms of the Bacteriological laboratory (Nalchik, Russia), in particular, *Esherishia coli*, *St. Aureus* and an pathogenic fungoid microflora of human *Candida alb*. The suspensions of the test-cultures of the *golden staphylococcus* prepared in a sterile physiological solution. The received suspension (10 ml) was filtered, dryed and incubated on a nutrient medium Endo (Fluka) at temperature 28^0C. After 24 hours of cultivation the results were calculated. The unmodified cellulose used as the control.

RESULTS

It was established, that the quantity of the implanted metacrylate guanidine is more at the samples processed by HIO_4 in comparison to H_2O_2. The samples received with use of cellulose were modifyied better, than cotton wool. At increase of an initial concentration of the monomer, the quantity of the implanted metacrylate guanidine varied from 4.5% up to 25% from celulose wight (table 1).

Table 1. The implantanted polymerization of the metacrylate guanidine and cellulose

sample	Concentration of the monomer in mother solution, %	Concentration of the implanted polymer, % (product wight)		
		Time, hours		
		5	10	20
Cellulose $+H_2O_2$	10 %	3,2	4,0	6,1
	25%	3,8	4,6	9,4
	50%	6,2	9,1	15,0
Cotton wool$+H_2O_2$	10 %	2,0	3,4	4,5
	25%	2,6	4,1	7,3
	50%	3,5	5,6	9,1
Celulose $+HJO_4$	10 %	5,8	7,4	12
	25%	6,9	12,6	17
	50%	13,5	18,2	25
Cotton wool$+HJO_4$	10 %	4,3	5,2	7,7
	25%	5,4	8,2	12,5
	50%	9,6	12,6	16,8

The effectiviness of a bactericidal action of cellulose modified with metacrylate guanidine determined quantitatively by the finding of the colony-forming units. The finding results shows that the optimal conditions for the receiving of the implanted copolymer of cellulose and metacrylate guanidine are the preactivation of cellulose with HIO_4 and processing with the high-concentrated monomer.

The amount of the metacrylate guanidine wich immobilizied on the cellulose surface also depends on duration of modifying. The increase in time of modifying from 1 up to 5 hours essentially does not influence on the change of the maintenance of a nitrogen in the samples of cellulose modified with the both 10% and 25% monomeric solutions,thus not on a degree of modifying. The essential increase of a nitrogen quantity in samples is observed at increase in duration of modyfying up to 20 h. of the time of the modification.

The concentration of an initial modifying solution considerably influences on the bactericidal activity of the modified cellulose. The bactericidal activity of cellulose modified by 25% solution of the metacrylate guanidine is ill-defined. At the increase in its concentration up to 50%, the bactericidal action increases.

In conclusion, the technique of modifying of the cellulose by metacrylate guanidine is developed. It is shown, that the optimum parameters of the modifying are the 50% concentration of a metacrylate guanidine and the 20 h duration of the modifying. The bactericidal properties of modified cellulose are investigated and it is established, that the efficiency of a bactericidal actions depends on the maintenance of the guanidine containing groups.

REFERENCES

[1] Sivov N.A., Martynenko A.I., Kabanova E.Yu., Popova N.I., Khashirova S.Yu., esmurziev A.M. // *Petrochemistry*. 2004, 44, №1. p. 47. Written in Russian.

[2] Zaikov G.E., Malkanduev Yu.A., Khashirova S.Yu., Esmurziev A.M., Martynenko A.I., Sivova L.I., Sivov N.A. // *Russian Polymer News*, 2003, V. 8, No. 4, p. 1.

[3] Zaikov G.E., Malkanduev Yu.A., Khashirova S.Yu., Esmurziev A.M., Martynenko A.I., Sivova L.I., Sivov N.A. // *J. Environ. Protect. And Eco*logy, 2003, V. 4, No. 4, p.863.

[4] Zaikov G.E., Malkanduev Yu.A., Khashirova S.Yu., Esmurziev A.M., Martynenko A.I., Sivova L.I., Sivov N.A. // J.*Appl.Pol.Sci.*, 2004, V.91, pp. 439.

[5] Khashirova S.Yu., Sivov N.I., Popova N.I., Kabanova E.Yu., Martynenko A.I., Topchiev D.A. // *Izvestiya vuzov, Sev-Kavk. Region, ser. Estestvennye nauki,* 2002,4, p. 45.

[6] Zaikov G.E., Malkanduev Yu.A., Khashirova S.Yu., Esmurziev A.M., Martynenko A.I., Sivova L.I., Sivov N.A., in *Biochemistry and Chemistry: Research and Developments*, Nova Science Publishers, New York, 2003, p.39.

[7] Gal'braikh L. S. // Cellulose and its derivatives. The educational magazine of Soros. 1996, 11.

[8] Rogovin Z.A. // Chemistry of cellulose. 1972; Moscow: "Chemistry"

In: Progress in Monomers, Oligomers, Polymers, Composites … ISBN 978-1-60456-878-3
Editors: R. A. Pethrick, G.E. Zaikov et al.

Chapter 9

Synthesis, Properties and Application Poliethersulphons and Polietherketons

R.A. Haraeva**, *O.R. Ashibokova*†, *G.B. Shustov*, *A.M. Haraev* and *A.A. Chaika

Kabardino - Balkarian the State University it. Berbekova H.M.
Nalchik city, Chemical faculty, Russian Federation

Abstract

The analysis of literary data for last 17 years on synthesis, properties and application poliethersulphons (PES) and polietherketons (PEK) is lead. Polyarilatsulphons synthesize in two perspective ways. On the basis of poliethersulphons heat-resistant shock-resistant materials have been received. It is synthesized polyarylenetherketone with lateral carboxylate group in each part. Introduction of this group is.

Keywords: poliethersulphone, polietherketone, polietheretherketone, bisthenole, polyaryletherketone, polyether- α-diketone, sopolyethersulfonketone, heat-resistant shock-resistant materials, polyarilatsulthone.

The chemistry of high-molecular connections is one of the major areas of a chemical science. Growth of volumes of manufacture such heat-proof polymers, as aromatic poliethersulphons (PES) and polietherketons (PEK) [1-11] every year proceeds.

The purpose of our researches was the analysis of the literature for last 17 years on synthesis, properties and application poliethersulphons and polietherketons.

The way of reception [12] aromatic PES on the basis of 4,4 '-dichlor (fluorine) diphenylsulfone and salts of alkaline metals corresponding bisphenols which receive interaction of these bisphenols with hydroxide metals with the subsequent dehydration the

* Correspondence to: Kabardino - Balkarian the state university it. Berbekova H.M., 360000, KBR, Nalchik city, Tchernyshev street 173, Chemical faculty, E-mail: ruzanna-haraeva@mail.ru

† Email to: ashib-ol@mail.ru

formed products distillation waters with azeotropisers-benzene, heptane, xylole, for toluene, benzene chloride, dichlorbenzene is known. The subsequent reaction heteropolycondensation spend in the environment of such solvents, as dimethylsulfoxide (DMSO) and sulpholane. Receive polymers of structures:

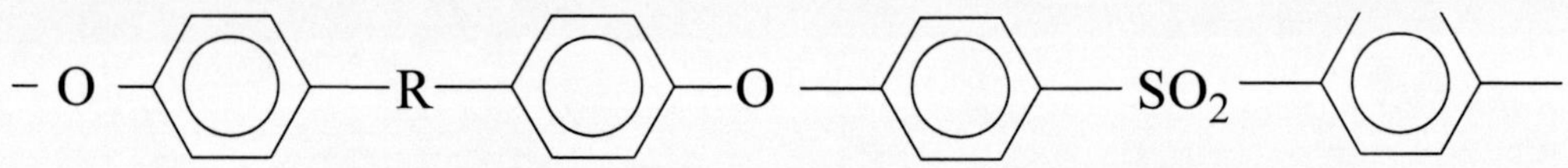

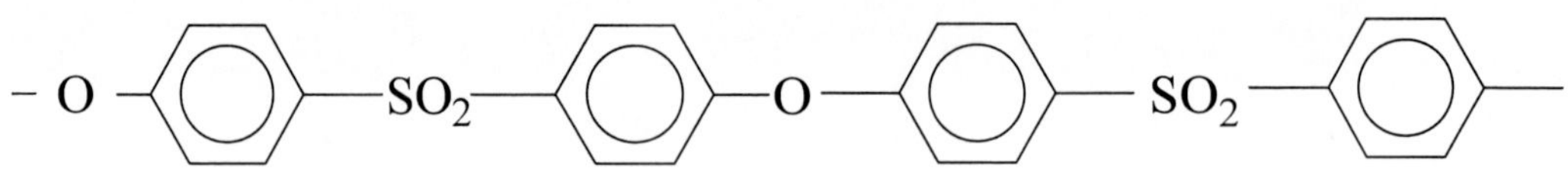

Where R=C $(CH3)_2$, etc.

The way of reception aromatic PES and sopolyethersulphonketons [13] is described. The essence of the invention consists in interaction bisphenole with humidity of 0,5-10,0 % with dihaloidarilensulphone and-or dihaloidarilenketone with humidity of 0,5-10,0 % in methyl sulphone environment with humidity of 1,0-20,0 % at presence crystallohydrate alkaline metal or bicarbonate of alkaline metal with humidity of 0,5-20,0 %.

Polyarilatsulphons synthesize in two perspective ways:

1) hightemperetion polycondensation at presence triethylamine;
2) lowtemperetion polycondensation at presence triethylamine, easily deleted in the form of hydrochloride after end of reaction.

On the basis of poliethersulphons heat-resistant shock-resistant materials [8,14,15] have been received.

In works [3,16] research of synthesis and properties aromatic PEK, synthesized by reaction nucleophilic replacements is executed. Synthesis kardov polyarilenftolidketons is carried out at use in polycondensation activated ethylene dihalide, containing kardove group.

For the first time [17] synthesis polyarilenftolidketons on reaction electrophilic substitution replacements has been carried out.

Methods dynamic mechanical, thermomechanical, thermogravimetric the analysis and differential scanning calorimetry study properties polyarylenetherketone with lateral carboxylate groups, the copolymers containing parts of it homopolymer, and their salts with a various degree of neutralization by alkaline metals. It is shown [18], that changing the nature of this metal, and also a degree of neutralization carboxylate groups homopolymer and the maintenance of such groups in a copolymer, it is possible to change properties of copolymers, in particular their ability to formation of ionic pairs, so also their temperature glass.

It is synthesized polyarylenetherketone with lateral carboxylate group in each part [19,20]:

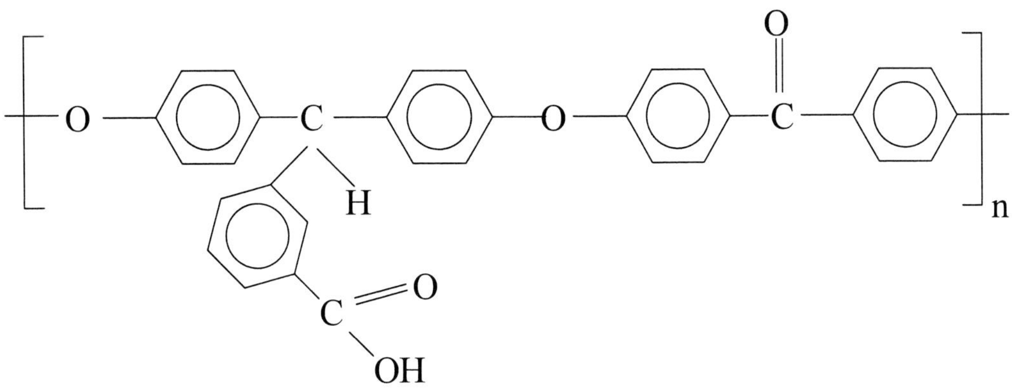

Introduction of this group is rather perspective as it is capable both to ionization, and to formation of hydrogen communications with various groups, that should give to the given polymers additional specific properties.

Within the limits of a single-phase condition in the studied polymers there is an association between cations metals and polyanione, found out in the diluted solutions a method sedimentation [21]. This association, naturally is kept and in a firm condition of polymers.

Four new polyaryletherketone, containing 1,4-naphthalenic parts, are received by [22] reaction nucleophilic replacements 1-chlorine-4 (4- chlorobenzoil) naphthalene with 1,4-hidroxynone, 4,4-isopropylidene diphenol, phenolphtalein and 4 (4-oxophenyl) (2H)-Phthalazin -1-this, accordingly. Polymers are characterized IKS with Fourie`s transformation, H-nuclear magnetic resonanse and broad-corner x-ray diffraction. The received polymers have a heat temperature glass and good heat stability. They are amorphous and soluble in some organic solvents.

Are synthesized new α-diketens on the basis of earlier not described diftoraromatic connections-2,2-bis-[4 (4- fluorophenil glyoxalic) phenyl] hexaftuoro- propane [23] and 4-fluorine-4 (p- fluorophenil glyoxalic) benzophenone [24] All of them are characterized by means of spectroscopy IK, a nuclear magnetic resonance 1 N, a nuclear magnetic resonance 13 S and a nuclear magnetic resonance 19 F, and also methods DSK, TMA and TGA.

Polyether -α- diketens are amorphous, their temperatures glass vary within the limits of 162-235°C, and temperatures 10 %-s' losses of weight in an interval 462-523°C. Values of a dielectric constant are within the limits of 2,90-3,25 at relative humidity of 0 %.

Polyethersulfone including sopolyethersulfonketons are used as constructional and electrical insulating materials in the automobile, aerospace industry, in the electrical engineer, electronics, medicine, and also are applied in the form of fibres, prepregs or at manufacturing membranes [4,7,10,12].

References

[1] D.S.Sharapov, V.V. Shaposhnikova, S.N.Salazkin. // *High-molecular Compounds.*, 2003, V.45, №1, p.113-116.

[2] G.I.Timofeeva, L.V.Dubrovina, T.P.Bragina, G.G.Nikiforova, V.G.Vasilev, L.I.Komarova, M.I.Buzin, A.N.Rjabev, V.V.Shaposhnikova, S.N.Salazkin, V.S.Papkov, L.Z.Rogovina. // *High-molecular compounds., A series*, 2005, V.47, №11, p.2005-2012.

[3] K.I.Donetsky, Synthesis and properties polyarylenetherketons with kardovs and lateral functional groups: *Autoabst. Ch. D., 2000*, 24 p.

[4] L.M.Bolotina, V.P.Chebotarev. // *Plastics, №11,* 2003.

[5] R.A.Haraeva, O.R.Ashibokova, Synthesis and some properties sopolyetherketons, the Collection of proceedings of young scientists, *Nalchik, 2005*, 138 p .

[6] A.M.Haraev, R.C.Bazheva, R.A.Haraeva, Z.L.Beslaneeva, E.V.Pampuha, E.B.Barokova, Reception polyetherketons on the basis of bisphenols a various structure, the Collection of proceedings of young scientists, *Nalchik, 2005*, p.44.

[7] R.A.Haraeva, O.R.Ashibokova, G.B.Shustov, A.M.Haraev, Synthesis poliethersulphons, polietherketons and copolymers on their basis, VII All-Russia scientific and technical conference « New chemical technologies: manufacture and application », *the Collection of clauses, Penza*, 2005, p.111.

[8] A.M.Haraev, R.C.Bazheva, F.K.Kazancheva, R.A.Haraeva, R.T.Bahov, E.R.Sablirova, A.A.Chajka, Aromatic polietherketons and polietheretherketons as perspective heat-resistant constructional materials, the Collection of proceedings of young scientists, *Nalchik, 2005*, p.68.

[9] K.B.Temiraev, Synthesis and properties of polyethers, poliformals and the block-sopoliethers on their basis: Autoabst. Ch. D., KBGU, *Nalchik,* 2000, 51 p.

[10] Mac Kinnon Sean M., Bender Timthj P., Wang Zhi Yuan (Department of Chemistrj, Carleton Universitj, 1125 Colonel Bj Drive, Ottawa, Ontario, Canada K 1 S 5B6). // *J.Poljm. Sci*., A, 2000, 38, №1, p.9-17.

[11] R.M.Kumykov, A.K.Vologirov, A.B.Ittiev, A.L.Rusanov, Simple aromatic polyethers and polietherketons on the basis of dinitroderivative chlorale. New polymeric composite materials: Materials of II All-Russia scientifically-practical conference, *Nalchik, 2005*, p.225-228.

[12] *The Stalemate. RU, 2005738, 5C 08G 75/23*

[13] *The Stalemate. RU, 2043370, 6 C 08 G 75/23*

[14] M.A.Gazaev, synthesis, properties and development of heat-resistant shock-resistant materials on the basis of polyarylatesulphones : *Autoabst. Ch. D*., 2001, 39 p.

[15] I.P.Storozhuk, the Block-copolymers polycondensation type on the basis of polyethers and of some other polymers: *Autoabst. Ch. D.,* 2002. 27 p.

[16] S.N.Salazkin, V.V.Shaposhnikova, K.I.Donetsky, G.V.Gorshkov, P.V.Petrovsky, L.I.Komarova, M.M.Genina, A.S.Tkachenko. // News of the Russian Academy of Science., *series chemistry., 2000*, 6.

[17] L.T.Iljasova, Research of synthesis and properties polyarilenftolidketons: *Autoabst. Ch.* C., UFA, 2003.

[18] G.G.Nikiforova, M.I.Buzin, V.G.Vasilev, M.N.Ilina, A.N.Rjabev, V.V.Shaposhnikova, S.N.Salazkin, L.Z.Rogovina, V.S.Papkov. // *High-molecular compounds, A series,* 2006, V.48, 1, p.66-73.

[19] L.Z.Rogovina, V.G.Vasilev, G.G.Nikiforova, L.V.Dubrovina, T.P.Bragina, L.I.Komarova, G.N.Timofeeva, M.I.Buzin, S.N.Salazkin, V.V.Shaposhnikova,

A.N.Rjabev, V.S.Papkov. // *High-molecular compounds, A series*, 2002, V.44, 8, p.1302.

[20] L.Z.Rogovina, G.G.Nikiforova, M.I.Buzin, V.G.Vasilev, G.I.Timofeeva, L.V.Dubrovina, T.P.Bragina, L.I.Komarova, A.N.Rjabev, V.V.Shaposhnikova, S.N.Salazkin, V.S.Papkov. // *High-molecular compounds, A series*, 2004, V.46, 12, p.2072.

[21] G.I.Timofeeva, L.V.Dubrovina, T.P.Bragina, G.G.Nikiforova, V.G.Vasilev, L.I.Komarova, M.I.Buzin, A.N.Rjabev, V.V.Shaposhnikova, S.N.Salazkin, V.S.Papkov, Л, Z.Rogovina. // *High-molecular compounds, A series,* 2005, V.47, №11, p.2005.

[22] Gao Ye, Jian Xi-gao (Dept.of Chem. Eng., Dalian Univ.of Technol., Dalian 116012, China). Synthesis and the characteristic polyaryletherketons, containing 1,4-naphthalenic parts. Dalian ligong daxue xuebao=*J.Dalian Univ. Technol.2001*.41, 1, p.56-58.

[23] M.L.Keshtov, A.L.Rusanov, S.V.Keshtova, A.N.ShChegolihin, P.V.Petrovsky. // *High-molecular compounds, 2001*, V.43, №12, p.2071-2080.

[24] M.L.Keshtov, A.L.Rusanov, S.V.Keshtova, P.V.Petrovsky, G.B.Sarkisjan. // *High-molecular compounds*, 2001, V.43, №12, p.2059-2070.

In: Progress in Monomers, Oligomers, Polymers, Composites ... ISBN 978-1-60456-878-3
Editors: R. A. Pethrick, G.E. Zaikov et al.

Chapter 10

THE RECEIPT OF POLYBUTYLENETEREPHTHALAT, WHICH IS STABLE FOR HIGH TEMPERATURES

Z.T. Alakaeva, T.A. Borukaev** *and A.K. Mikitaev
The Kabardino-Balkaria State University, KBR, Nalchik, Russian Federation

ABSTRACT

The given results of the investigation make it possible to come to the following conclusions. Firstly, the expediency of receiving PBT with different antioxidants is shown. Secondly, an effective systems of stabilizers are received, the usage of which enable to synthesize PBT which has high thermal characteristics.

Keywords: polybutylenterephthalate, synthesis, stabilization.

Polybutylenterephthalate (PBT) is one of the thermoplastic polymers, which is widely used in motor-car construction, electronics, electrical engineering, medicine, space, life and so on. [1, 2]. That is provided by a complex of valuable properties which PBT has [3]. But in the process of PBT synthesis under 200-270 °C the processes of thermal decomposition take place, which accelerating because of catalyst of interesterification and polycondensation [4]. In this work according to this the polymers- PBT with different stabilizers are synthesized and t5heir properties are investigated. As stabilizers and their mixtures the Irganox – 1010 (I-1010), Irganox B-561 (I-561), trinonilphenylphosphate (TNPHPH) and natrium hypophosphite $Na_2H_2PO_2$.

Synthesis of PBT with different stabilizers were held according to the method [5], then the properties of the received materials were investigated.

One of the affective methods of appraisal of the thermic steadiness of polymer flux is the comparison of the datas of flux fluidity (DFF) samples, received after 5- and 30- minutes ripening in the canal of measuring instrument [6]. For that in this work the samples were

* Correspondence to: The Kabardino-Balkaria State University, 360004, KBR, Nalchik, Chernishevskaya street 173, e-mail:boruk-chemical@mail.ru

ripened for 5, 20 and 30 minutes in the canal of IIPT-M instrument under the temperature 250 °C, then DFF was defined. The results of the measurings are shown in the table. As it is seen in the table with the increase of the time of flux ripening in canal of the instrument the DFF data increases. That prooves that the macromolecules PBT at 250 °C are exposed to decomposition, the molecular weight of polymers decrease, the flux fluidity increases.

Along with DFF, the maintenance of final carboxylic groups is an important characteristic of polyesters which influences their behaviour in the flux. In their turn the carboxylic groups may catalyse the hydrolytic dissociation of ester bonds in polyesters. According to this the maintenance of final –COOH –groups received DFF according to the method [7] is defined in the work. The results are also shown in the table. As it is seen in this table the maintenance of final –COOH –groups and the datas of thermal stability coefficient $K_{5/30}$=DFF_5/DFF_{30} and $K_{5/30}$=DFF_{30}/DFF_5 greately depend on applied in the synthesis of PBT heat setting system.

Table 1. Thermal characteristics of stabilized polybutyleneterephthalat flux

№	The composition of stabilizing system	DFF_5, g/(10 min)*	DFF_{30}, g/(10 min)**	ΔIITP, %***	$K_{30/5}$****	The composition of carboxylic groups, COOH·10^6, g-ekv/g
1	Irganox-1010 (I-1010) (0,5 %)	1,8	4,2	133	2,3	68
2	Irgafos (TNPHPH) (0,5%)	1,5	3,0	100	2,0	61
3	Natrium hypophosphite (0,5%)	8,2	13,9	70	1,7	50
4	0,3% I-1010+0,2% $Na_2H_2PO_2$	2,0	3,4	70	1,7	53
5	0,5 % I-1010 + 0,2% $Na_2H_2PO_2$	3,3	4,2	27	1,3	26
6	0,2%TNPHPH + 0,2% $Na_2H_2PO_2$	10,0	13,9	39	1,4	33
7	0,4% I-1010 + 0,1 % TNPHPH + 0,05 % $Na_2H_2PO_2$	6,3	7,6	21	1,2	19
8	0,3 % I-1010 + 0,1 % TNPHPH + 0,05 % $Na_2H_2PO_2$	15,5	18,0	16	1,2	18
9	0,2 % I-1010 + 0,1 % TNPHPH + 0,05 % $Na_2H_2PO_2$	18,4	23,0	25	1,3	23
10	Polybutyleneterephthalat sort «Celanex 2003» (Germany)	4,9	10,5	114	2,1	64

The given above results marginally proove the datas ΔDFF, % defined by the formula:

$$\Delta DFF = \frac{(DFF_{30} - DFF_5)}{DFF_5} \cdot 100\% .$$

For the systems investigated in the work the ΔDFF comes to 70÷133 %. Industrial PBT sort «Celanex 2003» of the firm «Tikona» (samples of table 10) has a high thermal stability of the flux, its data ΔDFF is 114 %.

The best result were received during the usage of stabilizing systems on the basis of TNPHPH/$Na_2H_2PO_2$ and TNPHPH/И-1010/$Na_2H_2PO_2$. The usage of these systems in the synthesis of PBT provided a low content of final –COOH –groups ($2,6 \cdot 10^{-5}$ и $1,9 \cdot 10^{-5}$ g-ekv/g, correspondingly) and an insignificant change of DFF in thermo-senescence in the canal IIRT-M (ΔDFF is 27 and 21 %, correspondingly) (table).

On the pic.1 the dependence of the change (increase) of DFF flux of PBT, synthesized with different thermally stabilized systems from the content of final –COOH –groups is shown. As it is seen on the picture the dependance has an exponential character – the higher the content of final –COOH –groups the lower the thermal stability of the flux when it is ripened for 30 min. The obtained for PBT dependance (pic.1) is not described in literature, but it is known [8], that –COOH –groups can catalyse the process of hydrolytic decomposition of ester bounds. According to this the reduction of the content of –COOH –groups in polymers is important. The results of our investigations showed that the content of –COOH –groups in PBT is minimal in case of receiving polymers with stabilizing systems, one of the components of which is $Na_2H_2PO_2$ (table). An increased thermal stability of these samples is explained not by inhibition of thermal-oxidative hypophosphite decomposition, but by partial prohibition of polycondensation catalyst, accelerating the thermic and thermal-oxidative decomposition of PBT. These datas are prooved by a synthesis, carried out with increased content of $Na_2H_2PO_2$ – 0,5 mass. %. The reaction doesn't pass, as the given quantity of $Na_2H_2PO_2$ is probably enougth for prohibition of catalyst, which is used in synthesis of PBT. Besides, we suppose that $Na_2H_2PO_2$ inhibits the reaction of cyclization of butanediol and it's transformation into tetrahydrofuran. We examine these estimates at present.

It should be mentioned that in the polymer oxidation reactions such radicals as $R^\bullet$, $RO_2^\bullet$ and ROOH –groups take place. The last one may dissociate with H_2O formation and such radicals as $RO^\bullet$ and $R^\bullet$.

It is known [9] that phosphoprous-containing stabilizers are effective as antioxidants, which prevent the formation of peroxide radicals and hydroperoxides. In their turn that confines water formation which causes additional decomposition of polyester in ester bonds. In this work TNPHPH is used as phosphoprous-containing stabilizer, which may stoichiometricaly restore hydroperoxides into corresponding alcohol according to the scheme:

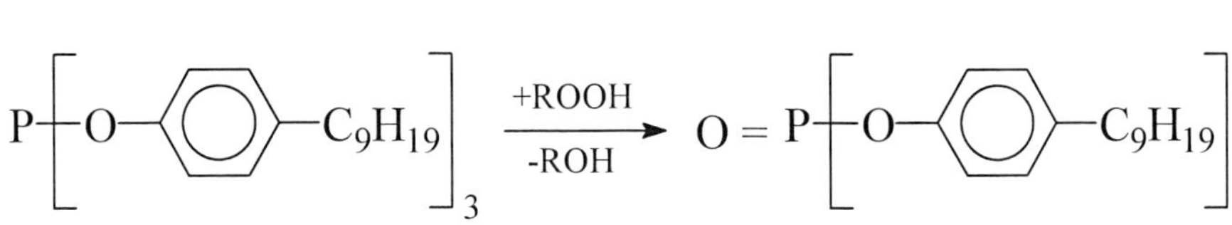

In their turn, the part of the Irganox – 1010 in temperature stabilized mixtures consist in deentrainment of active radicals, which initiating the process of polymer decomposition. The mechanism of its operation may be presented in the following way:

$$InH + {}^{\bullet}OR \rightarrow In^{\bullet} + ROH$$

Or

$$InH + {}^{\bullet}OOR \rightarrow In^{\bullet} + HOOR.$$

Thus, the results (pic.1 and table) make it possible to come to a conclusion that the best stabilizing effect is achieved only when the system of stabilizers on the basis of organophosphorus stabilizer TNPHPH, which neutralize the hydroperoxide groups of spatially hindered Irganox – 1010, which entraps active radicals of different construction, and natrium hypophosphite, which reduces the activity of catalyst in running incidental reactions and inhibitory cyclization of 1,4-butanediol with water isolation.

It should be mentioned that the individual usage of the mentioned stabilizers or their binary mixtures gives a lesser effect (pic.1 and table). That is why we may say that the system consisting of 0,4% Irganox – 1010, 0,1% TNPHPH and 0,5% $Na_2H_2PO_2$ has an synergetic effect and is more effective for practical usage.

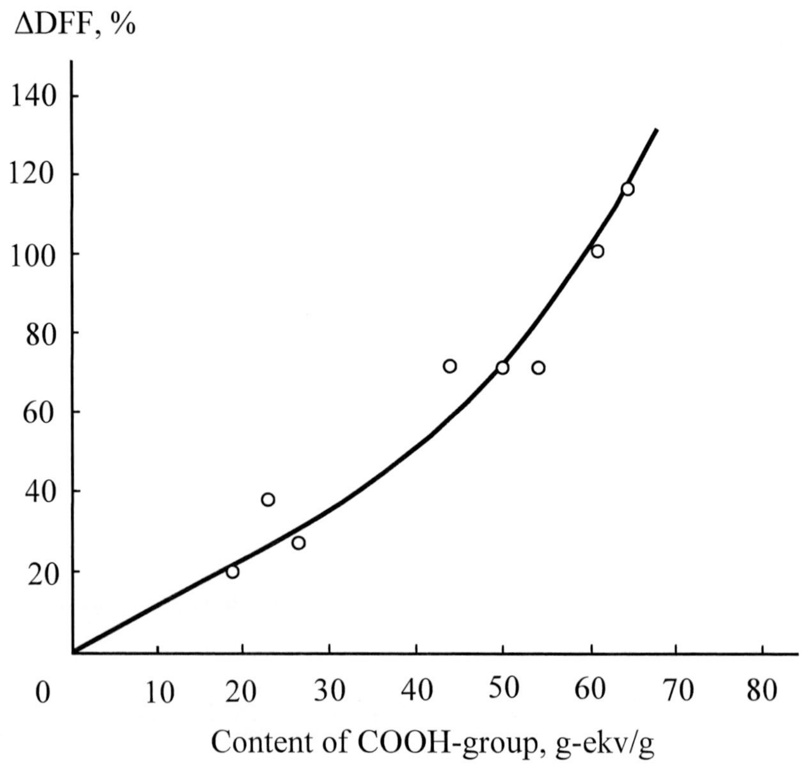

Picture 1. The dependance of DFF changing in 30 minutes from the content of carboxylic groups in PBT at T = 250 °C

To appreciate the effectiveness of the used heat setting a thermogravimetric analyses (TGA) of received polymers was held. The investigations were made on derivatograph grade Q-1500 firm "MOM" (Hungary); the heating rate of the samples in the air is – 2,5 grad./min. The results of TGA polymers (pic.2) showed that the thermal stability of PBT is mostly defined by the nature of the stabilizer. In the case when the stabilizing system consisted of two or more components, its effectiveness depended not only on the type of taken components but also on their correlation in the compound. At this it should be mentioned that one of the perspective ways of thermal stability of polymers and polyesters in particular is the switch from the usage of individual stabilizers to their compounds, which in most cases intensify their effect. Thus, the result of the introduction of stabilizers is that the effect of the antioxidant compounds turns out to be more effective than it could be expected on the bases of a simple summation of inhibitory abilities of some components. The received, as a result, effectiveness greatly exceed the effectiveness of each of individual antioxidants even with their higher content.

As it comes from datas of TGA and PTR measurings, the stabilizing systems turned out to be especially effective in the temperature range in which the refining of PBT is held: 240÷270 °C. At higher temperatures – 300 °C and more - the stabilizing effect is not so noticeable, though it remains till the temperatures at which the decomposition becomes inevitable by virtue of quite temperate energy of activation of decomposition of ester groups with the presence of initiators which are in great amount in PBT (-OH и – COOH, as terminal and formed in the process of thermal-oxidative decomposition of polymeric chain).

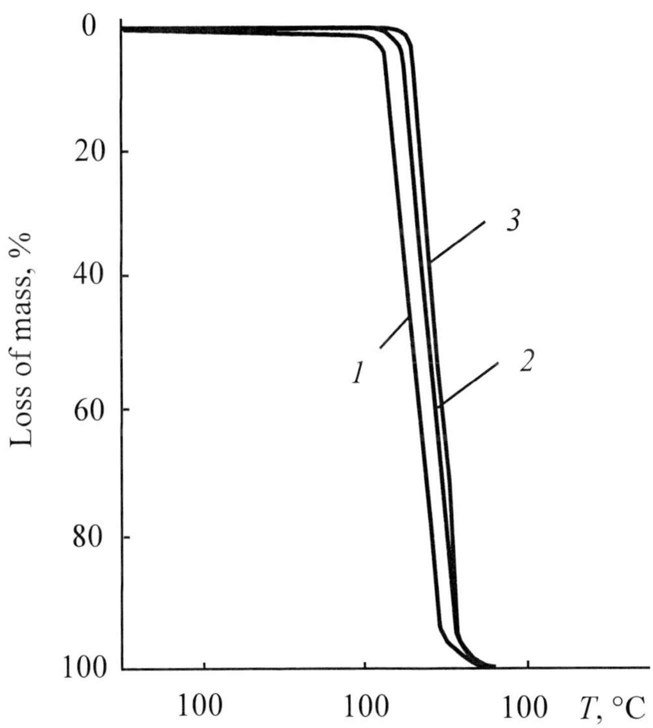

Picture 2. Curves of thermogravimetric analyses of PBT samples: unstabilized (1); PBT + 0,4 % I-1010 + 0,1 % TNPHPH + 0,05% NaH_2PO_2 (2); PBT + 0,2 % I-1010 + 0,2 % TNPHPH + 0,2 % NaH_2PO_2

An indirect confirmation of the received results is the fact that in the synthesis of PBT with stabilizers and especially with systems of stabilizers, the polymeric samples are white on the surface as well as in mass. Unlike that the samples of PBT synthesized without using the mentioned stabilizers had a light broun tint and at best of ivory colour. The last fact shous that in the process of PBT synthesis the macromolecules are exposed to destructive processes.

The given results of the investigation make it possible to come to the following conclusions. Firstly, the expediency of receiving PBT with different antioxidants is shown. Secondly, an effective systems of stabilizers are received, the usage of which enable to synthesize PBT which has high thermal characteristics.

References

[1] Matys V.//*Plasty a. kauc*. 1997. V.34. №7. P. 217-218.

[2] Blinne G. //*Kunststoffe. 1999*. V.89. №10. S. 107-110.

[3] Kore dzaire.//*Eng. Mater*. 1986. V.34. №9. P. 86-87.

[4] Patent № 5516879, USA //*RGCHem*. 20C 4513P. 1996.

[5] Kardanov A.A., Alakaeva Z.T., Storojuk I.P.//Thes. rep. IX conf. «Decomposition and stabilization of polymers». *M., 2001*. P. 77-78.

[6] Kalinchev A.L., Sakovceva M.B. Properties and processing of thermoplastic. L.: *Chemistry,* 1983. 288 p.

[7] Toropceva A.M., Belogorodskaya K.V., Bondarenko V.M. Laboratory practical work in chemistry and technology of high-polymeric compounds. L.: *Chemistry,* 1972. 415 p.

[8] Kovarskaya A.M., Бlumenfeld A.B., Levantovskaya I.I. Thermic stability of heterochain polymers. M.: *Chemistry,* 1977. 264 p.

[9] Matter M.R., Shukla S/R/, Sawant P.B.//*Polym. J.* 1996. V.28. № 3. P. 189-192.

The subscriptions to pictures of the article of authors Z.T. Alakaeva, T.A. Borukaeva, A.K. Mikitaeva "The investigation of thermal properties of polybutilenterephthalat received in presence of different heat settings".

In: Progress in Monomers, Oligomers, Polymers, Composites ... ISBN 978-1-60456-878-3
Editors: R. A. Pethrick, G.E. Zaikov et al.

Chapter 11

BIO- AND PHOTODESTROYED POLYMERIC COMPOSITIONS ON THE BASIS OF POLYTHENE OF HIGH DENSITY AND CORN STARCH

G.B. Shustov, M.L. Sherieva* and Z.L. Beslaneeva
The Kabardino-Balkaria State University, KBR, Nalchik, Russian Federation

ABSTRACT

The work presented nowadays, is devoted to results of creation of compositions on the basis of polyethylene and the starch, capable to keep initial physic mechanical and physical and chemical properties and biodegradable under action of microorganisms of ground and the UV-irradiation after expiry of the term of operation. As rheological characteristics values of a parameter of fluidity melt (PFM), the basic physic mechanical properties - on an example of a pressure and deformation of destruction were investigated at single-axle a stretching, and in some cases - the module of elasticity.

Keywords: biodecomposition, biodegradable, starch-natural polymer, "polyethylene + starch".

In connection with wide application of packing materials, polymeric plastic, polyethylene (polyolefin, copolymers ethylene) interest to creation of high-quality and economic bioplastics has increased.

Now there are many development and ideas for creation of such materials and three basic directions in this area are allocated: PEF hydroxyl carbonates acids (glycolic, dairy, valerians), plastics on the basis of reproduced polymers (water-soluble a film from a mix of starch and pectin, and also a mix of starch from PVC, PVS, compositions on the basis of cellulose, chitin), giving biodegradable to industrial high-molecular synthetic materials (polyethylene, PE, PVC, PS, PETP) [1].

* Correspondence to: The Kabardino-Balkaria State University, 360004, KBR, Nalchik, Chernishevskaya street 173, e-mail:zera14@yandex.ru

Most widely from of some natural connections in biodecomposed packing materials starch is used. Starch, as is known, is the most widespread material of plants [2].

The purpose of our researches is reception and research of some compositions on the basis of corn starch and polyethylene. Earlier we published works devoted to preparation of compositions on the basis of HDPE (B-273) and corn starch (GOST 12020-72), to reception from these compositions of test samples and to research of IR-spectra plastic initial polyethylene and the received compositions, physic mechanical properties of the pressed samples of compositions, dependence of a tangent of a corner of dielectric losses tg α from temperature T for film samples initial not stabilized HDPE (B-273) and compositions PE-273 + starch (the maintenance of starch from 1,5 % up to 15 %), and also results of the first attempts of studying of biodecomposition in ground of samples of the received compositions, change eventually them rheologicals and deformation-strength characteristics [3].

The work presented nowadays, is devoted to results of creation of compositions on the basis of polyethylene and the starch, capable to keep initial physic mechanical and physical and chemical properties and biodegradable under action of microorganisms of ground and the UV-irradiation after expiry of the term of operation [4].

For giving to polyethylene of high density biodegradable to it added natural polymer - starch. It is necessary to note, that, despite of preliminary preparation of starch during research, to prepare it was possible the mixes containing from 1,5 % up to 30 mass of % of starch, plasticized glycerin (on 100g to a mix – 10mil glycerin). The compositions were prepared of two parts and were buried in two different types of soil, the most typical for territory KBR (grey wood and light grey wood). As on key parameters these two types soil are very close, and at research rheologicals and physic mechanical parameters of samples before and after buried in ground of value practically coincide, for discussion the results received at a burial place of samples of compositions and initial polyethylene in grey wood ground are born. The received compositions after definition of a parameter of fluidity расплава represent firm white (sometimes grey or yellowish) a product with thin foam-figurative structure. From received extrusion samples at pressing it is formed strong elastic a film.

As rheological characteristics values of a parameter of fluidity melt (PFM), the basic physic mechanical properties - on an example of a pressure and deformation of destruction were investigated at single-axle a stretching, and in some cases - the module of elasticity. Results of the executed researches are presented in table 1.

The received results testify that to increase in percentage of starch in compositions there is an increase of a parameter of fluidity melt (PFM), and durability and lengthening at break decrease, i.e. compositions become more fragile. At the maintenance in compositions of 30 mass % of starch operational characteristics, and PFM accordingly above in comparison with PFM other structures are lost all. Most easily from the received compositions with the maintenance of starch of 10 and 20 mass %. As introduction already small amounts of starch up to 1,5 mass % leads to sharp falling of durability almost in 2 times and PFM thus increases in 3 and more times, it is possible to assume, that there is a distribution of starch between molecular formations, easing of communication between them and increase of their mobility. Simultaneously with increase in the maintenance of starch rigidity of samples raises, they become more fragile.

Table 1. Physic mechanical properties of the pressed samples of compositions on the basis of polyethylene and starch:

№	Structure of a composition, %		$PFM_{21.6}^{190}$	$PFM_{21.6}^{190}$	σ_{p_i}, MPa	ε ,% at break
	polyethylene	starch				
1	100	0	6,36	0,11	36,3	>500
2	98,5	1,5	17,57	0,30	17,7	35
3	97	3	34,87	0,58	17,7	53
4	95	5	45,93	0,77	17,7	27
5	93	7	37,94	0,63	15,1	15
6	90	10	31,5	0,53	10,8	9
7	85	15	17,06	0,28	16,7	12
8	80	20	36,36	0,61	12,3	19
9	70	30	43,02	0,71	*	

*-a composition so fragile, that collapses without external influence.

It is necessary to note, that research IR-spectra of initial polymer and the received mixes has shown, that during thermal processing there are some changes in the field of 1300 - 900 see $^{-1}$ Besides the researches executed with use roentgen structure of the analysis, electronic and scanning sound also have shown microscopy influence starch components on formation of morphological features of a composition "polyethylene + starch". The X-ray analysis has shown, that the degree crystallinity compositions at introduction of starch varies slightly and allows to conclude, that starch is not included into crystal areas of polyethylene. The data received by a method electronic and scanning sound of microscopy, allow to speak about heterogeneity of distribution starch phases in superficial layers of a composition.

At thermal processing the received compositions there are changes of dielectric properties. So, values tg for samples of initial polyethylene are constant up to 120^0C, values of a tangent tg in view of this frequency 10^4Hertz (10^{-3}-10^{-2}) correspond resulted in the literature. At temperature above starch 120^0C rise of dependence of a tangent tg from T with possible peak is observed at 190^0C. The specified temperature dependence essentially changes at introduction of starch.

The further increase the maintenance of starch up to 10 weights. % raises a parameter of fluidity melt and lowers breaking strength. At the maintenance of starch of 15 mass % of loss make no more than 0,15 mass %. Most likely, it is the composition most destroyed in the long term. The behavior of a composition from 20 % of starch is similar.

If to judge as a whole on initial rheological, dielectric and strength to characteristics the most suitable us both under operational characteristics, and by way of an opportunity of recycling are structures with 1,5; 3; 5; 10 mass % of starch.

Action of excited environments on received samples in accordance with GOST 12020-72 has been studied also. As excited environments were used: HCI-10 %-s solution, NaOH 10-th solution. Except for it the behavior of compositions was studied at contact to water.

Let's note, that PFM investigated compositions keeps higher values, than initial polyethylene after influence on them of excited environments and waters within 18 day. It is possible to assume, that as any ground is the certain chemical environment also the behavior of the received compositions at buried in ground can be analogous.

Influence of ultra-violet radiation on the received compositions was studied in the device for an irradiation (weatherometep) according to GOST 11279.2-83.

Samples of initial polythene are exposed to destruction already after 42 hours of an irradiation, at an irradiation within 96 hours lengthening increases, and loading at break decreases (figure 1, table 2).

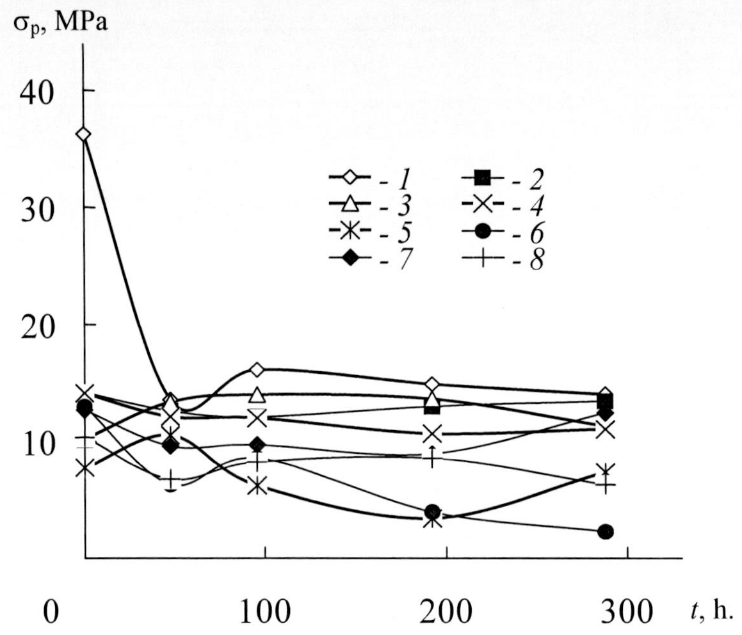

Figure1.Dependence of an explosive pressure of samples initial (1) and modified by starch PE from time of a photo irradiation. The maintenance of starch:
1,5 (2); 3 (3); 5 (4); 7 (5); 10 (6); 15 (7) and 20 (8) mass %

Table 2. Change of relative lengthening at break initial PE-273 and compositions on the basis of PE-273+starch at a photo irradiation:

N №	Structure of a composition, %		ε rel.,% initial	ε rel.,% 42h.	ε rel.,% 96h.	ε rel.,% 192h.	ε отrel.,% 288h.
	polyethylene	starch					
1	100	0	>600	30	46	38	35
2	98,5	1,5	23	23	25	21	17
3	97	3	29	28	28	23	18
4	95	5	34	23	23	23	20
5	93	7	20	21	18	16	17
6	90	10	35	16	18	18	11
7	85	15	31	24	20	18	18
8	80	20	28	23	17	15	14

In the beginning irradiations till 42 o'clock at the maintenance in compositions of 10 mass % of starch relative lengthening and explosive a pressure almost in 2 times decreases, and to the termination of an exposition (after 288 hours of an irradiation) the explosive pressure decreases in 6 times, and relative lengthening goes down in 3 times (table 2, figure 1). Whereas for initial polyethylene at an irradiation within 288 hours the explosive pressure decreases only in 2,5 times. A parameter of fluidity melts at the maintenance of 10 mass % of starch stably decreases. It is necessary to note, that for a composition with the maintenance of starch 1,5; 3; 5 mass % it is not observed such sharp decrease neither relative lengthening, nor durability. The parameter of fluidity melt for all compositions "polyethylene + starch" sharply decreases by the moment of 96 hours of an irradiation, and the further photo irradiation (288 hours) does not lead to significant changes of viscosity melt. From the received structures it is possible to allocate structure of containing 5 mass % of starch, as structure practically not losing the deformation-durability characteristics during a photo irradiation.

Biodecomposition in Ground of Compositions Polyethylene + Starch

Biodecomposition of polymer is complex process, speed and which completeness influences not only a structure and properties of polymer and polymeric a material, but also surrounding conditions. From surrounding conditions paramount influence humidity, temperature, pH render, light, and also such complex factor as contact to ground and type of ground. In turn the type of ground is a complex of corresponding factors and corresponding community of microorganisms.

Biodecomposition in ground was defined at keeping the received pressed samples in ground during 48 day. Type of ground: grey wood and light grey wood (the most widespread to territories KBR). Soil characteristics have preliminary been measured: pH (a water extract) = 6,5; pH (a salt extract) = 6; humus= 3,5 %; capacity of absorption of 25-30 mg-eguiv/100g ground. Then studying rheological and deformation-durability characteristics of the received compositions was spent. Results are presented in table 3,4 and on figure 2,3.

At biodegradable in ground the following is brightly expressed: an explosive pressure of the received compositions much more below an explosive pressure of samples of initial polyethylene (figure 3, table 4), which even after buried in ground not destructive (relative lengthening more than 600 %). Durability of samples of compositions " polyethylene + starch " after 14 day biodegradable in ground practically does not vary, and relative lengthening decreases for all received compositions, except for compositions with the maintenance of 1,5 and 3 weights. %. With increase in time biodegradable relative lengthening continues to decrease. For example, for a composition with the maintenance of starch 7 and 15 mass % relative lengthening decreases in 2,5 times, and durability in 2 times. It is necessary to note, that for compositions from 5, 7, 10, 15, 20 mass % of Starch at a finding in ground till 14 day biodegradable PFM decreases, viscosity melt the given compositions accordingly increases.

The subsequent stay of compositions in ground does not lead to sharp changes PFM though values remain above PFM initial polyethylene (figure 2, 3, table 3, 4). On operational properties and on ability to biodecomposition it is necessary to allocate structures with the maintenance of 5 and 10 mass % of starch.

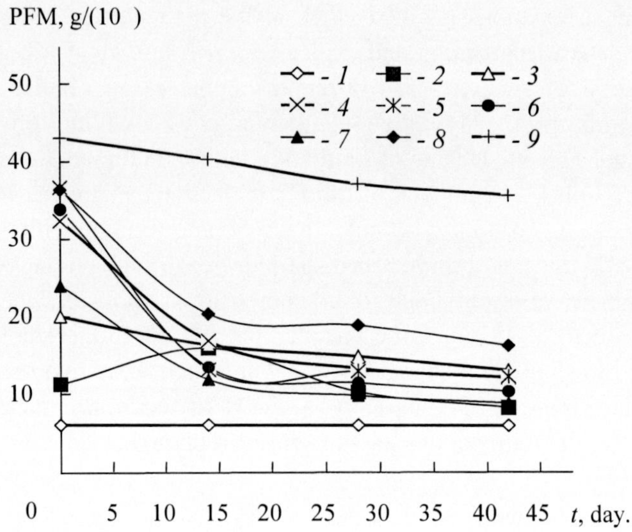

Figure2. Dependence explosive pressure of samples initial (1) and modified by starch PE - 273 from time. The maintenance of starch:
1,5 (2); 3 (3); 5 (4); 7 (5); 10 (6); 15 (7) and 20 (8) mass %

The analysis of the received results has shown that at biodegradable in ground the explosive pressure varies slightly, then as relative lengthening at break of samples decreases. It speaks that compositions at buried in ground become more rigid as there are structural changes in a matrix of polymer therefore compositions are exposed to greater destruction, than initial polyethylene.

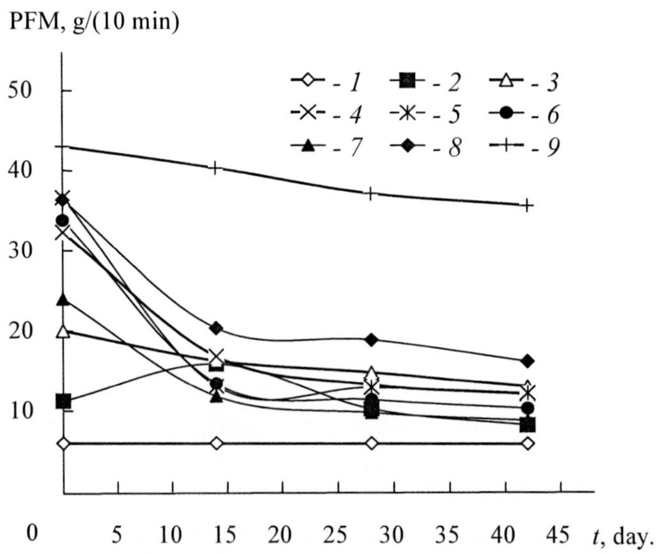

Figure3. Dependence of a parameter of fluidity расплава samples initial (1) and modified by starch PE from time biodegradable. The maintenance of starch:
1,5 (2); 3 (3); 5 (4); 7 (5); 10 (6); 15 (7) and 20 (8) weights.

Thus, introduction of starch as the additive to synthetic polymer allows to accelerate process destructive polymer under action of microorganisms and does not render thus significant influence on initial physical and chemical properties. It leads to that a number of the investigated mixes on the basis of polyethylene and local corn starch have improved rheology characteristics, good physic mechanical properties and ways to be exposed bio-and to photo destruction.

Table 3. Change of an explosive pressure of the pressed samples of compositions on the basis of polyethylene and starch at biodegradable:

№	Structure of a composition, %		σ_p, MPa Initial	σ_p, MPa 14 day	σ_p, MPa 28 day	σ_p, MPa 42 day
	polyethylene	starch				
1	100	0	36,3	36,2	36,2	35,3
2	98,5	1,5	14,2	13.5	13,0	10,6
3	97	3	10,2	15,8	14,5	13,7
4	95	5	14,1	10,4	8,8	8,0
5	93	7	7,8	7,1	3,4	3,0
6	90	10	13,0	12,4	9,0	8,3
7	85	15	12,7	12,3	7,9	6,1
8	80	20	10,4	10,8	8,1	7,9

Table 4. Change of relative lengthening at break of the pressed samples of compositions of polyethylene and starch at biodegradable:

№	Structure of a composition, %		$\varepsilon_{rel.}$,% initial	$\varepsilon_{rel.}$,% 14day	$\varepsilon_{rel.}$,% 28 day	$\varepsilon_{rel.}$,% 42 day
	polyethylene	starch				
1	100	0	>600	>600	>600	>600
2	98,5	1,5	23	27	28	28
3	97	3	29	38	38	39
4	95	5	34	18	11	10
5	93	7	20	18	8	7
6	90	10	35	25	17	8
7	85	15	31	16	15	12
8	80	20	28	24	22	18

REFERENCES

[1] Shustov G.B., Sherieva M.L., Mirzoev R.S., Kanametova I.K., Beshtoev B.Z. biologically utilized plastics: a condition and prospects // Mater . II All-Russian research-technical conference: new polymeric composite materials.- Nalchik. 2005. p. 34-38.

[2] 2. Sherieva M.L., Shustov G.B., Mirzoev R.S., Beshtoev B.Z., Kanametova I.K. reception and research of properties of the modified starch // Mater. II All-Russian

research-technical conference: New polymeric composite materials.- Nalchik. 2005. p. 117-120.

[3] Sherieva M.L., Shustov G.B., Shetov R.A., Beshtoev B.Z., Kanametova I.K. research of mixes on the basis of corn starch and polyethylene // Mater. II All-Russian research-technical conference: New polymeric composite materials.-Nalchik. 2005. p. 266-273.

[4] Sherieva M.L. Biodecompos of a composition on the basis of polyethylene of high density and starch // the Dissertation on competition of a scientific degree of Cand.Tech.Sci. Nalchik. 2005. p. 116.

In: Progress in Monomers, Oligomers, Polymers, Composites … ISBN 978-1-60456-878-3
Editors: R. A. Pethrick, G.E. Zaikov et al.

Chapter 12

THE STUDY OF INTERACTION BETWEEN THE BIOCIDE POLYELECTROLYTE AND THE BACTERIAL CELL WITH HELP OF SCANNING PROBE MICROSCOPY

S.Yu. Khashirova*, T.B. Guppoev, S.S. Pekar, R.A. Uzdenov and Y.A. Malkanduev
Kabardino – Balkarian State University, KBR, Nalchik, Russian Federation

ABSTRACT

The influence of the novel biocide guanidine – containing polyelectrolytes on the morphological properties of the St.Aureus bacteria is studied. The possible mechanism of the biocide influence of the polyelectrolyte is given. The process of the lysis of the bacterial cell caused by the guanidine polyacrylate has been detected by the atomic – force microscope Solver Pro – 47 in the Laboratory of the Nanoprobe Investigations of the Kabardino – Balkarian State University.

Keywords: guanidine – containing polyelectrolytes, atomic – force microscopy, biocides, bacterial cells, the morphology of the cell, guanidine polyacrylate

The modern methods of scanning probe microscopy (SPM) allow the scientists to deeper understand the morphology and properties of living bacteria. SPM gives the extra opportunities to study the growth, division, lysis, motion and aggregation of cells in their natural environments such as water and air.

The goals of this paper are: (1) the study of influence of novel biocide guanidine – containing polyelectrolytes on the morphology of bacterial cell St. aureus; (2) the uncovering

* Correspondence to: Russia, Kabardino – Balkarian State University, 360004, KBR, Nalchik, Chernishevskaya street, 173, e-mail:sveta_daova@mail.ru

of the biocide effect and (3) the registration of the elementary acts of lysis of bacterial cell under the influence of the polymer bacteriofagi with help of the SPM.

We applied the known technique of preparation of microbe cells [1,2] to visualize the surface of the bacteria. The cultures of golden – color staphylococci from the International Collection of Etalon Cultures were being transplanted on the nourishing yolky – saline agar according Chistovich, the agar being poured into the sterile Petri cups. Then the cups with crops were put into the incubator at 37°C for 18 hours. The "just – taken" bacteria were placed in the retorts filled with distilled water and 1% solution of the polymer with help of the loop. The acquired concentration of the bacteria was being reached visually by monitoring the turbidity of the solution. The drop of prepared solutions was put onto the mica substrates of size 6×6 square millimeters after 20 and 60 minutes after the solution preparing.

Preliminary observations with the microscope were being done after 10 – 15 minutes after the placing of the samples.

We point out that the placing of the microbe cells to the distilled water during sample preparing according our observations does not result in the lysis (destruction) of the cells in time range studied.

The visualization of the surface of bacteria cells has been performed in semi – contact regime of the atomic – force microscope Solver Pro 47 in the laboratory of nanoprobe investigations of the Kabardino – Balkarian State University. The probe used was the cantilever NGS-10 with the silicon nitride tip, the mechanical rigidity of which was 0.06 and 0.12 N/m. The force of the interaction between the tip and the focused surface was 10^{-9} N.

Fig. 1 represents the images of the initial St. aureus cultures and those treated by the polymer. Atomic – force microscopy shows significant distinctions in the morphology of the surfaces of St. aureus cultures before and after the treatment.

The sizes of separate bacteria were 0.81 to 0.90 microns in length and 0.7 to 0.8 microns in apparent width. The height of the bacteria was 0.05 to 0.2 microns what tells about some deformations of the bacteria during the placing onto the mica substrates. The use of distilled water allowed us to exclude the formation of the saline deposits appearing in great number at condensating the bacteria from the salty buffer.

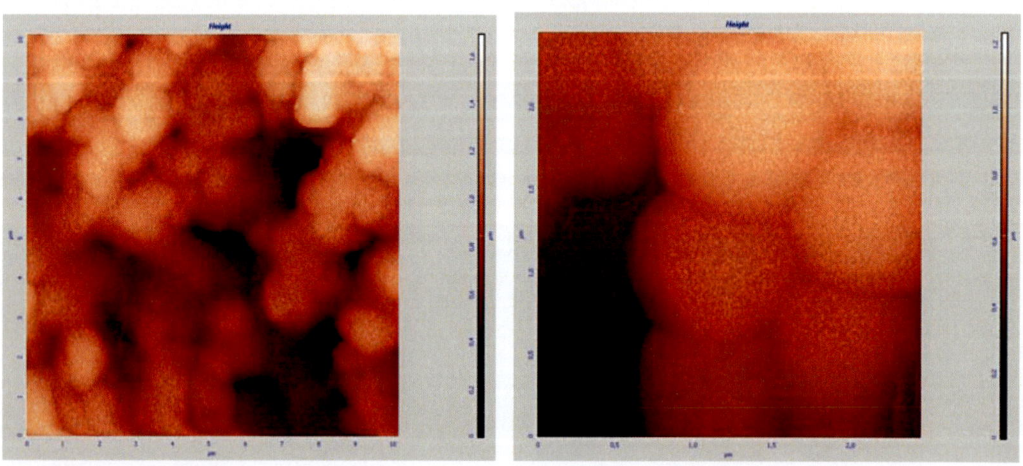

(a) (b)

Figure 1. The images of St. aureus cultures before (a) and after (b) polymer treatment.

The dimensions of the bacteria increase approximately two times (up to 1.6 to 1.8 microns) after the treatment, then bacterium becomes more spherical transforming itself to the spheroplast – cell with almost entirely removed walls – thus we observe the lysis.

Let us consider the cell membrane of gram – positive microorganisms to find out the mechanism of the biocide effect of guanidine – containing polymer on the bacterial cells St. aureus, see fig. 2.

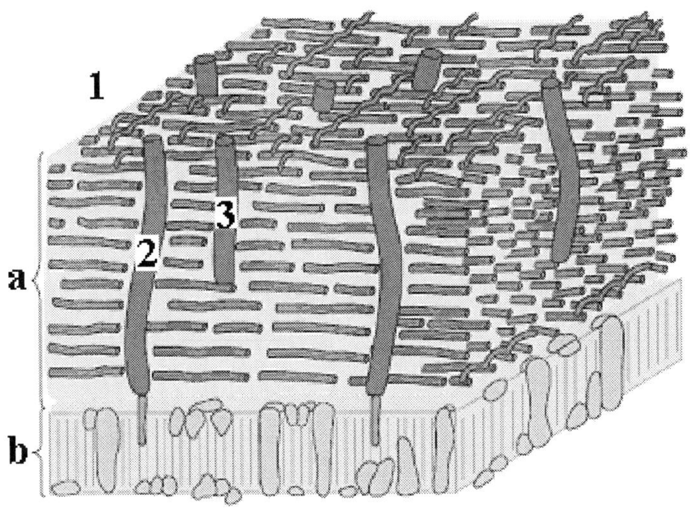

Figure 2. The schematic depiction of the cell membrane of gram – positive microorganisms: a) – cell membrane, b) – cytoplasmic membrane.
1 – peptidoglicane, 2 – lypoteichoic acids, 3 – teichoic acids.

The cell membrane of staphylococci consists of the multilayer structure of 70 to 80 nanometers in width named peptidoglicane. The peptidoglicane bag covering the cytoplasmic membrane of these cells is built up from the polysaccharide chains linked by the peptide bridges into the web. The cell membranes of studied bacteria includes the negatively – charge polymers, so called teichoic acids besides the peptidoglicane. Teichoic acids, due to the presence of the phosphorus acid in them, help the electronegative charge to be formed on the surface of the gram – positive bacteria.

The guanidine polyacrylate containing the negatively charged carboxyl groupings of acryl acid and positively charged groupings of the guanidinium's cation was used as polymeric biocide preparation in our observations.

The key point in the mechanism of the influence of the polyelectrolyte (guanidine polyacrylate) on the golden – color staphylococci cultures, probably, is the electrostatic interaction with the negatively charged teichoic acids, which play the role of "antennas" in linking the polymer and target cell. The consequences of this are the general neutralization of the charge of the membrane and the cell and also the change in the ratio of hydrophobic and electrostatic interactions stabilizing the membrane, see fig. 3.

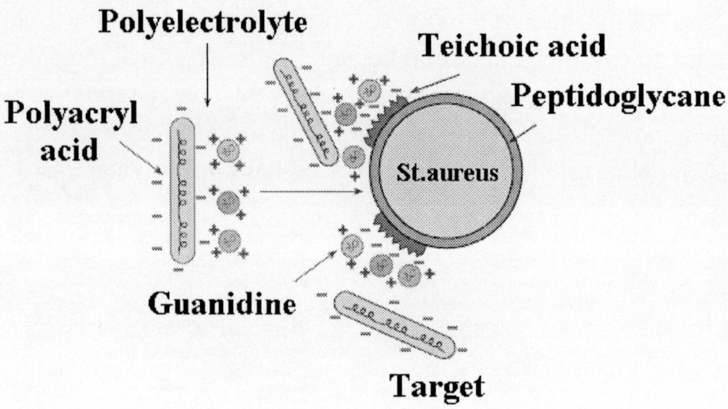

Figure 3. The hypothetic scenario of the interaction between the guanidine polyacrylate and the bacterium.

As literature says [3-9], the mechanism of the biocide influence of polyelectrolytes (and particularly guanidine – containing ones) is the following. Because the microorganisms usually possess total negative electric charge they are substantive to the cation of the bactericide preparation which, being in contact with the microorganism, adsorbs on the cell's membrane, destroys it and penetrates inside the cell. In there the preparation blocks the bioactivity of the ferments, stops the replication of the nucleic acids and oppresses the breathing system. Adding up all these effects kills the microorganism.

To conclude, we present the atomic – force observations of the morphology changes of the golden – color staphylococci cultures under the influence of the polymeric biocide preparation (guanidine polyacrylate) performed for the first time. Also we offer and explain the probable mechanism for the interaction between the guanidine polyacrylate and the surface of the bacteria cell.

ACKNOWLEDGEMENT

This research is financially supported by the Ministry of Education of the Russian Federation, grant R.N.P.2.2.2.3.4888, partially supported by the mini-grant Y3-C-04-05. The observations have been performed in the Laboratory of Nanoprobe Investigations of the Kabardino – Balkarian State University (Nalchik, Russia).

REFERENCES

[1] Yaminsky I.V., Demin V.V., Bondarenko V.M. // *Zhurnal mikrobiologii, epidemiologii i immunobiologii, 6* (1997) 15-18 [in russian]

[2] Yamisky I.V., Pyshkina O.A., Sergeev V.G., Semenov A.E / Vserossiyskoe soveschanie po zondovoi mikroskopii (*N.Novgorod: Institute of the Physics of the Microstructures of the Russian Academy of Sciences, 1997*) 124-127 [in russian]

[3] Plate N.A., Vasiliev A.E. Fiziologicheski aktivnie polimery (*Moscow: Khimia*, 1986) 296 p. [in russian]

[4] Plate N.A., Vasiliev A.E. // Vysokomol. *Soed A. 24*(4) (1982) 675 [in russian]

[5] Ryser H.J. // *Science 150* (1965) 501

[6] Ryser H.J. // *Biomembranes 2* (1971) 197

[7] Yaroslavov A.A., Kabanov V.A. / Materialy Vserossiyskogo Karginskogo Simpoziuma. *Abstracts 1 (2000*) 17 [in russian]

[8] Feldshtein M.M. / Sinteticheskie polimery medicinskogo naznachenia. *Proceedings of the 6th All – union Symposium* (Alma – Ata, 1983) 142 [in russian]

[9] *Otchet filiala №5* Instituta biofiziki Minzdrava SSSR po hozdogovoru na temu "Rezultaty issledovanij persektivnih soley PGMG (hronicheskii eksperiment) s cel'u vnedrenia ih v narodnoe hoziastvo i medicinu" (Angarsk, 1991)

In: Progress in Monomers, Oligomers, Polymers, Composites ... ISBN 978-1-60456-878-3
Editors: R. A. Pethrick, G.E. Zaikov et al.

Chapter 13

SYNTHESIS AND PROPERTIES OF PHOSPHORUS-CONTAINING POLYPEPTIDES FOR EFFECTIVE ION-EXCHANGE MATERIALS OF IMPROVED CAPACITY

Victor Kablov, Dmitry Kondrutsky* and Maria Sudnitsina
Volzhsky Polytechnic Institute (branch of) Volgograd State
Technical University, Volzhsky, Russian Federation

ABSTRACT

New phosphorus-containing polymers (polyproteinylmethenephosphinic acids) on base of peptides were synthesized through reaction phosphonomethylation by phosphinic acid and formaldehyde to the peptides with template ions. The product was characterized by IR spectroscopy and its ion-exchange properties toward Cu^{2+} cation was investigated. The material thermal behavior was studied by differential thermal analysis. The characterization results show that the templated phosphorus-containing polypeptides have improved capacity toward Cu^{2+} cation (2,9 mmol/g) and fireproof ability.

INTRODUCTION

The raw materials amounts reduction, such as oil and gas, and the environment pollutions in view of significant volume of the non biodegradable wastes pose an actual problem of use of the raw materials of renewed sources [1].

One of possible line of investigations is a synthesis of phosphorus-containing polymers of natural proteins. They possess sorption and ion-exchange activity, reduced combustibility, biodegradability etc.

* Correspondence to: Volzhsky Polytechnic Institute (branch of) Volgograd State Technical University, 42-a, Engels srt., Volzhsky 404120, Russia; dakondru@vlz.ru

In the present work results of phosphorus-containing polypeptides on the base of some substrates (peptides, gelatin and fish collagen) synthesis research are presented. The chemical structure of a product, sorption properties and thermal-oxidative degradation are investigated.

EXPERIMENTAL PART

The synthesis of peptides by enzymatic hydrolysis of proteins was carried out with gas-vortex gradientless bioreactor of type BIOK-022. As enzymes were used pepsin and trypsin. Initial concentration of the protein substrates was 20 % on weight. Temperature of the synthesis was 36-40 °C. Synthesis was carried out during 1 hour. Received peptides molecular weight was 800-1100 g/mol of titration method with formalin [2].

The synthesis phosphorus-containing polypeptides (polyproteinylmethenephosphinic acids) on base of peptides were carried out by the reaction phosphonomethylation. Sodium hypophosphite or phosphinic acid were added into 20 % by weight water solution of peptides under agitation at 45-60°C. Then 36.7 % formalin was added in the reaction solution. pH was controlled by litmus paper and one was supported in the range 3-4 by addition of hydrochloric acid. Molar ratio of peptides, formaldehyde and sodium hypophosphite (phosphinic acid) was varied in the range from 1:2:1 to 1:8:4 accordingly. Reaction was stopped by means of quick water dilution of the mixture and fast separation the insoluble products from the mixture by vacuum filter. Then the synthesized product was washed at first distilled water, then 0.1 normal solution of hydrochloric acid, again water, after that 0.1 normal solution of sodium hydroxide and again water until neutral pH. Remaining water in the product was removed by vacuum filter.

If the product was soluble that it was separated from the solution by sodium sulfate injection in the mixture with next double-ply processes soluble one in water and dissoluble in present of sodium sulfate for purpose to wash from reaction mixture leavings.

The synthesis phosphorus-containing polypeptides from fish collagen were carried out the same manner. Into 20 % by dry weight water suspension of fish collagen was added the known reagents (formaldehyde and sodium hypophosphite or phosphinic acid) in the molar relations to dry collagen from 1:8:3 to 1:10:20 accordingly. pH was supported in the range 3-4. The reaction was stopped by addition 10 ml of 2 normal solution of NaOH. The product was separated and in series washed in vacuum filter by means of 0.1 normal solution of hydrochloric acid, then water, after that 0.1 normal solution of sodium hydroxide and again water until neutral pH. Remaining water in the product was removed by vacuum filter.

The template synthesis of increase capacity toward cations Cu^{2+} sorbents were carried out by addition into water suspension of fish collagen 50 % cuprum sulfate water solution in mass relation of fish collage to $CuSO_4$ as 5:1. Then the synthesis was conducted in same previous manner. For more effective desorption residual cations Cu^{2+} from the polymeric die was applied double consecutive wash by 0.1 normal HCl solution, 0.1 normal NaOH solution and water until neutral pH.

A photocolorimeter FEC-56PM was used to quantitative chemical analysis to phosphorus (GOST 208512-75) [3]. A SPECORD-M82 was used to directly characterize the phosphorus-containing polypeptides. IR spectrum was performed for the product in form of thin suspension into vaseline.

Static ion-exchange capacity (SEC) toward cations Cu^{2+} versus pH was investigated for about 1 g. polymer samples, which had been consecutive washed 0.1 normal solution HCl, 0.1 normal solution NaOH and next water. This treatment converts the polymer into "Na-form". The samples were dipped into 30-repeated abundance by weight 0.1 normal solution cuprum sulfate and were got necessary meaning of pH by means of addition HCl or NaOH. The samples were kept at 25 ^{0}C for 3 days.

An electronic pH-meter was used to control pH of reactionary solutions. One included a glass electrode, an argentums chloride electrode and a thermo compensative electrode. They were plugged on an universal potentiometer EV-74.

The residual concentration of Cu^{2+} cation of solution was investigated after time of the sorption by the photocolorimeter FEC-56PM by use of ammonia method [4].

The SEC of polymer was calculated in agreement with expression:

$$SEC = \frac{\left(c_n^0\left(Cu^{2+}\right) - c_n^1\left(Cu^{2+}\right)\right)\cdot V_{sorb}}{m_{pol}}, mmol/g$$

where $c_n^0\left(Cu^{2+}\right)$ - initial normal concentration Cu^{2+} cation of solution before sorption, mol/l;

$c_n^1\left(Cu^{2+}\right)$ - final normal concentration Cu^{2+} cation of solution after sorption, mol/l;

V_{sorb} - volume of solution for sorption, ml;

m_{pol} - mass of phosphorus-containing polypeptides sample, g.

Water adsorption of initial phosphorus-containing polypeptides samples was calculated in agreement with expression:

$$\varphi = \frac{m_{wet} - m_{dry}}{m_{wet}}\cdot 100\ \%$$

where m_{wet} - mass of phosphorus-containing polypeptides sample ("Na-form") after consecutive wash by 0.1 normal solution HCl, then 0.1 normal solution NaOH and next water until neutral pH. Remaining water in the product was removed by vacuum filter;

m_{dry} – mass of phosphorus-containing polypeptides sample after exsiccation under standard condition.

Differential-thermal analysis (DTA) of product and initial peptides was carried out from 25 to 600 ^{o}C using a derivatograph "MOM", Hungary. The mass of samples was 100 mg. Rate of heating was 10 ^{o}C per minute. Sensitivity DTA was 1/5 ^{0}C.

Results and Discussion

Synthesis of Initial Peptides

To prepare the initial substances for the synthesis of phosphorus-containing polypeptides, the research of fermentative hydrolysis of gelatin to mixture of low-molecular oligomers (the peptides) was carried out. An action of series of hydrolytic enzymes such as pepsin and trypsin was investigated. These enzymes are high specific catalysts for selective hydrolysis of proteins. Trypsin splits bonds that have been built carboxylic group of lysine or arginine. Using trypsin conducts to sufficiently large variation of molecular mass and increases its value than in case of less specific pepsin. Pepsin requires acid condition (pH = 3-4). That pH was received by addition 0.1 normal solution HCl. Like receiving solution of peptides had acid pH and its neutralization by 0.1 normal solution of sodium hydroxide produced formation sodium chloride in reaction system that further required additional deionization. All of these facts decrease technological attraction of pepsin using.

In case of using trypsin the process was carried out in neutral or weakly basic conditions (pH=7,0-7,8). This fact presents significant advantage because the synthesis is able to carry out without using of other acid-basic regulators. Thus trypsin was selected as enzyme for hydrolysis.

The optimal concentration of gelatin for hydrolysis was 20 % by dry weight. In this condition the process was more technologic because medium viscosity wasn't too high and heat exchange was better. The optimal temperature of the process was 36-40 ^{0}C that well agree with optimum of enzyme catalytic activity.

The Phosphorus-Containing Polypeptides Reaction Synthesis Investigation

The phosphorus-containing polypeptides synthesis passes via few stages with series intermediate production. Further the intermediates lead to creation cross-linked high-branched polymer what includes derivation of phosphinic acid.

At first peptide methylol-derivations synthesize by adding formaldehyde as more reactionary substance to nucleophilic centers of peptides.

In general view schema of this reaction can be introduced as

peptid intermediate 1

At subsequent stage methylol-derivations of peptide react to phosphinic acid that conduct to form the product of phosphonomethylation. The scheme of this reaction can be introduced as

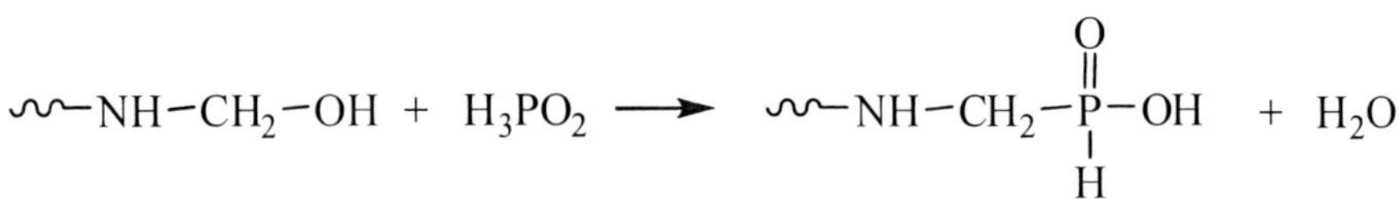

intermediate 1 product 1

At final stage of the synthesis a series of reactionary groups couple together and form cross-linked phosphorus-containing polypeptides. The scheme of such reaction can be introduced as

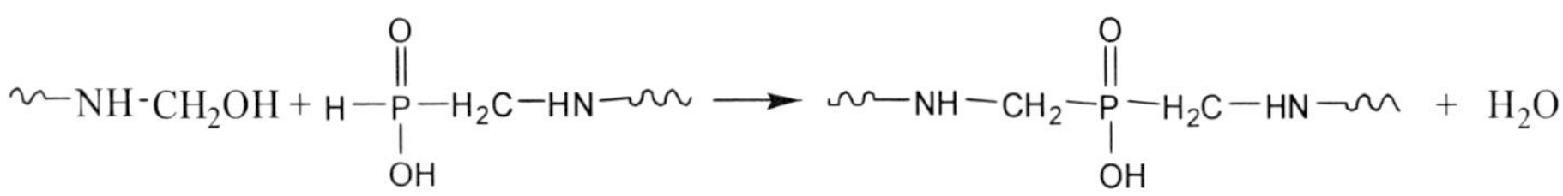

cross-linked product

Investigation of Synthesis Conditions Influence to Phosphorus-Containing Polypeptides Properties

Influence of different factors such as the relation of reagents, the medium pH and the temperature to quantity of phosphorus in the phosphorus-containing polypeptides was investigated

The experiment results of phosphorus-containing polypeptides synthesis at 21 ^{0}C for 24 h are presented in table 1. We can observe that the optimal conditions of synthesis are 3-repeated abundance of the reagents and acid medium.

Table 1. Investigation of influence of the reagent relations and the medium pH to quantity of phosphorus in the product

M_r peptides, g/mol	Molar ratio of the reagents			pH	ω(P), %	Comment
	peptide	CH_2O	H_3PO_2			
1400	1	2	1	7	0.04	The form of product is insoluble gel
	1	3	1,5		0.02	
	1	4	2		0.03	
	1	6	3		0.03	
	1	8	4		0.03	
1400	1	2	1	3	0.50	The form of product is viscous liquid
	1	3	1,5		0.75	
	1	4	2		0.95	
	1	6	3		1.13	
	1	8	4		1.07	

Table 1 shows that in neutral medium a good deal of phosphor can not be ingrafted to peptides.

In adding hydrochloric acid and decrease medium pH until 3 quantity of phosphorus in the product increased in a few orders and got at 1.1 % at molar relation of peptide:formaldehyde:phosphinic acid as 1:6:3. Like this the degree of phosphonomethylation was about 50 %. But such conditions weakness is soluble form of phosphorus-containing polypeptides that confuse them applications as ion-exchange materials.

Investigation of the temperature influences to quantity of phosphorus in the product showed that synthesis conducted at 50 ^{0}C under similar other conditions to cause to 2.3 % of phosphorus in the product. This is adequate ultimate value of the phosphonomethylation degree for peptides having similar molar weight. Using more high temperature is no purpose, because phosphinic acid is able to suffer disproportionation with elimination phosphine under heating.

The experiment results are shown in Table 2.

Table 2. Investigation of the temperature influences to quantity of phosphorus in the product

M_r peptides, g/mol	Molar ratio of the reagents			t, ^{0}C	pH	ω(P), %
	peptide	CH_2O	H_3PO_2			
1400	1	6	3	21	3	1.13
				50		2.3

For receiving of product in the insoluble form, which has high phosphorus quantity, fish collagen was investigated as an initial substrate. The fish collagen has high quantity amino groups, which include to functional lateral radicals of L-lysine and L-arginine.

The fish collagen swells limited in water and produces a suspension, which stables in acid medium. The concentration of amino group according to formalin titration in the fish collagen was 1.1 mol/g. The fish collagen contains in 10 % more amino group then the gelatin peptides.

To decrease hydrolysis and swelling the fish collagen during phosphonomethylation reaction sodium sulfate was added in reaction mixture as a graining reagent. The results of the experiment are presented in Table 3.

Table 3. Investigation of graining agent (Na_2SO_4) influence to phosphonomethylation reaction

M_r peptides, g/mol	Molar ratio of the reagents				t^0	τ, h	pH	ω(P), %
	Na_2SO_4	peptide	CH_2O	H_3PO_2				
900	1	1	6	3	50	1	3	1.2
	4							0.49
	6							0.34
	9							0.16

The experimental data analysis is showed that the product produces in a good spherical granule form under increasing molar ratio of sodium sulfate. But quantity of phosphorus in the product decreased in several times in comparison with ordinary synthesis. In that case application of abundance sodium hypophosphite is more perspective way to decrease the product losses all along of acid hydrolysis during phosphonomethylation. This way lets to obtain insoluble product having high phosphorus quantity.

IR-Spectroscopy of Phosphorus-Containing Polypeptides

Infrared spectroscopy was performed for confirmation scheme of the product synthesis.

The results of IR-spectroscopy are presented in Table 4. It was founded the characteristic vibration of groups and bounds, which build by atoms of phosphorus and carbon (the bond P-C) or atoms of phosphorus and oxygen (the bond P=O) in structure of the product.

At the same time signals of absorption for bound P-H disappeared and band for P-O-C was not detected. These facts show that the reaction passes via forming bounds P-C with taking bound P-H as it was shown in the scheme of phosphonomethylation reaction.

Table 4. The characteristic vibrations of IR-radiation absorption of reagents and product

Functional group	peptide	NaH_2PO_2	phosphorus-containing polypeptides
	υ, $см^{-1}$		
$-NH_3^+$	3130-3030	-	3130-3030
$-COO^-$	1454	-	1454
=C=O («Amide 1»)	1651	-	1651
=NH и ≡C-N= («Amide 2»)	1538	-	1538
≡C-NH-	1082	-	1082
P-C	-	-	920-930
P=O	-	1182	1153-1168
P-O-C	-	-	-
P-H	-	2320-2350	2320-2350
-OH (polyassociates)	-	-	3600-3200

Investigation of Sorption and Ion-Exchange Activity of Phosphorus-Containing Polypeptides

The main purposes of this paper were the investigation sorption activity of phosphorus-containing polypeptides, the determination of full (FEC) and static exchange capacity (SEC), the exploration of influence different factors to SEC, first of all the quantity of phosphorus and the water adsorption.

The value of FEC can be calculated as

$$FEC = \frac{k \cdot \nu(P)}{m_{pol}}, \text{ mmol/g}$$

where k – number of dentate of complexing groups. "k" shows number of ions to one reactionary centre

ν(P)- amount of substance ion-exchange group, mol

$$\nu(P) = \frac{m_{pol} \cdot \omega(P)}{M(P) \cdot 100\%}$$

m_{pol} – mass of phosphorus-containing polypeptides, g

ω(P) – phosphorus abundance, %

If we substitute ν(P) into expression for FEC that we take a new expression as

$$FEC = \frac{k \cdot \omega(P) \cdot 1000}{M(P) \cdot 100\%}$$

Thus, for phosphorus-containing polypeptides theoretical determining FEC value was 0.74 mmol/g at maximal possibly quantity of phosphorus about 2.3 % and k=1.

The results of experiments for investigation SEC depending on quantity of phosphorus and value of water adsorption are presented in Table 5.

Table 5. Investigation of influence quantity of phosphorus and water adsorption to SEC for phosphorus-containing polypeptides on base of the fish collagen

Mr peptides, g/mol	ω (P), %	FEC, mmol/g	φ, %	SEC, mmol/g (Cu)	k
900	0.49	0.16	84	1.87	11.7
	0.16	0.05	45	0.54	10.8

The experimental data confirm about increasing SEC under increase of quantity of phosphorus and water adsorption.

The experimental SEC value at 10-11 times more then calculated FEC at adequate ω (P) value. This fact indicates that series additional groups of residual peptides (carboxylic group, amide units and etc.) are involved in sorption process.

For increasing SEC value was investigated processes complexation and chelation of metal ions in the polymeric die to produce centers of complementary for these ions. In this case polyions and molecular sorption are able to pass with ultramicroscopic phase of salt formation into a polymeric die.

By means of the template synthesis at cation copper (II) was made an attempt to creation in the material die of the ligand sites complementary to this ion.

The using of template synthesis permitted to increase SEC at 2-3 times, but led to diminish quantity of phosphorus in the product in few times. It can be explained by inhibition of part of the reactionary centers (amino groups) of the fish collagen because the complexes of amino groups with cations Cu^{2+} form and reduction Cu^{2+} to Cu^{+} passes during the synthesis. These secondary processes determine different color changes of reactionary mixture and the product.

At pH change ion-contained polymeric die has to transform its conformation that ion-exchanges activity changes too. For phosphorus-containing polypeptides (Na-form) template synthesized with quantity of phosphorus 0.1 % decrease pH medium from 4.9 to 2.9 reduces SEC at 50 % from 2.9 to 1.9 mmol/g. The experiment results about investigation of influence pH to SEC are presented in Table 6.

Table 6. Investigation of □□ medium influence to SEC (□(□)=0,1 %) from $CuSO_4$ 0.1 normal solution

pH	SEC, mmol/g
2.94	1.88
4.41	2.84
4.88	2.90

The registered regularity presents important advantage for phosphorus-containing polypeptides in that it allows reversible to adsorb-desorb desirable ions from aqueous solution at pH medium regulation.

The research alternative ways of the product application or utilization the material waste present actual problems.

The phosphorus-containing polypeptides includes considerable number of reactionary centers (carboxylic group, phosphorus unit, amino group, amide group) which present in well-known self-extinguish material so it is very perspective to research thermal-oxidative material destruction.

Differential Thermal Analysis of Phosphorus-Containing Polypeptides

The results of differential thermal analysis of phosphorus-containing polypeptides are presented in Pictures 1, 2.

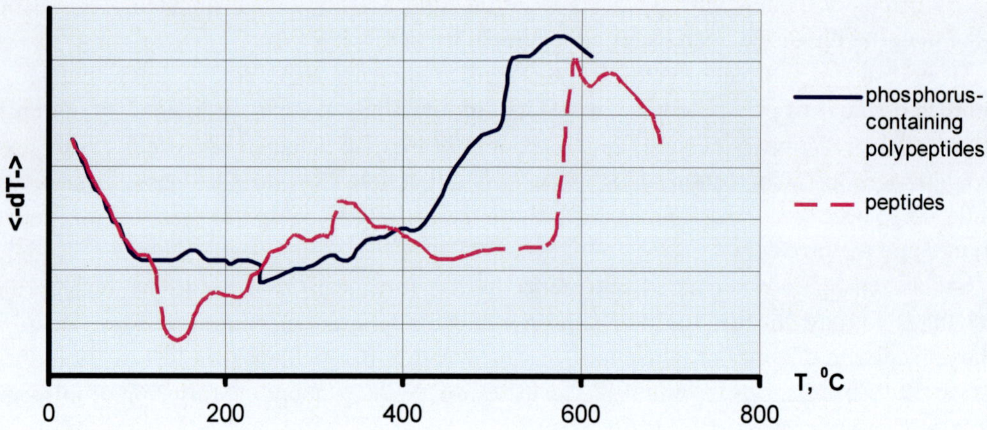

Picture 1. Curves of differential thermal analysis of phosphorus-containing polypeptides and initial peptides

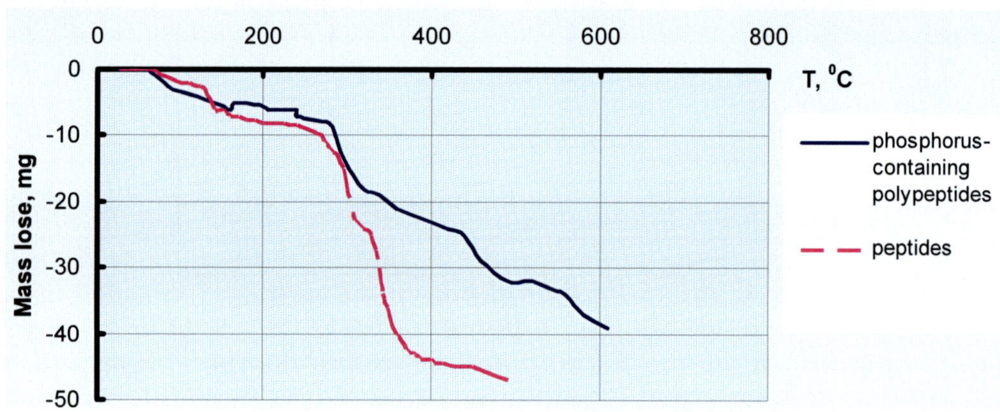

Picture 2. Thermo gravimetric curves of phosphorus-containing polypeptides and initial peptides

The phosphorus-containing polypeptides stable until 270 ^{0}C. Heat effect of structuring is only in that temperature range. Intensively thermal-oxidative destruction starts at 300 ^{0}C and passes until 600 ^{0}C. The phosphorus-containing polypeptides are more stable to thermal influence as experimental data show. That at 500 ^{0}C mass loss for peptides is 49 %, but for the phosphorus-containing polypeptides is only 31 %. In total phosphorus-containing polypeptides have reasonably stable to thermal-oxidative influences that high value of coking part mass confirms (about 69 %). This fact performs significant advantage and expands areas of application of these materials.

CONCLUSIONS

Completed experiments and calculations showed availability of using the phosphorus-containing polypeptides for adsorption cations Cu from technological mixtures and waste water with easy regeneration by way of pH medium regulation.

Application of template synthesis allows obtaining high SEC value. It opens possibility of manufacturing ion-selective materials with improved sorption activity. These future investigations present especial interest.

Developed material will be interesting in use first of all in hydrometallurgy and in processes of water treatment.

For waste of phosphorus-containing polypeptides are provided possibility conversion in the effective fireproof coverings.

References

[1] Ana Laura Martinez-Hernandez; Carlos Velasco-Santos; Miguel de Icaza, Victor M. Castano; *e-Polymers* 2003, *no. 016*, 1

[2] Shapiro D.K; *Practical work in biological chemistry*; High School: Moscow, 1976

[3] GOST 208512-75

[4] Korostelev P.P; *Photometric and complexonometric analysis in metallurgy*; Metallurgy: Moscow, 1984

In: Progress in Monomers, Oligomers, Polymers, Composites … ISBN 978-1-60456-878-3
Editors: R. A. Pethrick, G.E. Zaikov et al.

Chapter 14

PHYSICAL AND MECHANICAL PROPERTIES OF HIGHLY FILLED POLYMER COMPOSITES

O.A. Legonkova *

Moscow State University of Applied Biotechnology,
Moscow, Russian Federation

ABSTRACT

Physical and mechanical properties of the filled polymer composite materials (PCM) in dependence on the extent of filling, the rate of deformation were investigated. It was found out that structural properties of the filled composite materials are determined with the nature of polymer matrix, filling degree, nature of the fillers, structural organization of FCM, that is being formed in the process of receiving of the composite materials, and conditions of tests.

Key words: composite materials, physical and mechanical properties, strength.

Today the overwhelming majority of polymers are applied as composite materials (PCM). Properties PCM, as a rule, are not a sum of properties of the components, and are defined by the variety of chemical and physical processes as the result of interaction of components on the borders of phases. Introduction of fillers into polymers brings essential changes in the mobility of macromolecules on boundary layers, arises various kinds of interaction between polymers and the surface of fillers, influences on chemical structure of the fillers and polymers during reception and exploitation of PCM [1-5]. However insufficient attention is given to research of physical and mechanical properties of highly filled materials consisting of polymeric matrix and mixture of fillers of various nature, giving biodegrability of composition as a whole.

* Correspondence to: Moscow State University of Applied Biotechnology, Moscow, 109316, Talalikhina str., 33, email: OALegonkovaPB@mail.ru

At present work the following large-capacity polymers were taken as polymeric basis: acrylate-styrene carboxylated latex - Lentex A4 (TU 2241-001-47923137-01); copolymer of ethylene and vinyl acetate (cevilen, mark #11306-070; polyamide (PA, which is copolymer of hexamethylene diamine, adipate and sebacic acids with melting temperature 130^{0}C); thermoplastic polyurethane (PU), which is produced by "Agropolymer" firm (Russia, TU 5141-003-17823007), polyvinyl spirit (PVS) with the degree of saponification - 88 %. Waste products of thrashing grains(organic filler) of the third category (size of particles 63-240mcm, bulk density of 350 kg / m^3; humidity of 4 %), and water-soluble mineral fertilizer (inorganic filler with the following make-up (in %): $(NH_4)_2SO_4$- 35, $NH_4H_2PO_4$ - 6, KNO_3 - 32, $MgSO_4*7H_2O$ - 27) were taken as fillers. Depending on properties of polymeric matrix various ways of reception of composites were applied. Compositions on basis of water-soluble PVS and industrial Lentex A4 have been received via watering of water solutions consisting of organic and inorganic fillers. Compositions on polyurethane basis were received through hardening on air of the component mixture. For reception of samples of PCM on sevilen basis and PA matching of components were carried out by laminar mixture in melt, pressing, and also by making a mixture of components and pressing. Thus samples received as plates with thickness 500 microns.

It was noticed while investigating the physical and mechanical properties of PCM on PA basis and speed of deformation at 100 mm/min for two component systems, that at introduction of inorganic filler into composition up to 10 % the breaking point insignificantly falls. It could be explained by the following: the filler at its small concentration in PCM in absence of aggregation powdered particles, carries out role of concentrator of internal pressure and is a potential source of cracks' growth [1]. At increase of concentration inorganic filler up to 30 % durability of PCM increases in 2, 5 times and that is explained with the creation of more obstacles for development of cracks. Owing to this the breaking processes of destruction stops [1]. Reduction of durability with increase of contents of filler is typical for highly filled PCM. At concentration of inorganic filler in the quantity of 75 % samples become very fragile, have durability on order below, than the initial samples. Relative lengthening at break for all samples decreases, and finally, becomes in 2 times less, than at initial sample. For two componental PCM with organic filler similar law is traced, relative hardening system is observed at small concentration of filler (up to 10 %) which is not so significant as in two-componental system with inorganic filler (increase of durability - in 1,5 times).

At creation of two component systems on sevilen basis at monoaxial deformation 100mm/mines the «effect of temporary conversion of reinforcing action of fillers», for the first time described in the work [7], wasn't found: with increase of content of organic and inorganic fillers durability and relative deformation reduced, fig. 1,2. Samples become more rigid (module grows) in case of filling organic filler. At introduction of inorganic filler even at large concentrations (weights of 60 %) samples keep rather high plasticity (deformation at destruction is appr. 400 %).

A.

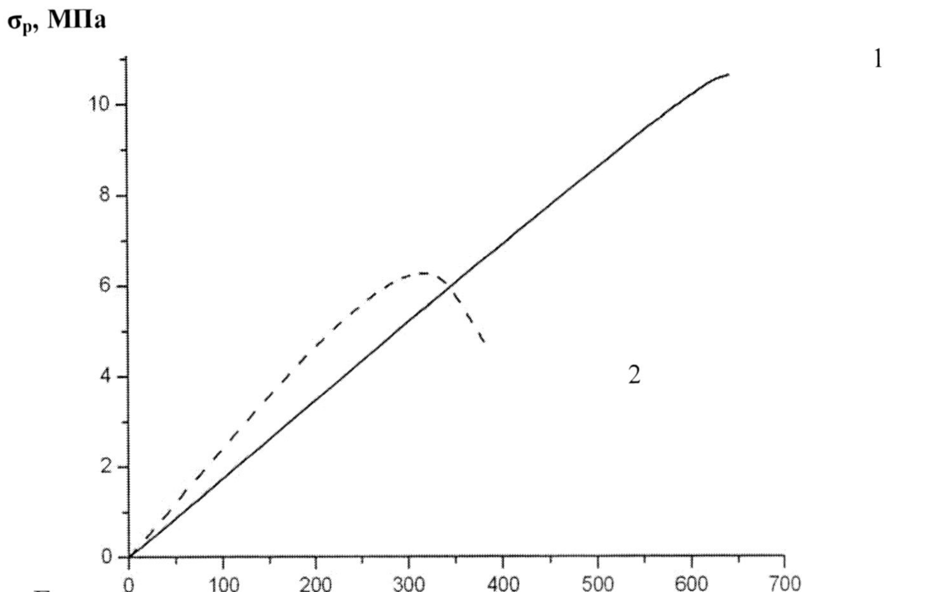

Б.

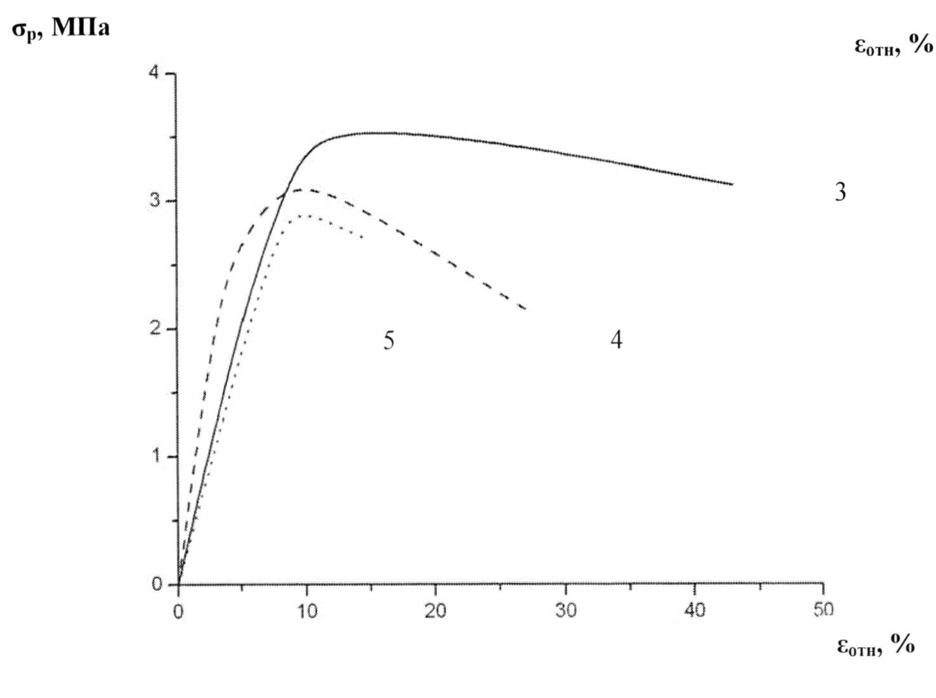

Figure 1. Physical and mechanical properties PCM on basis of sevilen. Content of organic filler, weight in %: 1 - 0 %; 2 - 20 %; 3 - 40 %; 4 - 50 %; 5 - 60 %. Speed of monoaxial deformation is 100 mm / min.

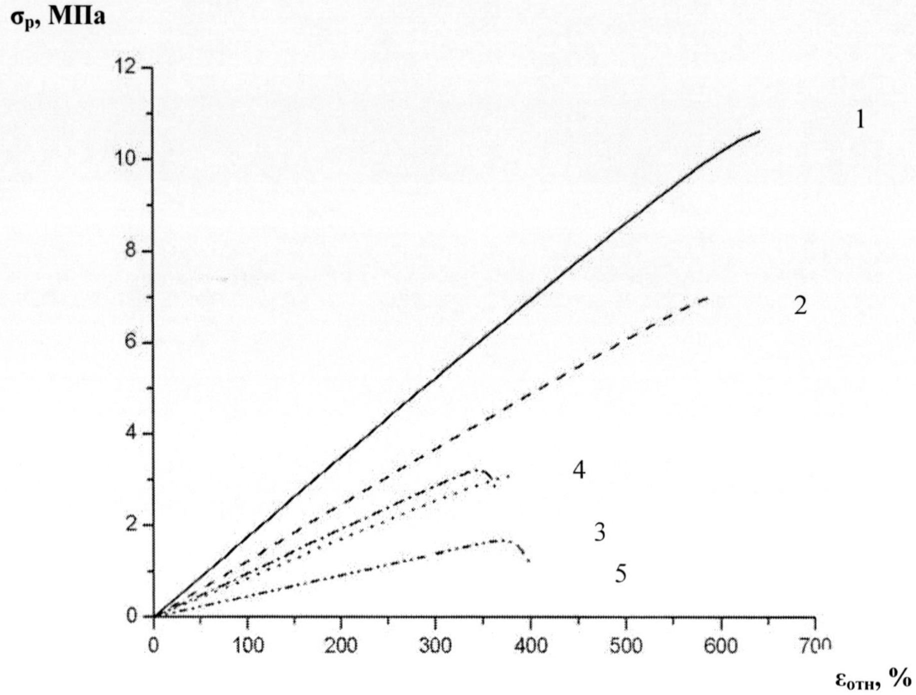

Figure 2. Physical and mechanical properties of two componental PCM on basis of sevilen. Content of inorganic of filler, weight in %: 1-0 %; 2 - 20 %; 3 - 40 %; 4 - 50 %; 5 - 60 %. Speed of monoaxial deformations - 100 mm / min.

According to IK-spectroscopy inorganic filler is inert in the relation to polymeric matrix because the identical spectra of absorption of initial unfilled polymer and the polymer filled with inorganic filler are the same. Hence, mechanical behavior of PCM is defined by structural organization of composite materials, its dependence on conditions of formation and test on PCM (in our case, speed of deformation).

Electronic- microscopic photos of PCM surface on basis of PA and sevilen are submitted on fig. 3, 4. It is visible, that systems heterogeneous. Introduction of organic filler results in reception of more homogeneous systems, than introduction of inorganic filler. Systems with inorganic filler include as crystallites of salts of filler as their more difficult formations. However distribution of inorganic filler in sevilen environment is distinct from its distribution in PA environment, fig. 3 (1-3). In PCM on basis of PA structure of composite at 20 % filling is more defective on all surface of spalling, creating obstacles for development of cracks. While in PCM systems on basis of sevilen defects "were pressed" in homogeneous structure of composite, therefore decrease in durability PCM occurs due to the reduction of maintenance of polymer in a composition and hardenings PCM is not observed at speed of deformation of 100 mm / min.

Physical and mechanical properties of two componental PCM on basis of PU, Lentex A4, PVS are submitted in tables 1-2, fig. 5. Introduction of organic filler resulting in decrease of elasticity of PCM and introduction of inorganic filler leading to preservation of enough elasticity of composite in comparison with PCM, filled with organic filler are general for all investigated systems.

1.

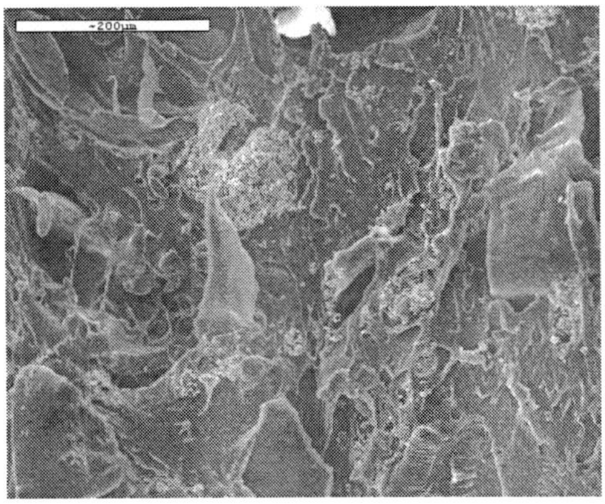

10% inorganic filler

4.

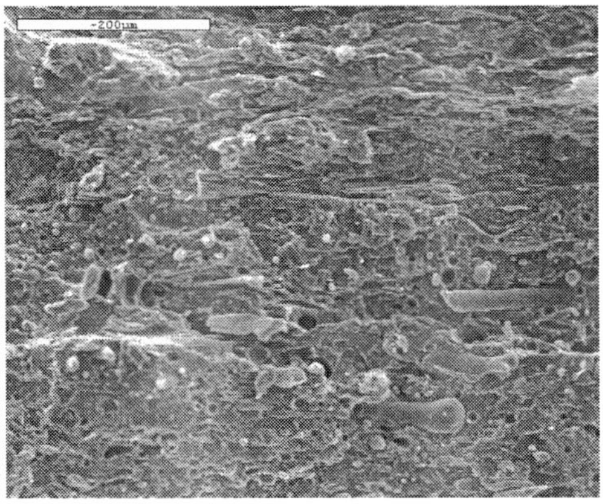

10% organic filler

2.

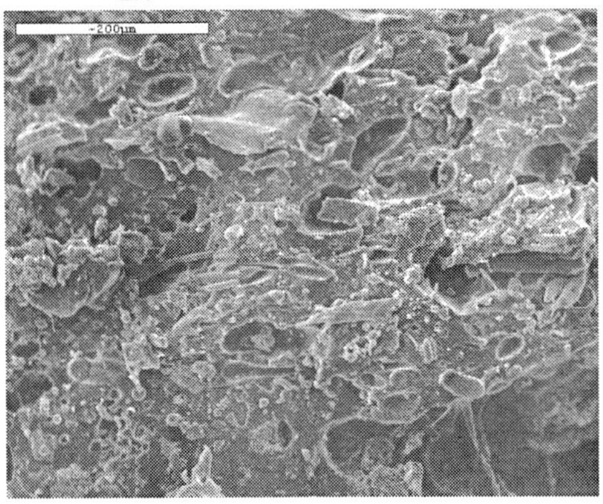

20% inorganic filler

5.

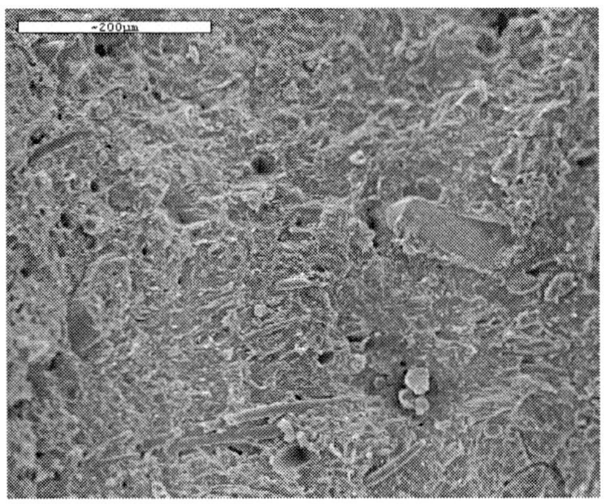

20% organic filler

3.

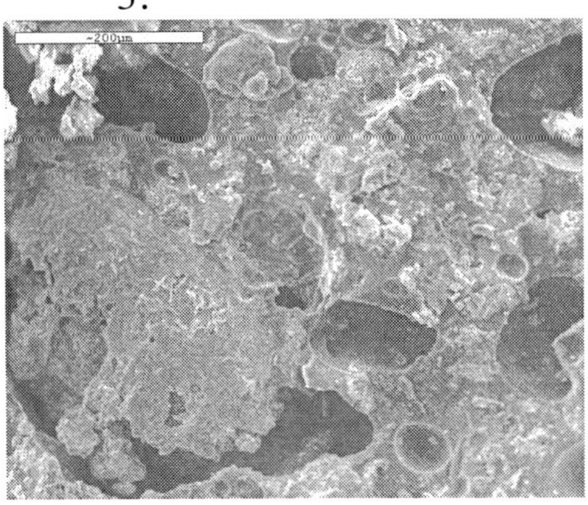

50% inorganic filler

6.

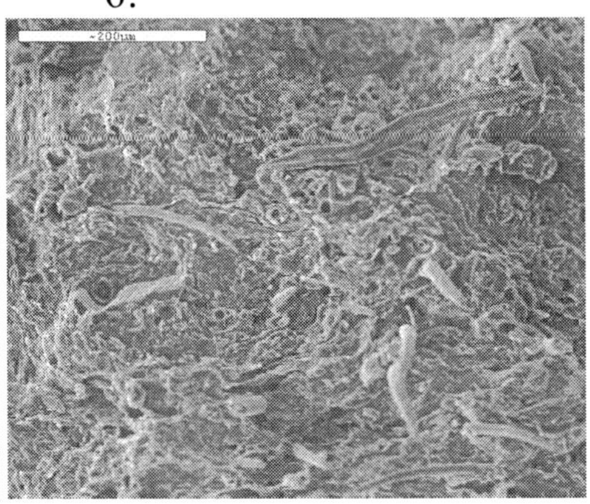

50% organic filler

Figure 3. Electronic - microscopic photos of chips of two component systems PCM on basis of PA and inorganic filler (1,2,3), and organic filler (4,5,6.) (Increase 200 microns)

1.

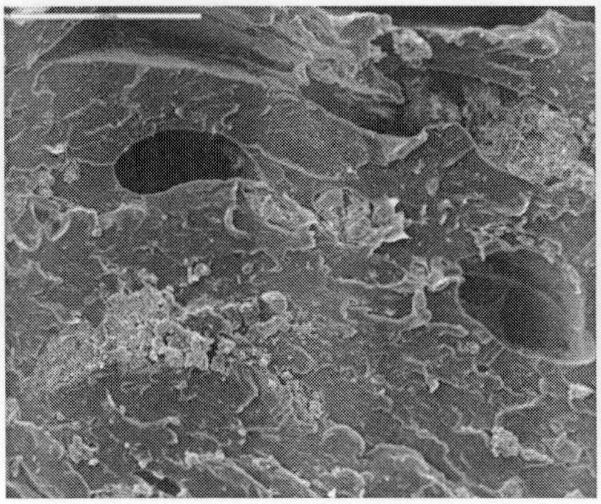

20% inorganic filler

4.

three componental PCM

2.

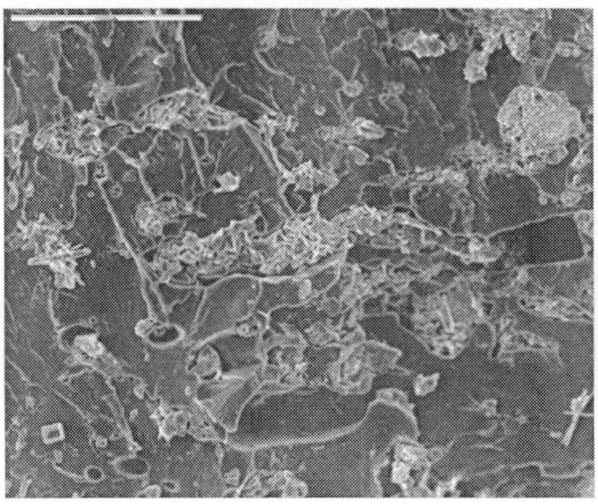

40% inorganic filler

5.

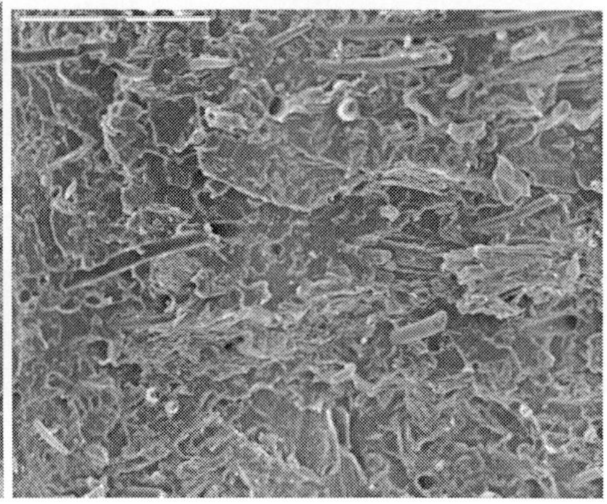

40% organic filler

3.

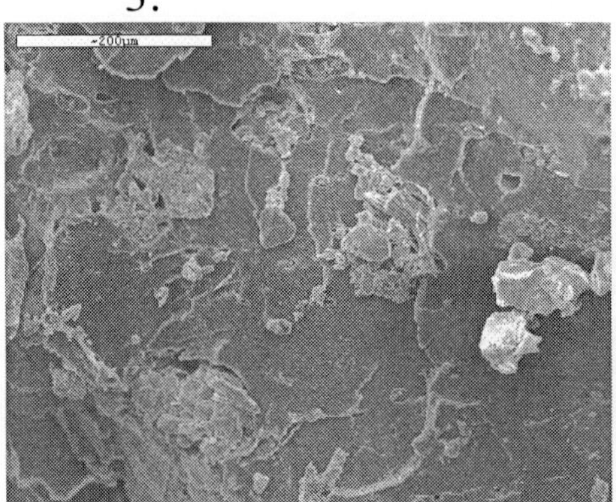

60% inorganic filler

6.

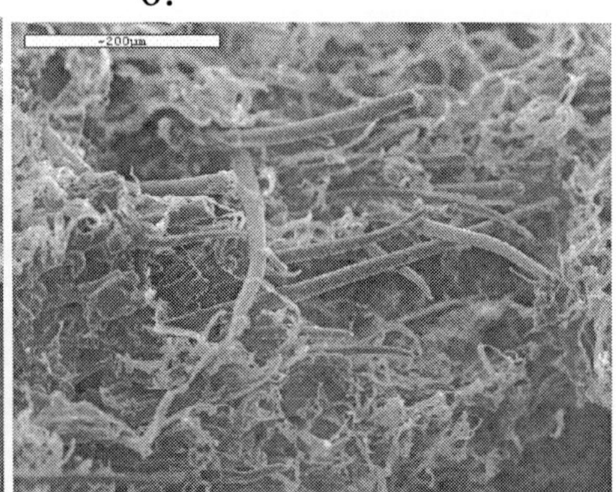

60% organic filler

Figure 4. Electronic - microscopic photos of chips of two component systems PCM on basis of sevilen and inorganic filler (1,2,3), and organic filler(5,6), three componental PCM: 50 % organic filler, 35 % inorganic filler. (Increase 200 microns)

«Effect of temporary conversion of reinforcing action of fillers» was found at filling with organic and inorganic fillers, both in two and three componental systems. Physical and mechanical properties of PCM on basis of PVS with concentration of filling inorganic filler 20 % at various speeds monoxial stretching are presented on fig. 6.

Table 1. Physical and mechanical properties of two component PCM on the basis of PU at stretching (speed of deformation is 100 mm / min)

Maintenance, weight %			durability, MPa	deformation, %
Полимер	organic filler	inorganic filler		
90	10	-	21,4	480
80	20	-	16,4	360
79	30	-	13,3	270
46	54	-	4,4	6,6
90	-	10	25,0	500
80	-	20	22,6	480
70	-	30	16,3	300
60	-	40	8,6	240

Table 2. Physical and mechanical properties of two componental PCM on basis Lentex 4 at stretching (speed of deformation is 100 mm / min)

Maintenance of Lentex A4, %	maintenance organic filler, %	maintenance inorganic filler, %	durability, MPa	Deformation of destruction, %
100	0	0	3,1	600
90	0	10	1,3	300
80	0	20	1,0	300
60	0	40	0,8	180
90	10	0	2,5	300
80	20	0	1,7	20
60	40	0	1,6	12

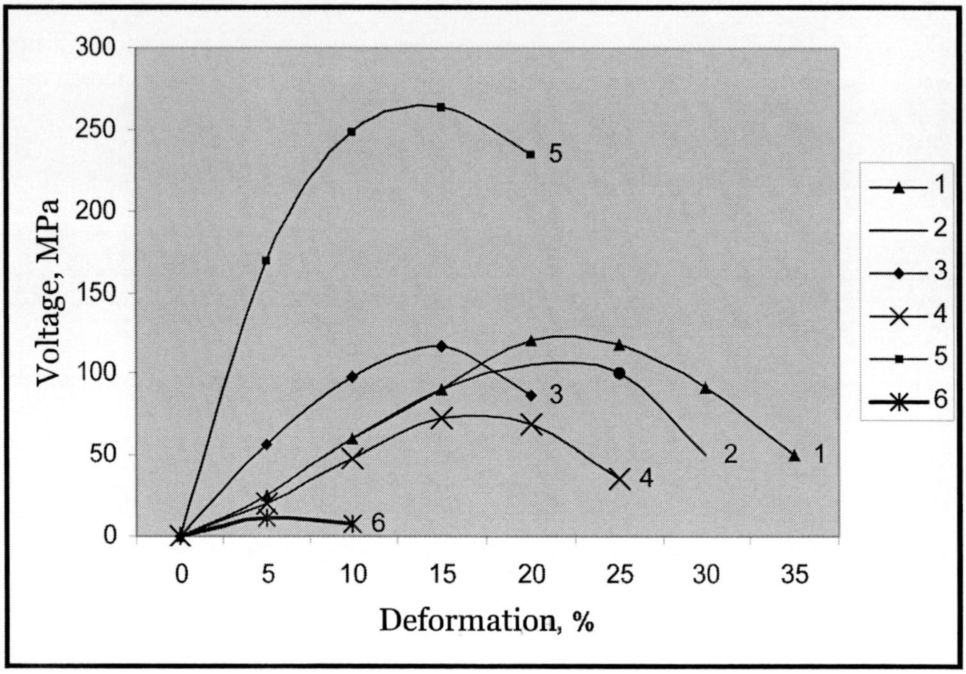

Figure 5. Physical and mechanical properties of PCM on the basis of PVS (monoaxial deformation 100 mm/min). Content of inorganic filler, weight in %: 1-0 %; 2 - 10 %; 3 - 20 %; 4 - 50 %; ratio of organic filler, weight in %: 5 - 20 %; 6 - a parity of polymer, organic filler and inorganic filler in three component PCM is 1:1:1.

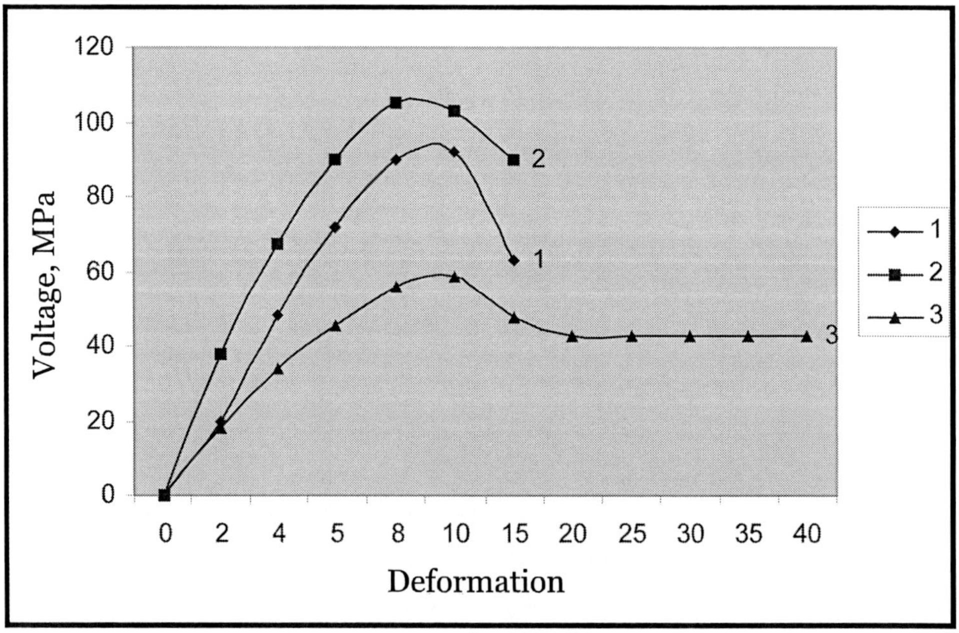

Figure 6. Physical and mechanical properties of PCM on the basis of PVS, filled inorganic filler (20 % of weights) at various speeds monoaxial deformation: 1-0,5 mm / min; 2 - 10 mm / min; 3 - 100 mm / min.

While investigating the three component PCM on basis of investigated polymers, fig. 6-7, tab. 3, 4, it should be noted that the basic contribution to durability of three componental PCM brings organic filler. Even in highly filled systems where the general content of fillers reaches 85 %, the given dependence is kept. Inorganic filler, as well as in case with two-component materials gives plasticity of PCM.

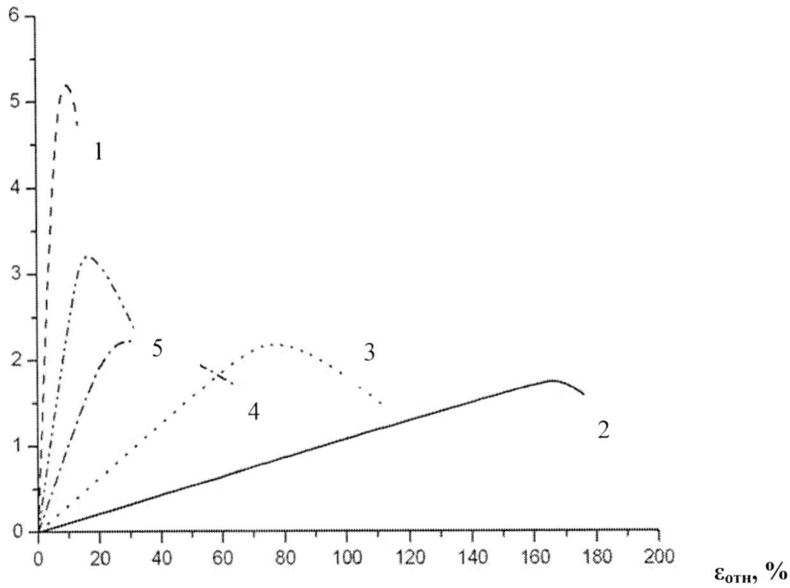

Figure 7. Physical and mechanical properties three component PCM on the basis of sevilen, weight in mass. %: 1 - sevilen/organic filler - 50/50; 2 - sevilen/ inorganic filler - 50/50; 3 - sevilen/ inorganic filler/ organic filler. - 50/37/13; 4 - sevilen/ inorganic filler/ organic filler.-50/25/25; 5 - sevilen/ inorganic filler/ organic filler.-50/13/37. The rate of stretching is 100 mm/ min.

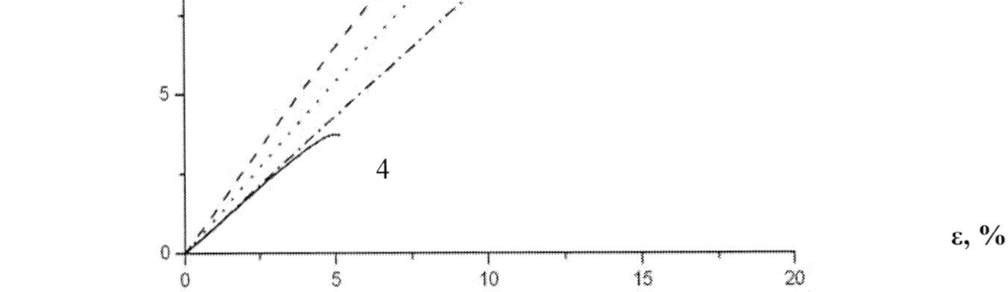

Figure 8. Physical and mechanical properties of three component PCM on the basis of PA. Content of polymer/organic filler/inorganic filler, weight in mass. %: 1 - 50/10/40; 2 - 50/20/30; 3 - 50/25/25; 4 - 50/40/10.

Table 3. Physical and mechanical properties of three componental PCM on basis PU at test for stretching (speed of deformation of 100 mm / min)

Content, %			Durability, MPa	Deformation, %
polymer	Organic filler	Inorganic filler		
80	10	10	7,3	110
72	14	14	7,1	94
60	20	20	6,5	63
50	25	25	6,1	26

At research of physical and mechanical properties of PCM received via pressing, all the above described dependences in changes of durability depending on degree of filling and conditions of test are kept.

Table 4. Physical and mechanical properties of the high filled PCM on the basis of sevilen (speed of deformation is 100 mm / min)

Content, %			durability, MPa	Deformation, %
polymer	Organic filler	Inorganic filler		
15	35	50	0,72	5,2
15	45	45	1,06	5,0
15	50	35	1,20	6,3
15	60	25	1,31	4,0
10	55	35	0,80	6,3

Being based on the above discussed results we can estimate that durability dependence on content of each type of the fillers (organic or inorganic) is linear:

$$f(x_j) = \alpha_{0j} + \alpha_{1j} \cdot x_j, \tag{1}$$

where j – factor number (filler), α_{0j}, α_{1j} – equation coefficient for the j-factor (j = 1, 2). For example, these dependences for samples based on sevilen are given in the fig. 8 (a, б).

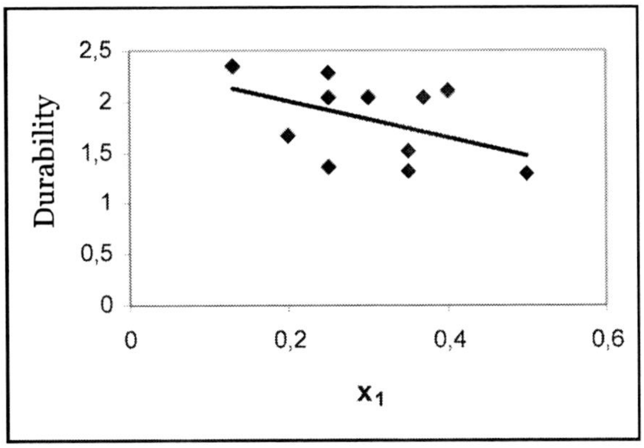

a)

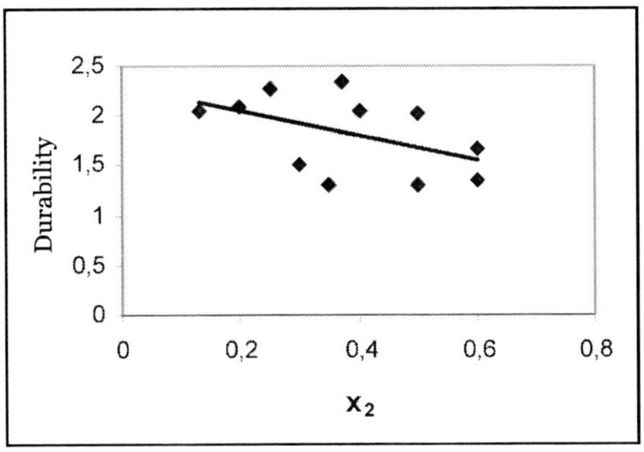

b)

Figure 8. Dependence durability of filler content: a – of inorganic filler (x_1); b – of organic filler (x_2).

Brandon's method was used to describe the behavior of three component PCM [7,8]. According to this method the approximate function has the following appearance:

$$y = \lambda \cdot f_1(x_1) \cdot f_2(x_2), \tag{2}$$

where y – durability of PCM; λ - constant; $f_1(x_1) = \alpha_{01} + \alpha_{11} \cdot x_1$, $f_2(x_2) = \alpha_{02} + \alpha_{12} \cdot x_2$ – dependence of PCM durability on filler content, correspondingly, inorganic (x_1) and organic (x_2) fillers.

The meaning of λ constant equals to the medium experimental meaning of the exit parameter

$$\lambda = \frac{1}{N} \sum_{i=1}^{N} y_i \text{ , i=1, 2, …, N.} \tag{3}$$

Where y_i – durability for PCM.

After normalization of experimental data via division y_i on λ according to the formula

$$y_i^* = \frac{y_i}{\lambda} \tag{4}$$

for two pairs of variables $\left(x_{1i}, y_i^*\right)$ with the help of method of minimum squares, constants of the first component in the regression equation $f_1(x_1) = \alpha_{01} + \alpha_{11} \cdot x_1$ were calculated.

Constants of the second component of the function $f_2(x_2) = \alpha_{02} + \alpha_{12} \cdot x_2$ for variable x_2 and remaining function y_{i1} were calculated via excluding the meaning of the first component of the function $f_1(x_1)$ out of the normalized meanings of the exit parameter y_i^* according to the formula

$$y_{i1} = \frac{y_i^*}{f_1(x_1)}. \tag{5}$$

After determination of $f_1(x_1)$ и $f_2(x_2)$ the common formula of plural regression was built.

As a result, the following regression equations for three component PCM were received:

- for PCM based on sevilen

$$y = 1{,}82 \cdot (1{,}29 - 0{,}96x_1) \cdot (1{,}36 - 0{,}95x_2) \tag{6}$$

- for PCM based on PA

$$y = 8{,}41 \cdot (2{,}07 - 4{,}08x_1) \cdot (0{,}96 - 0{,}15x_2) \tag{7}$$

- for PCM based on PU

$$y = 6{,}75 \cdot (1{,}21 - 1{,}23x_1) \cdot (1{,}01 - 0{,}03x_2) \tag{8}$$

- for PCM based on Lentex

$$y = 1{,}40 \cdot (2{,}01 + 9{,}64x_1) \cdot (1{,}27 - 1{,}01x_2) \tag{9}$$

- for PCM based on PVS

$$y = 39{,}63 \cdot (4{,}36 - 11{,}40x_1) \cdot (2{,}54 - 6{,}76x_2) \quad (10)$$

The quality control of approximation according to Fisher criteria revealed that equations (6-10) sufficiently reflect the behavior of three component PCM at $p<0{,}1$ (the level of the meaning).

Thus, strength characteristics of PCM are defined with nature of polymeric matrix, degree of filling and nature of fillers, that structural organization which is formed during reception PCM and test specifications (as response PCM to external influence).

Fillers bring corrective amendments into mechanical properties of composite material independently from each other, namely: presence of organic filler results in increase of durability, presence of inorganic filler keeps plasticity of systems almost for all investigated materials.

For each composite there is the rate of deformation at which "the effect of strengthening action of the filler" realizes.

The equation of regression for three component composite materials based on sevilen, PA, PU, Lentex, PVS were received that let us to forecast the durability properties of PCM and finally properties of industrial articles.

REFERENCES

[1] Lipatov U.S. Physical and chemical of basis of filling of polymers. *M.: Chemistry*. 1991.260 p.

[2] 2 Simonov-Emeljanov I.D., Kuleznev V.N. Basis of creation of composite materials. *M.: МИХМ*, 1986, 86 p.

[3] L.Nilsen. Mechanical properties of polymers and polymeric compositions. M., *Chemistry*, 1978, 310 p.

[4] Physical chemistry of polymeric compositions. collection of articles, « *Naukova Dumka* », Kiev, 1974, 183 p.

[5] V.E.Gul. Structure and properties of polymers. M., *Chemistry,* 1978, 300 p.

[6] SolomkoV.P. Filled crystallized polymers. Kiev, *Naukova Dumka*, 1980, 263 p.

[7] 7.Bondar A.G. Mathematical modeling in chemical technology. Kiev, *Vuscha shkola*, 1973, 280 p.

[8] 8. Podvaljnii S.L. Modelling of industrial processes of polymerization. Moscow, "*Chemisty*", 1979, 256 p.

In: Progress in Monomers, Oligomers, Polymers, Composites … ISBN 978-1-60456-878-3
Editors: R. A. Pethrick, G.E. Zaikov et al.

Chapter 15

QUANTUM-CHEMICAL CALCULATION OF ISOOLEFINES

A.V. Babkin[1,*], D.A. Frolov[1], U.A. Sangalov, E.S. Titova[2,†], V.U. Dmitriev[1] and G.E. Zaikov[3,‡]

[1]Volgograd State Architect-Build University Sebrykov Departament, Michailovka, Volgograd region, Russian federation

[2]Volgograd State Technical University, Volgograd; Russian Federation

[3]Institute of Biochemical Physics Russian Academy of Sciences; Moscow, Russian Federation

ABSTRACT

Quantum-chemical calculation of isoolefines: isobutylene, 2-methyllbutene-1, 2-methylbutene-2, 2-methylpentene-1, 2-ethylbuten-1, 2,3-dimethylbutene-1, 2,3,3-trimethylbutene-1, 2,4,4-trimethylpentene-1, 2-isopropyl-3-butene-1, 2-tret.-butyl-3,3-dimethylbutene-1 was made by classical method CNDO/2 in parametrization of Santri-Poppla-Segala. Electronic and geometric structure of these isoolefiines was calculated. Acid power of these compounds was determined. All under study isoolefines possess low acid power and pertain to class very weak H-acids (pKa >14). Exists the row an isoolefiines more capable to cation polymerization, than isobutylene

Keywords: cation polymerization, isoolefines, acid power, quantum-chemical calculation, method CNDO/2.

* Correspondence to: e-mail organic@vstu.ru

† Correspondence to: e-mail chembio@sky.chph.ras.ru

‡ Correspondence to; e-mail sfi@reg.avtlg.ru

AIMS AND BACKGROUNDS

Polymerization (oligomerization, telomerization) of isoolefines (isobutylene, 2-methylbutene-1, 2-methylbutene-2, 2-methylpentene-1, 2,3-dimethylbutene-1, 2-ethylbutene-1, 2,3,3-trimethylbutene-1, 2,4,4-trimethylpentene-1, 2-propiy-3-methylbutene-1) was studied [1-3]. The Full review of industrial important cation processes isobutylene there is in work [4]. Dates by chemistries of oligomers of isobutylene has got Butlerov [5]. The studies of conditions of polymerization of isobutylene are presented in monograph [1-2]. Polymers of 2-methylbutene-1, 2- methylbutene-2, 2-methylpentene-1, 2,3-dimethylbutene-1 have got the Danish researchers, used catalysts $AlCl_3$, $AlBr_3$, $TiCl_4$ (solvents-ethyl chloride, vynil chloride T=-175 ÷ -78 ^{0}C). Polymerization of 2,3,3-trimethylbutene-1 and 2,4,4-trimethylpentene-1 goes with formation of oligomers [6]. Mechanisms of initiation, growing and breakaway of circuit were studied are only for isobutylene [1-3]). The Geometric structure of enumerated monomer, influence of acid power of isoolefines on process of polymerization not studied.

The aim of this work is an estimation of acid power of isoolefines by quantum-chemical methods. Calculations will help to understand the mechanisms of initiation, growing and breakaway circuit processes of катионной polymerization and oligomerization.

METHODICAL PART

Quantum-chemical calculation of isoolefines was made by classical method CNDO/2 in parametrization of Santri-Popla-Segala with optimization of the geometries by all parameter by standard gradient method [7]. Acid power was calculated by formula pKa = 43,7 – 190,3 · $q_{H^+}^{max}$ (1) (where pKa - an universal factor to acidity, $q_{H^+}^{max}$ - a maximum positive charge on atom of the hydrogen)[3].

RESULTS AND DISCUSSION

Results of quantum-chemical calculation geometric and electronic structure, the general energy - E_0, dipole moment got by method CNDO/2 are presented in tabl. 1-10, on fig. 1 - 9.

From tabl. 1-10, the draw. 1-9 is seen that all under investigation isoolefines possess alike (a mistake of the method CNDO/2 ± 0.02) maximum charge on atom of the hydrogen. The universal factors of acidity were calculated by formula (1). All under study isoolefines possess low acid power and this is class of very weak H-acids (pKa >14). The charge on α-atom of carbon of isoolephynes defines the ability of the monomer to cation polymerization. The Charge increases, olefine easier enters in reaction [1-3]. Variety of isoolefines hase equal or even more prone to cation polymerization, than isobutylene (table 10).

Molecules with more branched radical of β-atom of carbon form the more stable active centers (AC). Polymers with big MM are got [ ]. This regularity is exists only for asymmetric isoolefines (the degree of branch of radical of C_β— atom of carbon increases for one branch only. The second branch is methyl. The row of 2-methylbutene-1, 2-methylpentene-1 and other MM increases accordingly 322000, 336000) [1].

Table 1.

Coordinate	Length of bond R, Å	Coordinate	Valency corners, Grad
C(1)-C(2)	1.32	C(3)-C(2)-C(1)	118
C(2)-C(3)	1.46	C(4)-C(2)-C(1)	125
C(2)-C(4)	1.47	C(5)-C(4)-C(2)	120
C(4)-C(5)	1.45	H(6)-C(1)-C(2)	125
C(1)-H(6)	1.12	H(7)-C(1)-C(2)	127
C(1)-H(7)	1.12	H(8)-C(3)-C(2)	114
C(3)-H(8)	1.12	H(9)-C(3)-C(2)	112
C(3)-H(9)	1.12	H(10)-C(3)-C(2)	112
C(3)-H(10)	1.12	H(11)-C(4)-C(2)	109
C(4)-H(11)	1.12	H(12)-C(4)-C(2)	117
C(4)-H(12)	1.12	H(13)-C(5)-C(4)	111
C(5)-H(13)	1.12	H(14)-C(5)-C(4)	110
C(5)-H(14)	1.12	H(15)-C(5)-C(4)	116
C(5)-H(15)	1.11		

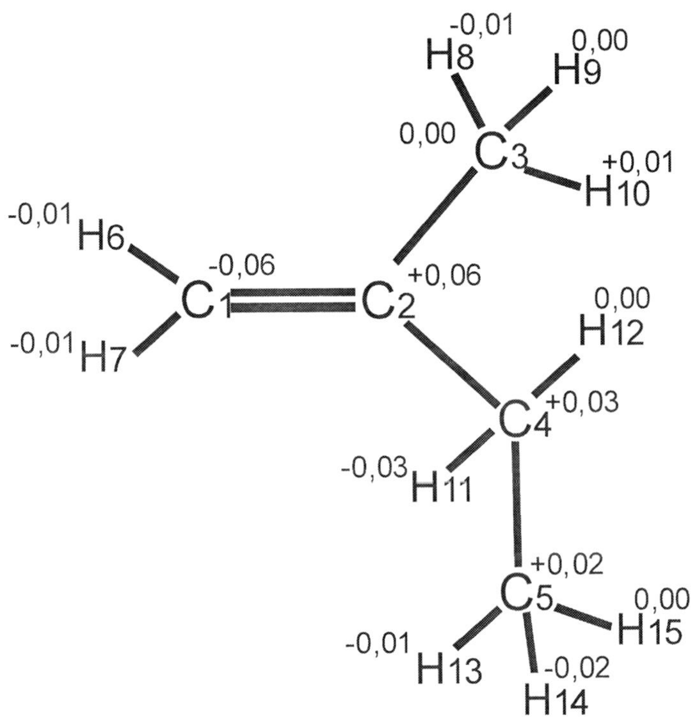

Figure 1. Geometric and electronic structure of 2-methylbutene-1.

Table 2.

Coordinate	Length of bond R, Å	Coordinate	Length of bond R, Å
C(1)-C(2)	1.35	C(3)-C(2)-C(1)	127
C(2)-C(3)	1.45	C(4)-C(1)-C(2)	118
C(1)-C(4)	1.46	C(5)-C(1)-C(2)	124
C(1)-C(5)	1.46	H(6)-C(2)-C(1)	117
C(2)-H(6)	1.12	H(7)-C(3)-C(2)	113
C(3)-H(7)	1.12	H(8)-C(3)-C(2)	113
C(3)-H(8)	1.12	H(9)-C(3)-C(2)	113
C(3)-H(9)	1.12	H(10)-C(4)-C(1)	113
C(4)-H(10)	1.12	H(11)-C(4)-C(1)	112
C(4)-H(11)	1.12	H(12)-C(4)-C(1)	112
C(4)-H(12)	1.12	H(13)-C(5)-C(1)	112
C(5)-H(13)	1.12	H(14)-C(5)-C(1)	112
C(5)-H(14)	1.12	H(15)-C(5)-C(1)	112
C(5)-H(15)	1.12		

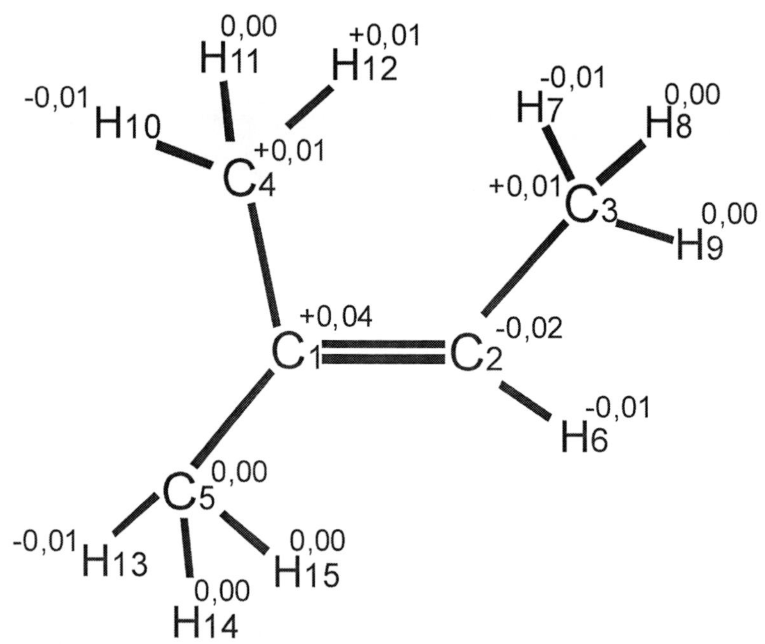

Figure 2. Geometric and electronic structure of 2-methylbutene-2.

Table 3.

Coordinate	Length of bond R, Å	Coordinate	Length of bond R, Å
C(1)-C(2)	1.33	C(3)-C(2)-C(1)	119
C(2)-C(3)	1.46	C(4)-C(2)-C(1)	118
C(2)-C(4)	1.46	C(5)-C(4)-C(2)	117
C(4)-C(5)	1.49	C(6)-C(5)-C(4)	118
C(5)-C(6)	1.46	H(7)-C(1)-C(2)	125
C(1)-H(7)	1.12	H(8)-C(1)-C(2)	124
C(1)-H(8)	1.12	H(9)-C(3)-C(2)	113
C(3)-H(9)	1.12	H(10)-C(3)-C(2)	113
C(3)-H(10)	1.12	H(11)-C(3)-C(2)	114
C(3)-H(11)	1.12	H(12)-C(4)-C(2)	109
C(4)-H(12)	1.12	H(13)-C(4)-C(2)	110
C(4)-H(13)	1.12	H(14)-C(5)-C(4)	109
C(5)-H(14)	1.12	H(15)-C(5)-C(4)	109
C(5)-H(15)	1.12	H(16)-C(6)-C(5)	112
C(6)-H(16)	1.12	H(17)-C(6)-C(5)	113
C(6)-H(17)	1.12	H(18)-C(6)-C(5)	110
C(6)-H(18)	1.12		

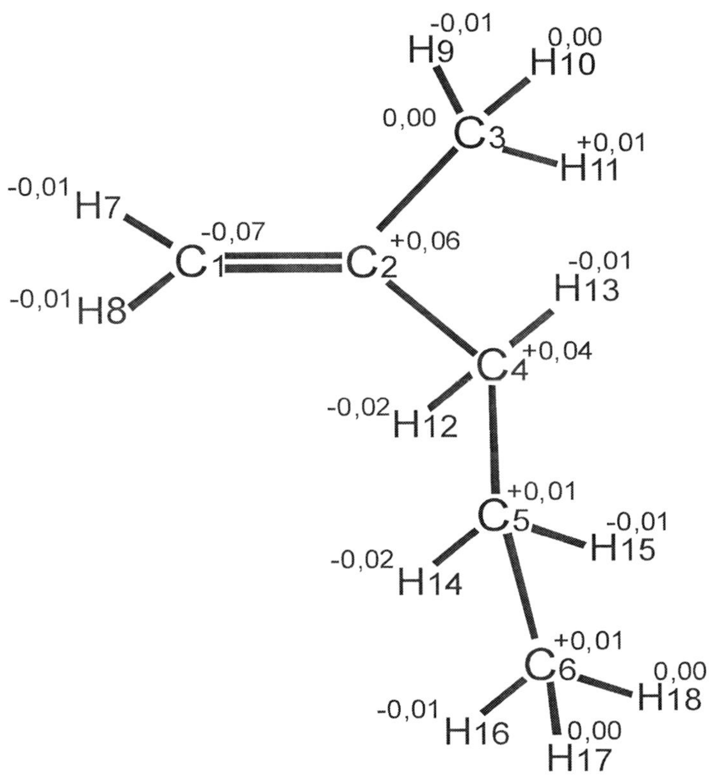

Figure 3. Geometric and electronic structure of 2-methylpentene-1.

Table 4.

Coordinate	Length of bond R, Å	Coordinate	Length of bond R, Å
C(1)-C(2)	1.34	C(3)-C(2)-C(1)	123
C(2)-C(3)	1.46	C(4)-C(3)-C(2)	113
C(3)-C(4)	1.46	C(5)-C(2)-C(1)	123
C(2)-C(5)	1.46	C(6)-C(5)-C(2)	113
C(5)-C(6)	1.48	H(7)-C(1)-C(2)	121
C(1)-H(7)	1.12	H(8)-C(1)-C(2)	121
C(1)-H(8)	1.12	H(9)-C(3)-C(2)	112
C(3)-H(9)	1.12	H(10)-C(3)-C(2)	111
C(3)-H(10)	1.12	H(11)-C(4)-C(3)	112
C(4)-H(11)	1.12	H(12)-C(4)-C(3)	112
C(4)-H(12)	1.12	H(13)-C(4)-C(3)	113
C(4)-H(13)	1.12	H(14)-C(5)-C(2)	112
C(5)-H(14)	1.12	H(15)-C(5)-C(2)	111
C(5)-H(15)	1.12	H(16)-C(6)-C(5)	112
C(6)-H(16)	1.12	H(17)-C(6)-C(5)	112
C(6)-H(17)	1.12	H(18)-C(6)-C(5)	109
C(6)-H(18)	1.12		

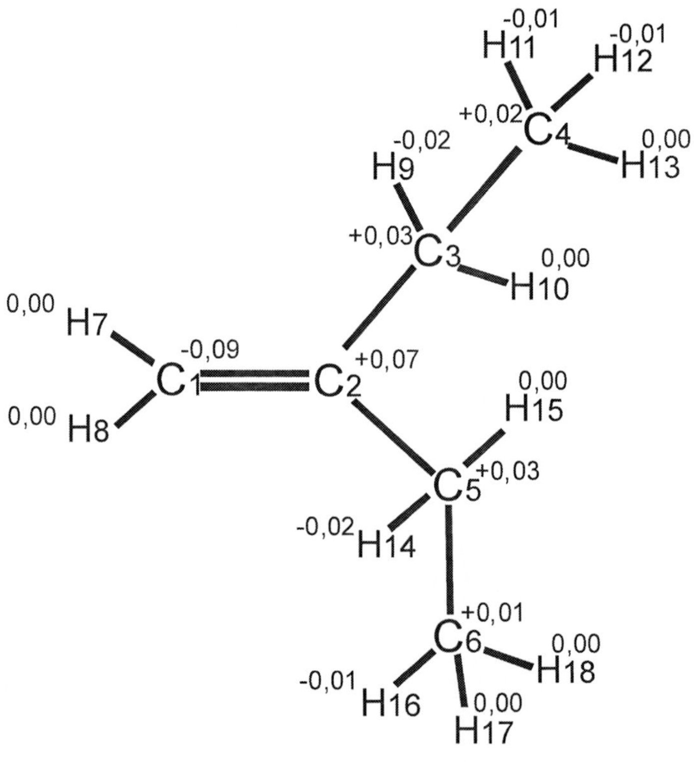

Figure 4. Geometric and electronic structure of 2-ethylbutene-1 C_6H_{12}**.

Table 5.

Coordinate	Length of bond R, Å	Coordinate	Length of bond R, Å
C(1)-C(2)	1.34	C(3)-C(2)-C(1)	119
C(2)-C(3)	1.48	C(4)-C(2)-C(1)	124
C(2)-C(4)	1.48	C(5)-C(4)-C(2)	112
C(4)-C(5)	1.48	C(6)-C(4)-C(2)	113
C(4)-C(6)	1.48	H(7)-C(1)-C(2)	120
C(1)-H(7)	1.12	H(8)-C(1)-C(2)	121
C(1)-H(8)	1.12	H(9)-C(3)-C(2)	113
C(3)-H(9)	1.12	H(10)-C(3)-C(2)	112
C(3)-H(10)	1.12	H(11)-C(3)-C(2)	112
C(3)-H(11)	1.12	H(12)-C(4)-C(2)	110
C(4)-H(12)	1.12	H(13)-C(5)-C(2)	113
C(5)-H(13)	1.12	H(14)-C(5)-C(2)	112
C(5)-H(14)	1.12	H(15)-C(5)-C(2)	112
C(5)-H(15)	1.12	H(16)-C(6)-C(5)	112
C(6)-H(16)	1.12	H(17)-C(6)-C(5)	112
C(6)-H(17)	1.12	H(18)-C(6)-C(5)	109
C(6)-H(18)	1.12		

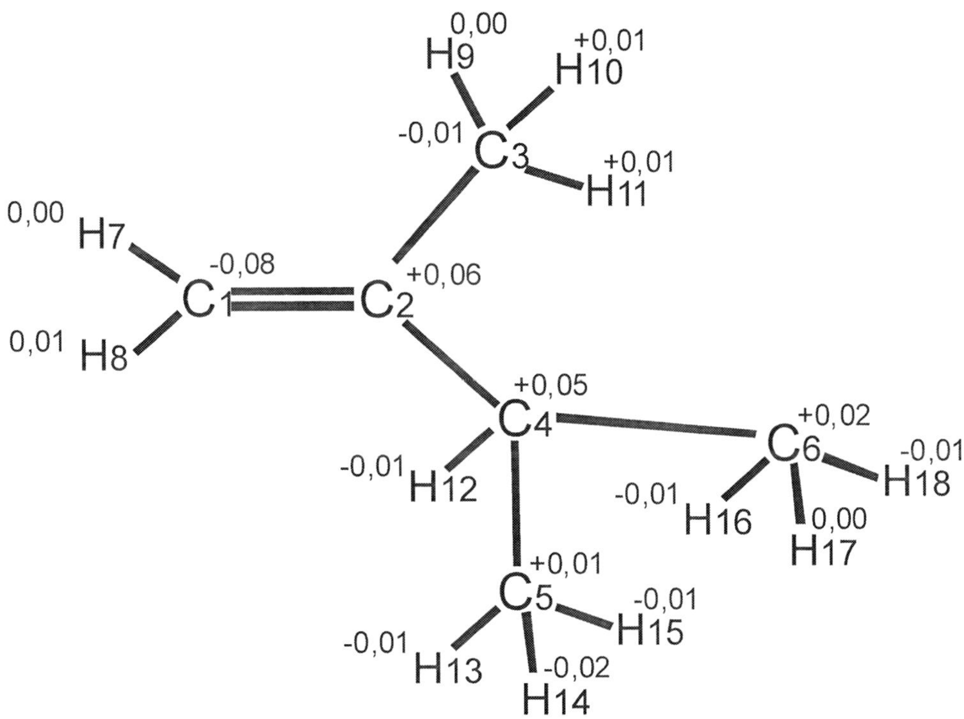

Figure 5. Geometric and electronic structure of 2,3-dimethylbutene-1 C_6H_{12}***

Table 6.

Coordinate	Length of bond R, Å	Coordinate	Length of bond R, Å
C(1)-C(2)	1.34	C(3)-C(2)-C(1)	119
C(2)-C(3)	1.46	C(4)-C(2)-C(1)	124
C(2)-C(4)	1.48	C(5)-C(4)-C(2)	111
C(4)-C(5)	1.48	C(6)-C(4)-C(2)	111
C(4)-C(6)	1.48	C(7)-C(2)-C(2)	111
C(4)-C(7)	1.48	H(8)-C(1)-C(2)	119
C(1)-H(8)	1.12	H(9)-C(1)-C(2)	121
C(1)-H(9)	1.12	H(10)-C(3)-C(2)	113
C(3)-H(10)	1.12	H(11)-C(3)-C(2)	112
C(3)-H(11)	1.12	H(12)-C(3)-C(2)	112
C(3)-H(12)	1.12	H(13)-C(5)-C(4)	112
C(5)-H(13)	1.12	H(14)-C(5)-C(4)	112
C(5)-H(14)	1.12	H(15)-C(5)-C(4)	112
C(5)-H(15)	1.12	H(16)-C(6)-C(4)	112
C(6)-H(16)	1.12	H(17)-C(6)-C(4)	112
C(6)-H(17)	1.12	H(18)-C(6)-C(4)	109
C(6)-H(18)	1.12	H(19)-C(7)-C(4)	109
C(7)-H(19)	1.12	H(20)-C(7)-C(4)	109
C(7)-H(20)	1.12	H(21)-C(7)-C(4)	109
C(7)-H(21)	1.12		

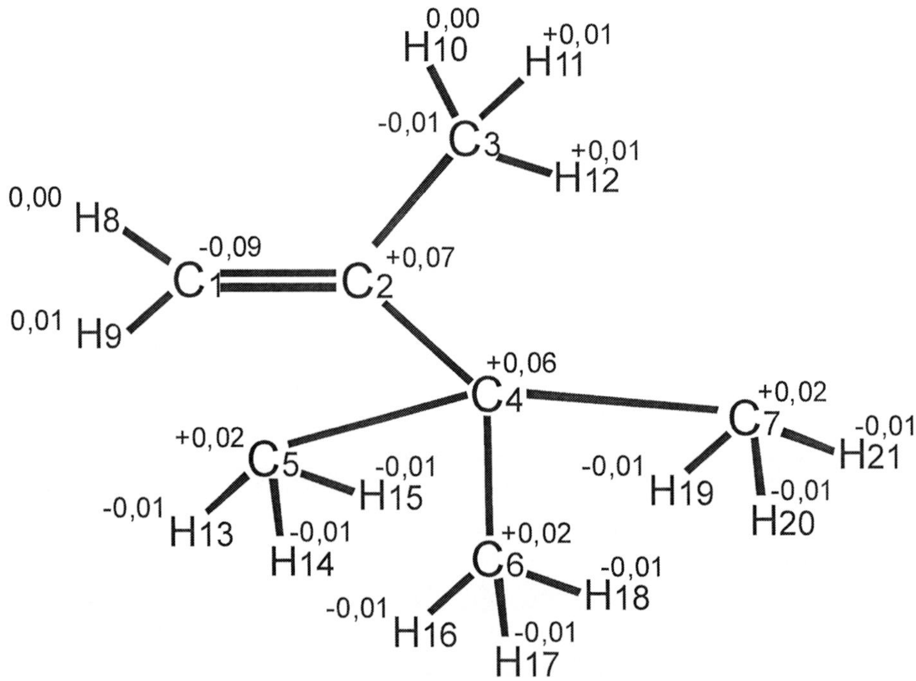

Figure 6. Geometric and electronic structure of 2,3,3-trimethylbutene-1 C7H14

Table 7.

Coordinate	Length of bond R, Å	Coordinate	Length of bond R, Å
C(1)-C(2)	1.33	C(3)-C(2)-C(1)	121
C(2)-C(3)	1.46	C(4)-C(2)-C(1)	123
C(2)-C(4)	1.47	C(5)-C(4)-C(2)	115
C(4)-C(5)	1.51	C(6)-C(5)-C(4)	109
C(5)-C(6)	1.48	C(7)-C(5)-C(4)	109
C(5)-C(7)	1.48	C(8)-C(5)-C(4)	115
C(5)-C(8)	1.49	H(9)-C(1)-C(2)	123
C(1)-H(9)	1.12	H(10)-C(1)-C(2)	124
C(1)-H(10)	1.12	H(11)-C(3)-C(2)	114
C(3)-H(11)	1.12	H(12)-C(3)-C(2)	113
C(3)-H(12)	1.12	H(13)-C(3)-C(2)	113
C(3)-H(13)	1.12	H(14)-C(4)-C(2)	109
C(4)-H(14)	1.12	H(15)-C(4)-C(2)	108
C(4)-H(15)	1.12	H(16)-C(6)-C(5)	114
C(6)-H(16)	1.12	H(17)-C(6)-C(5)	114
C(6)-H(17)	1.12	H(18)-C(6)-C(5)	109
C(6)-H(18)	1.12	H(19)-C(7)-C(5)	109
C(7)-H(19)	1.12	H(20)-C(7)-C(5)	109
C(7)-H(20)	1.12	H(21)-C(7)-C(5)	109
C(7)-H(21)	1.12	H(22)-C(8)-C(5)	109
C(8)-H(22)	1.12	H(23)-C(8)-C(5)	109
C(8)-H(23)	1.12	H(24)-C(8)-C(5)	109
C(8)-H(24)	1.12		

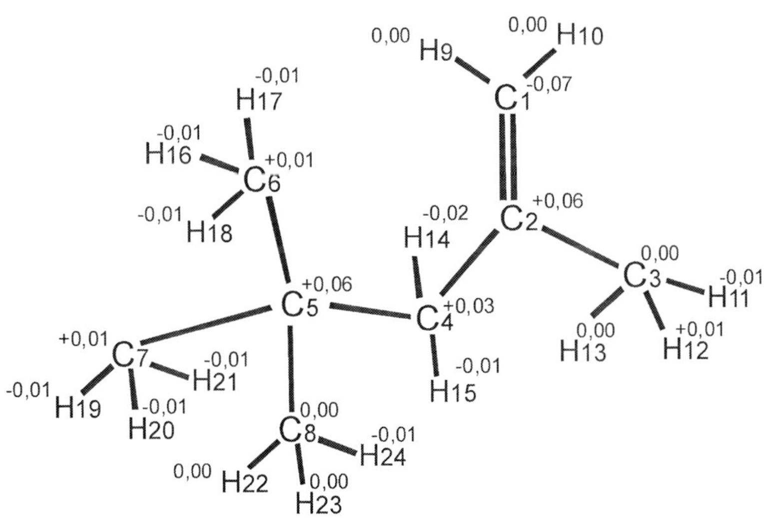

Figure 7. Geometric and electronic structure of 2,4,4-trimethylpentene-1 C8H16

Table 8.

Coordinate	Length of bond R, Å	Coordinate	Length of bond R, Å
C(1)-C(2)	1.34	C(3)-C(2)-C(1)	122
C(2)-C(3)	1.48	C(4)-C(3)-C(2)	112
C(3)-C(4)	1.48	C(5)-C(3)-C(2)	112
C(3)-C(5)	1.48	C(6)-C(2)-C(1)	120
C(2)-C(6)	1.48	C(7)-C(6)-C(2)	111
C(6)-C(7)	1.48	C(8)-C(6)-C(2)	112
C(6)-C(8)	1.48	H(9)-C(1)-C(2)	119
C(1)-H(9)	1.12	H(10)-C(1)-C(2)	118
C(1)-H(10)	1.12	H(11)-C(3)-C(2)	111
C(3)-H(11)	1.14	H(12)-C(4)-C(3)	112
C(4)-H(12)	1.12	H(13)-C(4)-C(3)	112
C(4)-H(13)	1.12	H(14)-C(4)-C(3)	112
C(4)-H(14)	1.12	H(15)-C(5)-C(3)	112
C(5)-H(15)	1.12	H(16)-C(5)-C(3)	112
C(5)-H(16)	1.12	H(17)-C(5)-C(3)	112
C(5)-H(17)	1.12	H(18)-C(7)-C(6)	109
C(7)-H(18)	1.12	H(19)-C(7)-C(6)	109
C(7)-H(19)	1.12	H(20)-C(7)-C(6)	109
C(7)-H(20)	1.12	H(21)-C(8)-C(6)	109
C(8)-H(21)	1.12	H(22)-C(8)-C(6)	109
C(8)-H(22)	1.12	H(23)-C(8)-C(6)	109
C(8)-H(23)	1.12	H(24)-C(6)-C(2)	109
C(6)-H(24)	1.12		

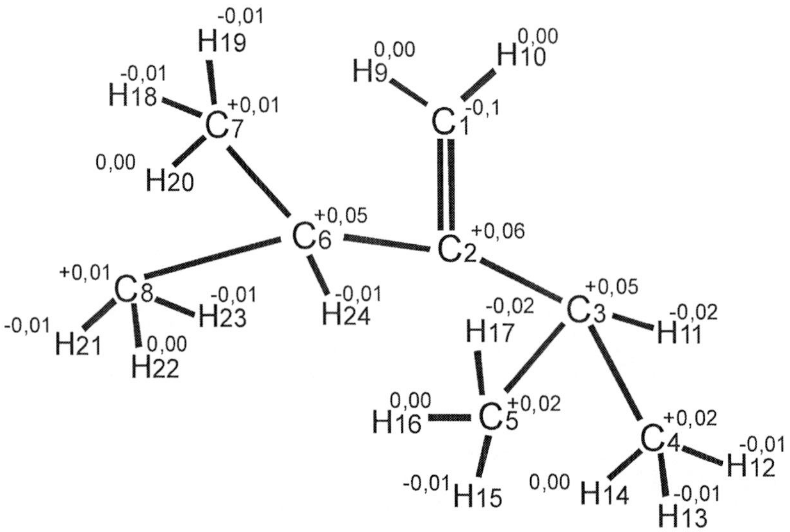

Figure8. Geometric and electronic structure of 2-isopropyl-3-butene-1 C8H16*

Table 9.

Coordinate	Length of bond R, Å	Coordinate	Length of bond R, Å
C(1)-C(2)	1.34	C(3)-C(2)-C(1)	121
C(2)-C(3)	1.50	C(4)-C(3)-C(2)	113
C(3)-C(4)	1.48	C(5)-C(3)-C(2)	109
C(3)-C(5)	1.49	C(6)-C(3)-C(2)	109
C(3)-C(6)	1.47	C(7)-C(2)-C(1)	119
C(2)-C(7)	1.51	C(8)-C(7)-C(2)	113
C(7)-C(8)	1.51	C(9)-C(7)-C(2)	105
C(7)-C(9)	1.50	C(10)-C(7)-C(2)	109
C(7)-C(10)	1.48	H(11)-C(1)-C(2)	123
C(1)-H(11)	1.12	H(12)-C(1)-C(2)	122
C(1)-H(12)	1.12	H(13)-C(4)-C(3)	112
C(4)-H(13)	1.12	H(14)-C(4)-C(3)	112
C(4)-H(14)	1.12	H(15)-C(4)-C(3)	114
C(4)-H(15)	1.12	H(16)-C(5)-C(3)	113
C(5)-H(16)	1.12	H(17)-C(5)-C(3)	113
C(5)-H(17)	1.12	H(18)-C(5)-C(3)	113
C(5)-H(18)	1.12	H(19)-C(6)-C(3)	113
C(6)-H(19)	1.12	H(20)-C(6)-C(3)	114
C(6)-H(20)	1.12	H(21)-C(6)-C(3)	112
C(6)-H(21)	1.12	H(22)-C(8)-C(7)	88
C(8)-H(22)	1.12	H(23)-C(8)-C(7)	140
C(8)-H(23)	1.14	H(24)-C(8)-C(7)	110
C(8)-H(24)	1.12	H(25)-C(9)-C(7)	77
C(9)-H(25)	1.12	H(26)-C(9)-C(7)	96
C(9)-H(26)	1.12	H(27)-C(9)-C(7)	147
C(9)-H(27)	1.13	H(28)-C(10)-C(7)	114
C(10)-H(28)	1.12	H(29)-C(10)-C(7)	113
C(10)-H(29)	1.12	H(30)-C(10)-C(7)	111
C(10)-H(30)	1.12		

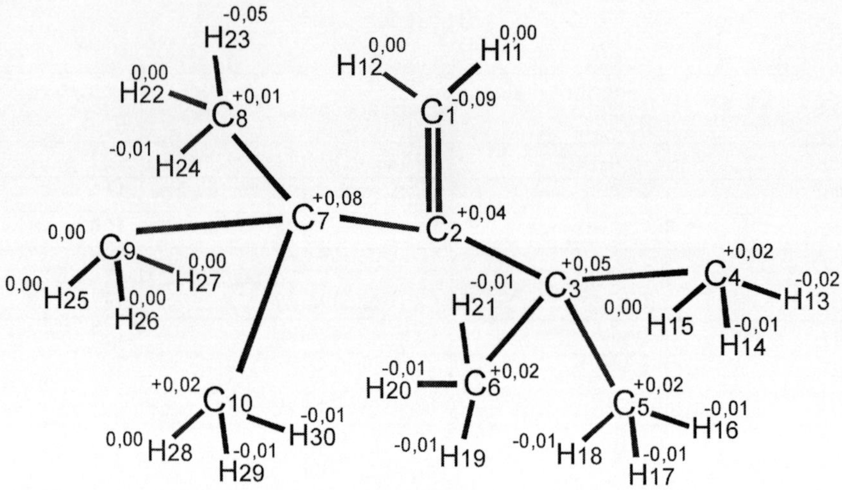

Figure 9. Geometric and electronic structure of 2-tret.-butylbutene-1 C10H20.

Table 10.

Isoolefines	E_O A.U.	$E_{of\ bond}$ A.U.	D dB	q_C	$q_{C\beta}$	q_{H^+}	pKa
1. isobutylene C_4H_8	-39.49	-4.72	0.55	-0.07	+0.07	+0.01	41.79
2. 2-methylbutene-1 C_5H_{10}	-43.17	-5.96	0.44	-0.06	+0.06	+0.01	41.79
3. 2-methylbutene-2 C_5H_{10}*	-43.20	-5.99	0.23	+0.04	-0.02	+0.01	41.79
4. 2-methypentene-1 C_6H_{12}	-51.88	-7.23	0.48	-0.07	+0.06	+0.01	41.79
5. 2-ethylbutene-1 C_6H_{12}*	-51.89	-7.23	0.67	-0.09	+0.07	+0.00	43.70
6. 2,3-dimethylbutene-1 C_6H_{12}**	-51.87	-7.22	0.53	-0.08	+0.06	+0.01	41.79
7. 2,3,3-trimethylbutene-1 C_7H_{14}	-60.56	-8.46	0.60	-0.09	+0.07	+0.01	41.79
8. 22,4,4-trimethylpentene-1 C_8H_{16}	-69.26	-9.72	0.55	-0.07	+0.06	+0.01	41.79
9. 2-isopropyl-3-butene-1 C_8H_{16}	-69.26	-9.72	0.64	-0.10	+0.06	+0.00	43.70
10. 2-tret.-butylbutene-1 $C_{10}H_{20}$	-86.57	-12.15	0.55	-0.09	+0.04	+0.00	43.70

MM of symmetrical isoolefines: isobutylene, 2-methylbutene-1, 2-propylbutene-1, 2-tret-butylbutene-1 decreases. Such isoolefines do not form the polymers, form oligomers.

Ions with radical more than methyl in symmetrical isoolepfines, can not provide the efficient growing a circuit because of steric compression in connecting condition [1]. Steric factor in asymmetric isoolefines is not dominant. Stability of carbcation increases with increase degree of branched of radical of β-carbon atom in composition AC. MM of polymers grows.

The formation of polymers with big MM for asymmetric isoolefines needs for pieces of evidence (in presence patented catalyst: mixture initiator $Fe(O\text{-}H\text{-}B)_3$ and BF_3 [8-11], $Al(O\text{-}F\text{-}C_4H_9)_3$ /BF_3 / $TiCl_4$ [12], $MgCl_2$ [3]). Analysis of the literature [1] and quantum-chemical calculation of isoolefines prove the possibility to get the polymers 2-methylbutene-1, 2-methylpentene-1 and the other isoolefines with MM more, than polymer of isobutylene (1303000) [1].

We have done quantum-chemical calculation of row known isoolefines (tabl.10) by classical method CNDO/2 in parametrization of Santri-Poppla-Segala. Geometric and electronic structure of these compounds was determined. Their acid power was theoretically evaluated. All under investigation isoolefines possess low acid power and pertain to class very weak H-acids. Variety of asymmetric isoolefines with equal or greater ability to cation polymerization than isobutylene exists.

REFERENCES

[1] Kennedy Dzh. Cation polymerization of isoolefines. 1978, g. Moscow, 431c.

[2] Sangalov YU.A., Minsker K.S. The Polymers and copolymers of isobutylene. 2001, g. Ufa, 383c.

[3] Babkin V.A. The Structure of active centre, mechanism of initiation and growing circuit under cation polymerization of olephyns in presence complex catalyst. *The Doctoral thesis.* The Institute of chemical physics The Laureate of Nobel prize of the academician N.N. Semyonova,1999,214s.

[4] Guterbock. Polyisobutylen, *Springer,* Berlin, 1959.

[5] Butlerov A.M., Liebigs. *Ann. Chem.*, 189, 47, 1877.

[6] Von Lohuizen O.E., De Vries K.S.J. *Polymer Sci.* p.16, 3943 (1968)

[7] Sigal Dzh. The Semitheoretical methods of calculation electronic structure. - *M.: The World*, 1980. 327c.

[8] Miyoshi M., Uemyra S., Tsuchiya S., Kato O. *the stalemate.* USA 3402146 (1968).

[9] *Pat. 1056730* (1967)

[10] *Pat. 1183118* (1970)

[11] *Pat. 1233557* (1971)

[12] ImanishiY., Yamamoto R., Higeshimura T., *Kobunshi Kagaku*, 24, 397, 1967.

[13] Addecott K.S.B., Mayor L., Turton C.N., *Europen Polymer* J., 3, 601.1956.

[14] Pople I.A. In *Approximate Molecular Orbit Theory*| *Pople* I.A. D. Beveridze Y,: Me. Grew-Hill, 1970, 214c.

In: Progress in Monomers, Oligomers, Polymers, Composites … ISBN 978-1-60456-878-3
Editors: R. A. Pethrick, G.E. Zaikov et al.

Chapter 16

RADICAL COPOLYMERISATION OF STYRENE WITH C 2-METHYL-2-VINYL-1,1-DICHLOROCYCLOPROPANE

***S.V. Kolesov*[1], *A.I. Vorobyeva*[1], *S.S. Zlotsky*[2], *A.R. Khamidullina*[2], *E.A. Brusentsova*[2], *R.R. Muslukhov*[1], *L.V. Spirikhin*[2], *G.E. Zaikov*[3,*]**

[1]Bashkirian State University, Ufa, Russian Federation
[2]Ufa State Technological Oil University, Ufa, Russian Federation
[3]N.M. Emanuel Institute of Biochemical Physics, Moscow, Russian Federation

ABSTRACT

In the present research radical copolymerization of styrene with 2-methyl-2-vinyl-1,1-dichlorocyclopropane has been investigated. It is shown that the copolymer has a unit structure and is enriched with styrene units.

Keywords: radical copolymerization, 2-metyl-2-vinyl-1,1-dichlorocyclopropane, copolymerization constants.

We have showed earlier that radical copolymerization of vinyl-hem.-dichlorocyclopropane involves ring-opening and polyunsaturated macromolecules are formed [1].

In the present paper we have investigated radical copolymerization of 2-metyl-2-vinyl-1,1-dichlorocyclopropane I with styrene II. Comparison of the parameters of nuclear magnetic resonance spectra of the copolymer that is formed in this case (Table. 1) with analogous spectral characteristics of the homopolymer [1] and with estimated values of chemical shifts proves that addition of units of the monomer II by copolymerization proceeds with *cis-/trans-* units of block copolymer I (Table. 2).

---

* Correspondence to: N.M. Emanuel Institute of Biochemical Physics, 4 Kosygin str., Moscow 119334, Russian Federation, Chembio@sky.chph.ras.ru

Table 1. Chemical shifts and multiplicity of signals of ^{13}CNMR (δ, ppm., (MHz), C_6D_6, TMS, +30°C).

Compound	C_1	C_2	C_3	C_4	C_5	C_6	C_7	C_8	C_9	C_{10}
Homopolymer **I** [1] 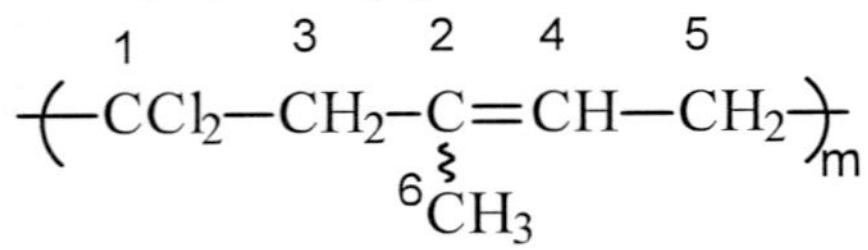	93.1 s	134.2* 134.1** s	56.6* 49.6** t	126.3* 125.6** d	46.1* 47.2** t	19.2* 26.3** q				
Copolymer **I** with **II** 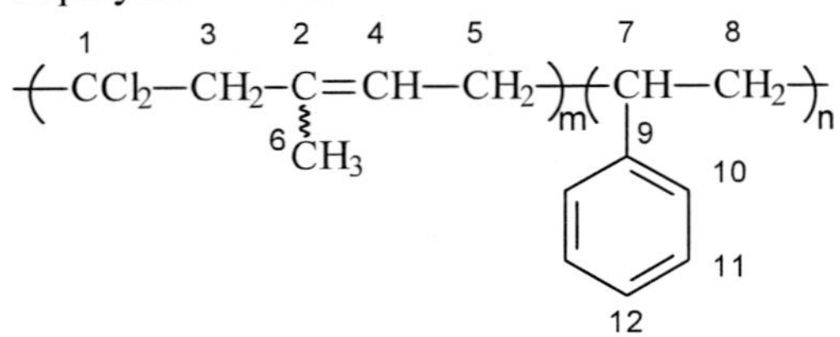	89.9 s	133.5 s	54.0* 52.3** t	126.0 d	46.4 t	19.2* 26.3** q	40.1 d	44.2 t	145.3 s	127.4 d

Note: in the structure of copolymer I with II C_{11} – doublet in 127.4 ppm.; C_{12} – doublet in 125.5 ppm.; * – trans, ** – cis.

Table 2. Copolymerization of 2-vinyl-2-methyl-1,1-dichlorocyclopropane (I) with styrene II. ([I] : [II] = 1 : 1), 75 °C

Azoisobutiro-nitrile, mol.%	Reaction time, h	Yield, %	Structure of copolymer, mol.%	
			I	**II**
3	14	41*	100	0
1	25	23	12	88

Note: * – homopolymerization of olefin **I** [1]

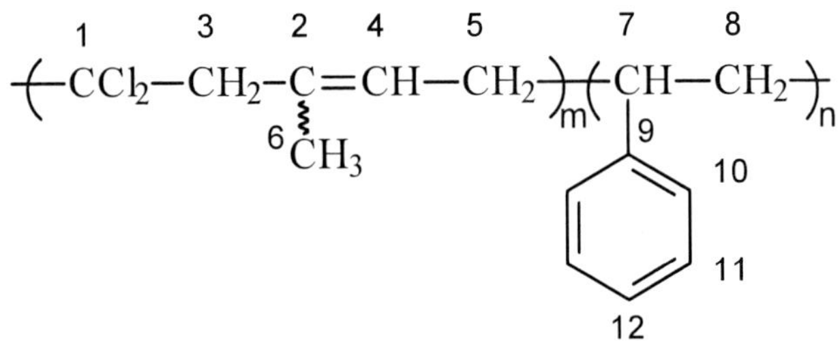

Judging by the dependence of the composition of the copolymer on the concentration of the monomer I (Fig. 1), the copolymer is enriched with olefine units II.

Using the Feineman-Ross method, we have estimated the values of effective constants of copolymerization of compounds I and II, and they turned out to be r_1=0,1 and r_2=16,6. If the concentration of vinyl-cyclopropane I increases, in the original mix of reagents the speed of copolymerization of decreases. (Fig.2). The high value of product of copolymerization constants $r_1 * r_2$=1,6 shows that susceptibility to alternation of comonomer links is low, and the copolymer mostly has a unit structure.

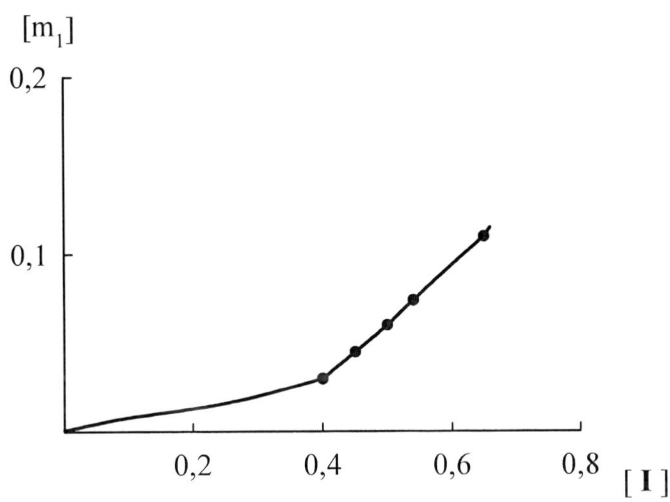

Figure 1. Relationship of constitution of copolymer of olefin I with styrene II from initial mixture in the polymerization. [I] и m_1 – mole fraction I in the initial mixture and in the copolymer correspondingly. [IBN] = 3 mas.%, T = 75 °C.

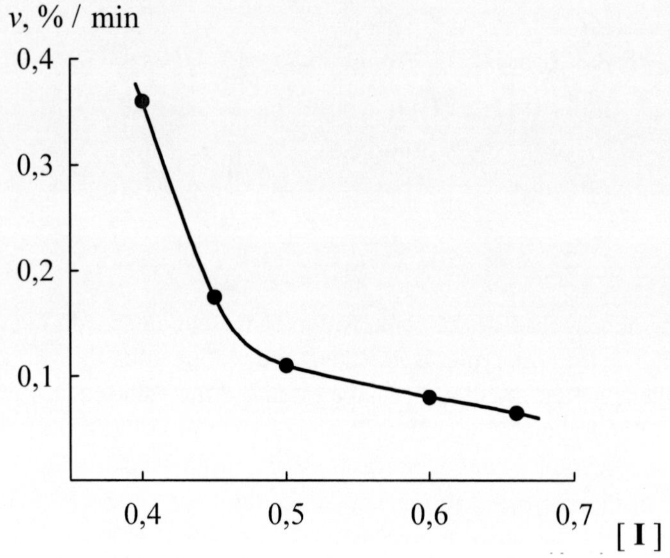

Figure 2. Relationship of initial speed (v) of copolymerization of olefin I with styrene II from the composition of initial mixture of monomers. [IBN] = 3 mas.%, T = 75 °C.

Experimental Part

2-vinyl-2-metyl-1,1-dichlorocyclopropane (I):

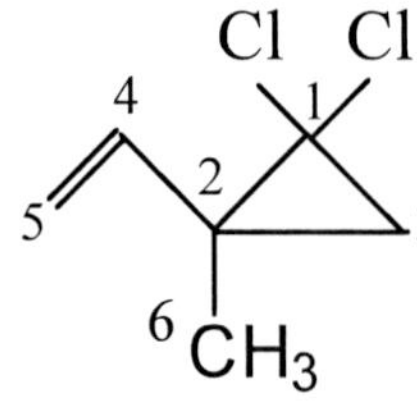

I

was synthesized through carbonization of isoprene using method [2].

Styrene (II) of chemical pure qualification was cleansed of the stabilizing element through triple sluicing with a 7% sodium alkali solution, was scoured with distilled water until neutral reaction, was dehydrated with incinerated calcium chloride and was ran twice. The monomers, initiating agents and dissolvents used in the experiment after cleansing with standard methods had properties that correspond to the data in the literature.

The copolymerization of vinyl-hem.-dichlorocyclopropane I with styrene II was carried out in bulk in vacuum under the effect of azoisobutyronitrile. When the required extent of conversion was reached, the reaction was interrupted through cooling and further precipitation of the copolymer. Cleansing of the copolymer was made through triple reprecipitation from benzene solution into methanol and vacuum drying at 50 °C up to a constant mass. The composition of the copolymer was determined using the results of elemental analysis.

Nuclear magnetic resonance spectra ^{13}C and ^{1}H were registered on the spectrometer "Bruker AM-300" with operating frequency of 75.5 and 300 MHz. Nuclear magnetic resonance spectra ^{13}C were recorded in regimes with broadband suppression with protons and JMOD. Solvent benzol-d_6, internal standard – tetramethylsilane.

REFERENCES

[1] *S.V. Kolesov, A.I. Vorobyeva, S.S. Zlotsky, A.R. Khamidullina, R.R. Maslukhov, L.V. Spirikhin.* The structure of homopolymers of vynil-hem.- cyclopropane dichloride // Doklady Akademii Nauk (ib Russian) (in print).

[2] *A.P. Molchanov, S.A. Kalyamin, R.R. Kostikov* // Zhurnal organicheskoi khimii. 1992. Vol. 28. Issue. 1. P. 122. (in Russian)

In: Progress in Monomers, Oligomers, Polymers, Composites ... ISBN 978-1-60456-878-3
Editors: R. A. Pethrick, G.E. Zaikov et al.

Chapter 17

STABILIZING INFLUENCE OF THE POLYMERS CONTAINING OXIMATES GROUPS ON POLYALKILENETHEREPHTALATE

Yu.I. Musaev, E.B. Musaeva, V.A. Kvashin**[*]**, ***A.B. Zhekamukhov and R.V. Pak***

H.M. Berbekov Kabardino-Balkarian State University, Nalchik, Russian Federation

ABSTRACT

The comparative analysis of the basic physical and chemical properties of not stabilized polyalkilenetherephtalates and their compositions with polymers containing oximates groups is submitted. It is shown, that use of new polyformaloximates (PFO) and polyarylateoximates (PAO) as modifying additives allows improving operational properties and technical characteristics of polyalkilenetherephtalate compositions.

Keywords: polyethylene therephtalate, polybutylene therephtalate, polyformaloximates, polyarylate oximates, compositions, thermostability, thermooxystability.

INTRODUCTION

Polyalkilenetherephtalates possess a number of unique properties: almost zero permeability for water паров and паров organic solvents for oxygen and nitrogen, small water absorption, high resistance fats, oils, the organic solvents, the diluted acids and alkalis, high stability to action light, x-ray and UV - beams, biological inertness. Besides polyalkilenetherephtalates possess fine physicomechanical properties. Compositions polyalkilenetherephtalates find wide application as constructional materials.

[*] Correspondence to: H.M. Berbekov Kabardino-Balkarian State University, Chernyshevsky Str., 173, Nalchik 360004, Russian Federation, email: ,kva78@mail.ru

With the purpose of increase it is warm also to thermostability, thermooxystability to stability, increases of dielectric characteristics, expansions of a temperature interval of processing of compositions and a temperature interval of products on a basis polybutylene therephtalate (PBT) and polyethylene therephtalate (PET), as modifiers into compositions entered polyformaloximate-1 (PFO-1) on a basis of diacetophenyloxid diketoxime (1); polyarylate oximate (PAO-1) on a basis of diacetophenyloxid diketoxime and dichloride anhydrides of iso-phtalic and terephtalic acids (2); polyarylate oximate (PAO-2) on a basis phenylketoxime, containing phtalude-grouping (3).

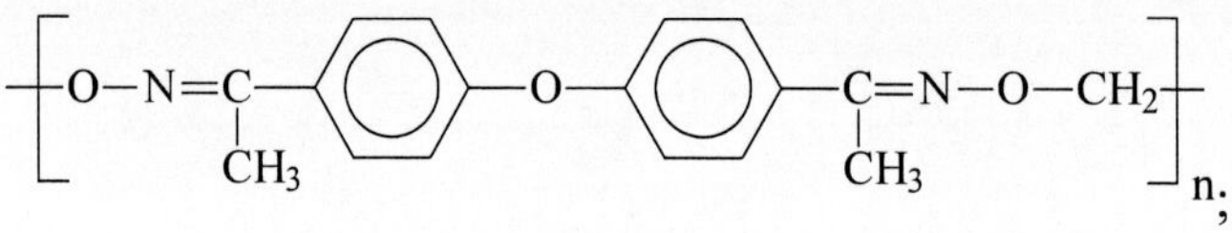

(1)

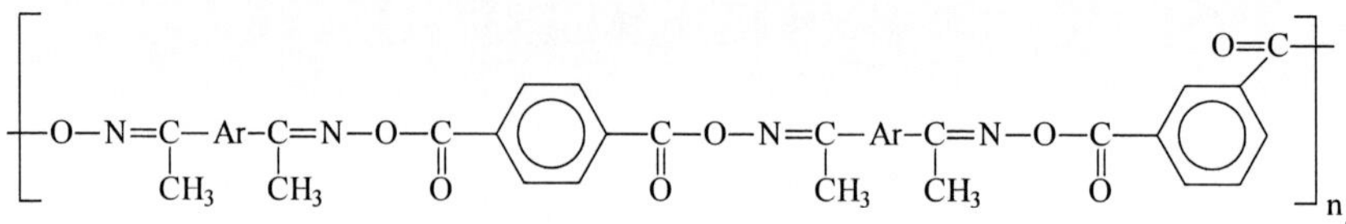

(2)

где Ar = 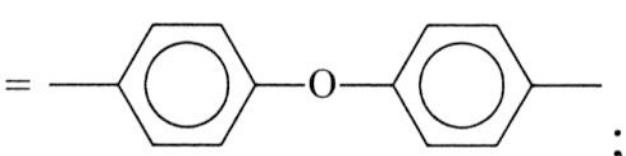

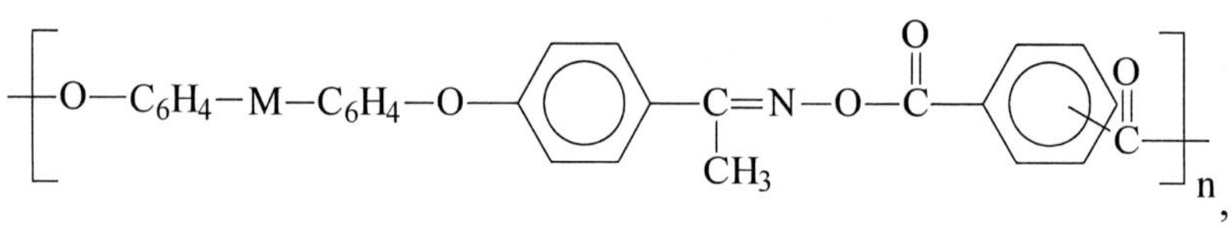

ãäå Ì = 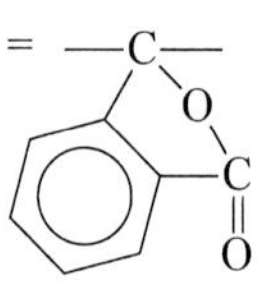

Experimental Part

Preparation of PBT (PET) Blends

Polymers – modifiers have been synthesized by us by that methods, which were described in [1]. Compositions of polybutylene therethtalate with PFO-1 have prepared by extruding of composition of granular PBT with addition of definite quantity of 1 % - solution of PFO-1 in chloroform previously dried up in vacuum at 100° C for 2 hours [2].

For getting of composition of polyalkilentherephtalates with PAO-1 and PAO-2 in composite of thetrachloroethane:phenol (3:1) dissolved by the composition of PBT (PET), which contained a definite quantity of PAO-1(PAO-2). After a small interfusion the received solution had been precipitated in isopropanol, filtered and washed by distillated water many times. The received polymeric composition was dried up in vacuum at 70 °C [3].

Instrumental Methods

Physical properties of the prepared samples were investigated by various analytical methods. Thermal behavior was analyzed under air and argon by TGA instruments "MOM" (Hungary), the heating rate of samples was kept at 5 °C/min, the of a sample was 25 mg. The melt flow index (MFI), which determines the processing method for thermoplastics, was measured by standard method of GOST 11645-73. For the estimation of MFI values the instrument IIRT-M2 was used. Electric measurements were carried out with the help of the bridge by applying alternating current of 10^3 Hz and digital readout R-5058 in a temperature interval from 20 to 250 °C. The error in measurements of a loss-angle tangent did not exceed 5 %.

Results and Discussion

The comparative a valuation of physical and chemical properties of non-stabilized polyalkilen therephtalates and composition was lead with using of accelerated methods. It is Known that thermal analysis of значения of mark of fusion's fluidity (ПТР) and behaviour of polymeric materials in electric поле are the most informative at the avaluation of the operating properties and also technical characteristics of the polymeric compositions.

The results of our investigation show that the compositions of polyalkilen therethtalates containing PFO-1, PAO-1, PAO-2 as a modifying additives, by their basic physical and chemical properties are over of analogous properties of well-known compositions in a wide interval of their temperature.

Figure 1 represent the thermal analysis for njn-stabilized PBT (curve 5), PBT industrially stabilized (curve 6) and PBT modified by PFO-1 and PAO-1.

It is shown on fig.1a at increasing of the mass part of PFO-1 (curves 2 and 3 -0,5 mass%, curve 4 – 1,0 mass%) and the time of thermal aging (curve 3 – thermal aging 30 minutes at T= 250 ° C), the mass part coconut rest increases too.

DTA curves 5 and 6 (fig. 1b) showed that thermo-oxidation degradation process of non-stabilized and stabilized industrial PBT began at 260°C and 325°C accordingly. For compositions PBT/PFO-1 containing 0,5 mass % (curve 2), it after 30 minutes thermal aging at T=250°C (curve 3) and 1% mass% (curve 4) temperature of beginning of oxidation equally 350°C. Herewith, the parts of curves of DTA active thermo-oxidation destruction of composition have 2 peaks, which area and position depends on the PFO content. With the increasing of content of PFO-1 in parallel increasing of thermo destruction, the crosslinking increases too, what is in keeping with the curves of TG.

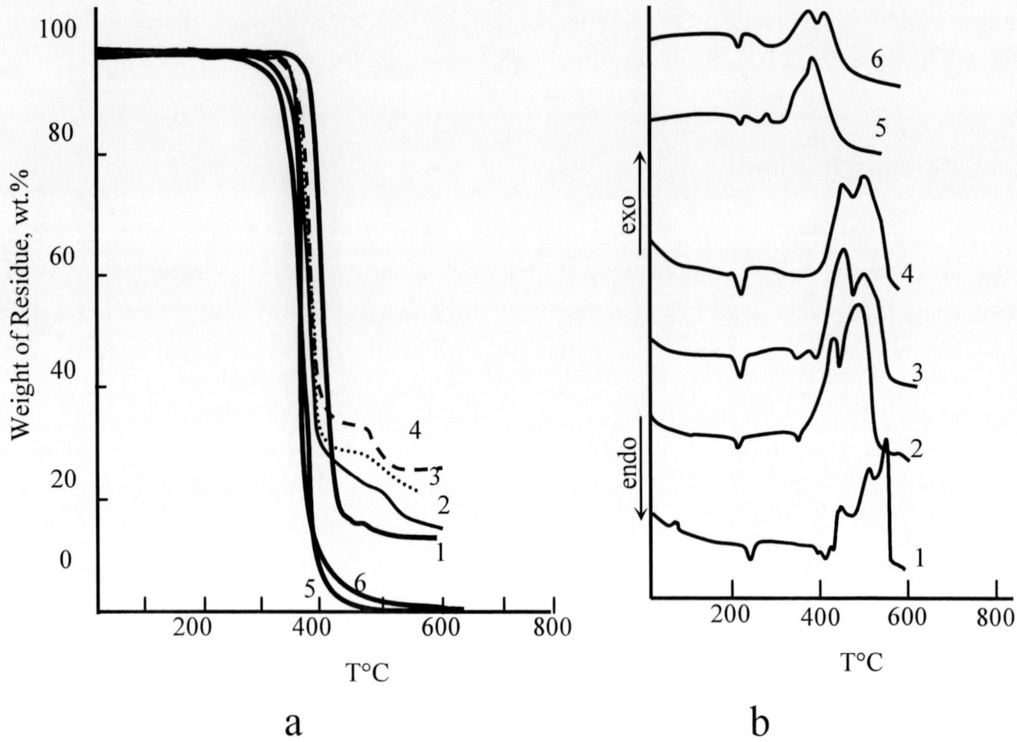

Figure 1. Thermal analysis data for PBT and PBT/PFO(PAO) blends (a – TGA, b – DTA)

On the graph of DTA (curves 5, 6) one can see that the peak of thermo-oxidation destruction have a crest in the interval of the temperature between 462 -466 ºC. At the addition of PFO-1 in composition the character of graphs of DTA and DTG is changing. At the fig. 1, one can see that the crosslinking becomes the main process.

And besides, if the composition with PBT contain 0,5 % of PFO-1 (without of thermal aging) there is absent the difference in the character of curves DTA for non-stabilized PBT and composition of PBT/PFO-1 in the area of 460 - 470 ºC, for composition PBT with 0,5% of PFO-1, after 30 minutes exposure at 250 °C, one can see essential stabilizing influence of PFO-1 (the 1-st peak which is in congruence with the processes of oxidation decreases, the 2-nd peak which is in congruence with the processes of crosslinking - increases). In this regard the most optimum results we received for the compositions with 1% PFO 1 on curve 4 DTA one can see, that the oxidation processes are absent till 390 ° C, but in the area of 460-470 ° C the crosslinking of composite passes; the 1-st peak, which is reply for thermo-oxidation destruction, almost fully disappears.

On the 1-st curve (a, b) on the figure 1 presented the data of the thermal analysis of composition PBT with 1 % of PAO-1. On the grounds of the data received from curves TG we can conclude that the processes of sewing, oxidation, assignation and other reaction passes at the higher temperature than in the case of industrial PBT and composition of PBT with PFO-1. The using of polyarylate oximates as stabilized additives allows increasing the temperature of loss 2 % in the weight about 357 ° C. Besides data of TG shows that in the composition of of PBT with PAO-1 appears coconut rest which increases with the increasing of mass part of PAO-1 in composition, in the case with PBT non-stabilized, industrial stabilized coconut rest is absent.

So, in the case of PBT +1% of PAO, coconut rest on air made up 10%, but in argon 15%.

On curves of DTA it is shown, that endothermic peak, which characterizes the temperature of melting, in the case of PBT+1% PAO (curve1) is higher in 35°C than non-stabilized PBT (curve 5) and industrial one (curve 6). The increasing of the temperature of melting of the composition PBT+PAO-1 contributes to expansion of the upper border line of the temperature interval of operation of product of polybutylene therethtalate. For composition of PBT/PAO-1 (curve 1) the temperature of oxidation is 425 ° C, which is in 165 ° C higher than the temperature of oxidation of non-stabilized PBT, and in 100 °C higher than the temperature of oxidation of stabilized industrial PBT. These facts point to the increasing of the temperature interval of processing of compositions on a basis PBT and PAO-1(the interval between the temperature of melting and temperature of oxidation of thermo-oxidation processes).

It is known, that in using of stabilizators on the curves of DTA (air) appears several exothermic peaks, which area and position depends on the content of stabilizing additions. These peaks are characteristically for the processes of and oxidation, herewith, as it notes in literature, the speed of oxidation processes higher than the speed of thermal crosslinking. On the figure 1b of DTA (curve 5) for non-stabilized PBT we can see only one exothermic peak at 457 °C, while on the figure 1b of DTA for stabilized industrial PBT (curve 6) are two peaks in interval of 455-456 °C. On the curve of DTA of the composition PBT/PAO-1 (curve 1) appears already three exothermic peaks of active destruction, the first of them is in congruence with the processes of oxidation, appears at 467 °C, the second and the third, which are in congruence with the processes of thermal crosslinking, appear at 500 ° C and 520 ° C, accordingly.

In modification of PET by the polyarylate oximate-2 (PAO-2) the most optimum results were achieved in the case of composition of PET with 1 % of PAO-2. The results of TGA are presented on the fig. 2 (curve 1) and the fig. 3.

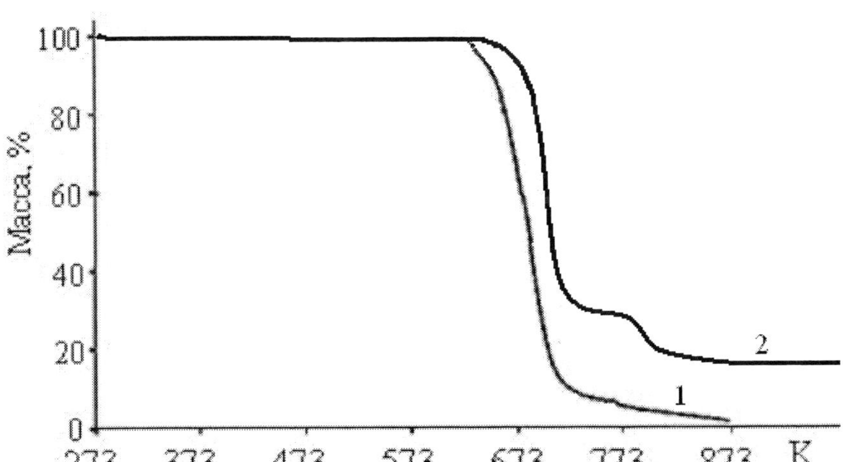

Figure 2. TG of non-stabilized PET (1), of composition of PET with 1 % of PAO-2 (2)

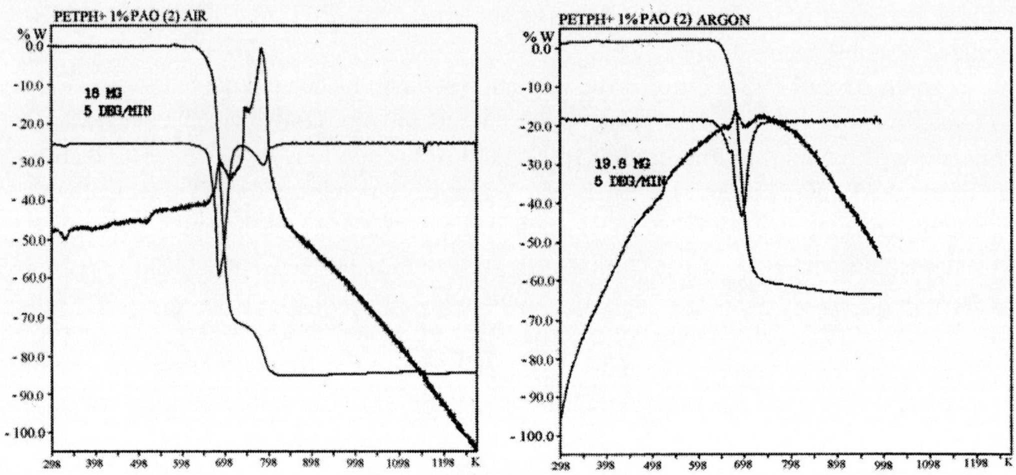

Figure 3. The data of thermal analysis of composition of PAT with 1 % of PAO-2 on air and argon.

On the figure 2 of TG for compositions PET with of PAO-2 it is shown that the loss at the weight begins at 358 °C (fig. 2, curve 2), while for the non-stabilized PET this process begins at 340 °C (fig. 2, curve 1). TGA non-stabilized PET showed, that the loss at the weight begins at 270 °C, in the case of PET with of PAO-2 the loss at the weight begins 358 °C. The temperature of the full destruction of non-stabilized PET on air is 600 ° C. It doesn't observe the full destruction of PAT with 1 % of PAO-2, the coconut rest on air made up 25% (in argon-40%) at the 600 °C, which is argumentative of expansion of temperature interval of operation of products on a basis polyethylene therephtalate. The received data of TGA are argumentative of the high thermostability of received composition.

The melt flow index (MFI) is the only standard reological characteristic of the melt. The MFI of polybutylene therethtalate compositions measured at 250 ° C and 5kg, the MFI of polyethylene therephtalate compositions measured at 260 ° C and 5kg. The results of the measure are presented in the table 1.

Table 1. MFI of PBT(PET)/PFO(PAO) blends

Containing of modifying additive, %	**0**	**0,05**	**0,1**	**0,2**	**0,5**	**1**
MFI of PBT/PFO-1	6,51	2,81	1,24	1,34	3,25	3,09
MFI of PBT/PAO-1	6,5	3,42	3,22	2,55	2,34	1,83
MFI of PBT/PAO-2	8,1	5,06	4,92	4,02	3,91	3,6

So, depending on the content of modifying additive MFI of compositions PBT with of PFO-1 decreases in 2,4 times; of compositions PBT with of PAO-1 in 1,3 – 3,3 times; of compositions PET with of PAO-2 in 1,6 – 2,2 times.

MFI, measured in the same way for 15 minutes thermal aging in the capillary of the tester, are differ from the datum not more than 1,5 – 2,0 %, what is argumentative of well thermostability of polymeric compositions.

The additive PFO-1 and PAO-1 to polybutylene therethtalate and PAO-2 to polyethylene therephtalate allows making better dielectric properties of modifying polymers. The temperature of beginning through conduction in the case with PBT non-stabilized is 125 °C, of composition of PBT+1% of PFO-1 is 16 °C, of PBT+1% PAO-1 is 185 °C,. The temperature of beginning through conduction of non-stabilized PET is 150 °C, for the compositions of PET+PAO-2 it is 200 ° C.

CONCLUSONS

So, the using of PFO-1, PAO-1 and PAO-2 as a modifying additives in compositions with polyalkilentherephtalates allows to increase the temperature of melting of compositions, it also allows to increase the temperature of beginning of chemical destruction, therefore it allows: to increase the warm- and thermostability, to increase the thermo-oxystability, to expansion the temperature interval of operation with the products on basis polyethylene therephtalate, to increase the temperature of beginning through conduction, and also it allows to change the composition of MFI to that size, which will be easy for conversion.

REFERENCES

[1] Musaev Yu.I., Musaeva E.B., Kvashin V.A., Zhekamuhov A.B., Photov R.B. Synthesis and some properties new hydroxyl-containing aromatic monomers, *Materials of II-nd of the international scientific-practical conference* "Fundamental and applied researches in an education system". Tambov, 2004. - p. 97-100.

[2] *RF Patent № 2261878 IPC C 08 L 67/02 2005.* Polymer composition. Musaev Yu.I., Mashukov N.I., Musaeva E.B., Kvashin V.A., Mikitaev M.A.

[3] *RF Patent № 2292366 IPC C 08 L 67/02 2006.* The polymeric composition used as a constructional material Musaev Yu.I., Musaeva E.B., Kvashin V.A., Mikitaev M.A.

In: Progress in Monomers, Oligomers, Polymers, Composites ... ISBN 978-1-60456-878-3
Editors: R. A. Pethrick, G.E. Zaikov et al.

Chapter 18

STUDYING OF CHARACTER OF THERMODYNAMICS, KINETIC AND THE MECHANISM OF HIGH-TEMPERATURE INTERACTION OF COMPONENTS OF SYSTEMS ME_2 WO_4-WO_3 - ME CL (ME - LI; NA, K, RB, CS) AND SEARCH OF WAYS OF PRACTICAL APPLICATION OF RESULTS OF RESEARCHES

I.A. Shogenov*, G.K. Shurdumov and Z.A.Cherkesov
H.M. Berbekov Kabardino-Balkarian State University, Nalchik, Russian Federation

ABSTRACT

The purpose of the present work studying interaction of components in the heterogeneous environment in systems of type Me_2WO_4 - WO_3-MeCl. (Me-Li, Na, K, Rb, Cs) and development on the basis of received is stated at, it of the given electrolytes for synthesis oxides tungsten bronze alkaline metals with adjustable physical characteristics.

Study thermal stability of low-melting structures, enriched with tungsten and acidic tungsten (VI)

Keywords: thermodynamic, interaction of components, kinetic

The chemistry of rare elements, in particular molybdenum and tungsten, concern to number of the most important for modern ethnics of chemical elements. Especially they have great value in metallurgy, mechanical engineering, radio electronics, and the aviation industry, without them it is difficult to present satisfaction of needs of defense. The modern science and ethnics extremely requires materials on the basis of these elements possessing a

* Correspondence to: H.M. Berbekov Kabardino-Balkarian State University, Nalchik, Chernyshevskaya str. 173, Russian Federation, Shogenov.islam.@mail.ru

complex of valuable physical and chemical properties: catalytic, high corrosion stability, half - conductor, superconductivity, etc.

One of the main primary goals of modern chemistry is reception of new substances and materials with the set of structurally-target properties to which number, certainly, it is possible to carry oxides tungsten bronze (OVB) various metals. The important role in the decision of the specified problem takes place with phase diagrams multicomponent salt and oxides - salt systems, data on which make a basis of development of electrolytes with optimum physical and physical and chemical properties, suitable for the synthetic purposes.

In connection with, the purpose of the present work studying interaction of components in the heterogeneous environment in systems of type Me_2WO_4 - WO_3-MeCl. (Me-Li, Na, K, Rb, Cs) and development on the basis of received is stated at, it of the given electrolytes for synthesis oxides tungsten bronze alkaline metals with adjustable physical characteristics.

Binary components of studied systems Me_2 WO_4 - WO_3 and Me_2WO_4-MeCl are characterized by formation in them isopolxtungetic structure Me_2WO_{3n+1}. As if to their third component MeCl-WO_3 as our data and data of other authors show, in it chemical reaction of type 2MeCl + 2WO_3 = Me_2WO_4 + WO_2C1_2 which reliability proceeds is confirmed by us kinetic, x-ray phase and chemical methods of the analysis.

Its impossibility un deer standard conditions is proved thermodynamics However, as show calculations T<298,15 °K. This process it is quite possible. It finds experimental acknowledgement in data of the thermodynamic analysis, in particular, in data to cuts VI-XI which specify presence of a boundary line between fields of crystallization of chloride and oxides tungsten (VI). Its extrapolation in view of temperatures of fusion of adjoining phases and that the studied system concerns to type threefold mutual provided that oxides tungsten (VI) and chloride of cesium with WO_2C1_2 forms eutectic (probability of it is sufficient greater), its surface of crystallization can be transferred figure. Studying of cuts I-VII and IX-XI has allowed revealing in its stable areas of a field of crystallization tungetic cesium, congress ditungetic and incongruous three tungsten cesium. It is important to emphasize here, that studying of the specified sections and an establishment of areas of crystallization of the specified phases, has allowed to reveal a lot of the low-melting electrolytes representing practical interest, for example, for synthesis OVB of high dispersive ness and cleanliness. Also binary components of systems Li (Na, K, Rb) CI-WO_3 similarly behave.

From stated follows, that liquids the studied systems as have shown experimental data, consist of fields of crystallization of initial components, connections with various characters of fusion. Lines joint allocation of components of systems Me_2WO_4 - WO_3-MeCl converge in four nonvarient points (lithium and sodium) and three nonvarient points (potassium, rubidic, cesium) various character of fusion. By way of the practical appendix of results of the lead researches we a kinetic method study thermal stability of low-melting structures, enriched with tungsten and acidic tungsten (VI) corresponding data are presented by chloride of cesium in figures 2.62-2.67 at temperature 500°C. Alongside with it on equations of Arreniouse, Yerofeev-Kolmogorov and Sakovich the constant of speed and energy of activation of reaction MeCl with WO_3 for the structures adjoining to diagonal section WO3-MeCl are calculated. In general quality standard of data of the table shows, that with promotion aside Me_2WO_4 thermal melts stability sharply increases. It proves to be true also gaugings of a flying product-WO_2C1_2 for one and too time at the same temperature. Corresponding data are resulted below in the table:

Table 2.57

Structure of an initial mix	**Temperature, °C**	**Time, in minutes**	**m WO_2CI_2 gr.**
5% Cs_2WO_4+ 60% CsCl +35% WO_3	500	40	1,856
10% Cs_2WO_4+ 40% CsCl +50% WO_3	500	40	0,942

The nature of a sublimating phase have qualitatively established methods of the chemical analysis and RFA. Both of them specify presence at it of tungsten and chlorine. Besides as it is visible from data RFA, in it presence of cesium - obviously result of partial sublimation of chloride of cesium is revealed also.

Table 2.58 Structure (a pier. %) and the average size (and 10» 6) particles (OVB), received from melts systems Me2WO4 - WO3-MeCI (T=550-650°C, i = 0,025 □/sm)

Li_2WO_4	**LiCl**	**$W0_3$**	**X**	**Na_2W0_4**	**NaCl**	**$W0_3$**	**X**	**K_2W0_4**	**KC1**	**WO_3**	**X**
22	60	18	0.58	80	15	5	0.29	60	5	35	0.57
17	70	13	0.58	75	15	10	0.29	85	5	10	0.57
14	75	11	0.58	85	5	10	0.29	60	15	25	0.57
Rb_2WO_4	**RbCl**	**WO_3**	**X**	**Cs_2WO_4**	**CsCl**	**WO_3**	**X**	**Dimensions,m**			
70	5	25	0.27	30	40	30	0.20	$1 \cdot 10^{-6} \cdot 5 \cdot 10\text{-}^{7}$			
70	15	15	0.27	25	50	25	0.20	$1 \cdot 10^{-6} \cdot 5 \cdot 10\text{-}^{7}$			
75	5	20	0.27	15	67	18	0.20	$1 \cdot 10^{-6} \cdot 5 \cdot 10\text{-}^{7}$			

It is established, that coarse grains OVB depends on temperature, density and structure of electro lit. Thus the sizes of particles bronze decrease with increasing density of a current, downturn of temperature and growth of concentration of chloride in electro lit, and the greatest on change of a degree of dispersive ness OVB renders of melts structure .

For today we have executed:

1. 1 .Visually - the polyhedral method leads the thermal analysis of system Me_2WO_4 – WO_3-MeCl in a wide interval of structures and temperatures.

2. On the basis of data thermal, chemical and the Russian Federation of analyses for the first time are given characteristics of morphology liquids threefold mutual systems Me, WO_2 // Cl, WO_4, one of which subsystems are systems Me_2WO_4-WO_3-MeCl studied in work (Me-Li, Na, K, Rb, Cs).
3. It is studied kinetic interactions of components of systems Me_2WO_4-WO_3 + MeCl depending on the maintenance of alkaline metals which data are presented for cesium system in threefold mixes. It is shown, that with increase in the maintenance of the last in a mix, thermal stability corresponding melts sharply raises.
4. A number of structures from which electrochemical and chemical by methods are synthesized OVB alkaline metals, with the sizes of particles from 1-10 up to 5-10 m, depending on maintenance MeCI in electro lit is developed.

REFERENCES

[1] Mohosoev M.V., Bazarova Z.G. / Complex oxides molybdenum and tungsten with elements I - IV M.: *the Science*, 1990p : p. 256 .

[2] Mohosoev M.V., Alekseev F.P., Lucik V.I.diagram of a condition molybdenum and tungsten system .-*Novosibirsk: the Science*, 1978, p. 320.

[3] Salmon R., Ceillet P. *Bull. Soc. Chim. France* 1969, № 5, p. 1569-1573.

[4] Rean J.-M., Fonassier C.L. *Bull. Soc. Chim. France* 1971, № 2, p. 398-402.

[5] Beljaev I.N., Doroshenko A.K., Nesterov A.A.Zhurn. inorganic, *chemistry, 1971*, т. 16, 9, p. 2604.

[6] DrobishevaSo-called, Bogoduhova N.A., Rabkina I.G/ И.Г. // proceedings. AH the USSR. *Inorganic materials*. 1975, т. And, 6, p. 1056.

In: Progress in Monomers, Oligomers, Polymers, Composites ... ISBN 978-1-60456-878-3
Editors: R. A. Pethrick, G.E. Zaikov et al.

Chapter 19

THE ELECTROCHEMICAL SYNTHESIS OF THE TUNGSTEN CARBIDE NANOPOWDERS AND CARBON NANOTUBES

Kh.B. Kushkhov
Kh.M. Berbekov Kabardino – Balkarian State University, Nalchik, Russian Federation

The principally new methods of electrochemical synthesis of the tungsten carbide nanopowders and carbon nanotubes have been developed in the centre of the collective use "The X-ray diagnosis of the materials" of the Kabardino – Balkarian State University. The research is a part of the priority project "The industry of the nanosystems and nanomaterials".

The electrolysis of the oxide and oxide – halogenide ionic solutions is employed for obtaining the tungsten carbide nanopowders. The basis of the electrochemical synthesis of the tungsten carbide is the simultaneous electrodeposition of atoms of both the tungsten and carbon on the cathode which may later join to form the nanoparticles.

The developed scheme is closed, ecologically safe and makes it possible to produce the nanodispersed powders of high purity and gives the opportunity to regulate the structure of the material. Comparing to the other known methods of manufacture, the high – temperature electrochemical synthesis allows to reduce the energy expenses and to simplify the process of production, because it consists of two stages only. The first is the very electrolysis while the separation of the acquired product from the electrolyte happens on the second one. The tungsten carbide nanopowders obtained with the electrochemical method could be pressed and sintered without introducing the linking additives. The exclusion of cementing agents results in the harder ceramics of the same viscosity. Besides, the tungsten carbide nanopowders possess the high electrocatalytic activity in the reactions of the electroemission and electrooxidizing of the hydrogen and can be used in various electrochemical devices instead of platinum (batteries and fuel cells, hydrogen energetics). Our tungsten carbide nanopowder can be characterized by: the specific area up to 20 square meters per gram, the particle size less than 200 nanometres, the impurity content varying in the limits of 0.01 to 0.001%.

The electrochemical synthesis of the carbon nanotubes is processed in two regimes. In first one the graphite cathode is being electrochemically eroded during the electrolysis of the individual chloride melts of alkali and alkali – earth metals and their mixes. Nanotubes are being produced after intercalating of the alkali and alkali – earth metals into the graphite. In the second one the chloride melts of alkali metals containing the lithium carbonate are being electrolysed at the excess pressure of the carbon dioxide (up to 15 atmospheres). The "building" unit of the nanotubes in the last case is the carbon after the electroreduction of the carbon dioxide dissolved in the melt.

In: Progress in Monomers, Oligomers, Polymers, Composites ... ISBN 978-1-60456-878-3
Editors: R. A. Pethrick, G.E. Zaikov et al.

Chapter 20

The Adsorption of the Polyurethane on the Surface of the Iron Oxide Particles

M.Kh. Ligidov*[1,*], S.Yu. Khashirova*[1]***, S.I. Pakhomov*[2]

[1]Kabardino – Balkarian State University, Nalchik, Russian Federation
[2]Moscow Academy of Jurisdiction, Moscow, Russian Federation

The influence of the surface – active agent on the thermoplastic polymer adsorbed on the surface of the magnetic powder is studied.

Keywords: polyurethane, adsorption, magnetic powder, iron oxide

The dispersing of the magnetic lacquer in the film – forming solution results in the following processes [1,2]: (1) the destruction of the powder aggregates and agglomerates caused by the external force, (2) the melting of the powder particles and the removing of the adsorbed gases and liquid from the solid surface, (3) the stabilization of the suspension resulted from the adsorptional blocking of the majority of the coagulated active centers by the linking components.

The particles of the magnetic powder are the initial needle – like crystals of the size 0.3-0.4 microns and the crystalline joints of the agglomerates containing the "bridges" of the solid phase. (The latter joints were formed during the thermal treatment). The separate initial crystals are also aggregated as a result of the Van-der-Waals interaction and the surface of the particles is contaminated with the gases and water besides that.

The dispersion of the magnetic powder is being processed in the multicomponent system consisting of the mixture of the solvents, antistatic, linking, dispersing, melting, peptizing and other functional agents.

The two stages of the melting and stabilizing of the suspension of the magnetic powder are mainly determined by the compounding and the order of the placing the agents into the

* Correspondence to: Russia, 360004, Nalchik, Shernyshevskogo street 173,E-mail: ligidov@inbox.ru

ferrolacquer while the stage of destruction (separation) of the aggregates and agglomerates is associated with the considerable expenditure of energy.

The wettability of the powder materials is dependant from the value of the specific surface, the microrelief and the porosity of the filler.

The surface active agents or dispersants, SAA for brevity, are being added into the suspension for improving the wetting of the surface of the filler by the liquid phase and to simplify the processes of dispersion and stabilization.

The introducing of the SAAs prevents the sticking of the particles together in the filled systems and increases their sedimentational and aggregational stability. The SAAs used were the L-α-lecytine and the KD-6 oxytos, the former possessing the higher affinity to the particles of the magnetic powder. The large values of the adsorption measured by the thermogravimetric (TG) method confirm that high affinity.

The value of the adsorption is the important parameter which defines the order of introducing the components into the solution [3]. The modified polyurethane adsorbed on γ-Fe_2O_3 was studied by the method of differential – thermal analysis (DTA). The ratio of the polymer mass (in milligrams) to that of the powder (in grams) was taken for quantitative estimation of the adsorption:

$$K = \frac{m_{polymer}}{m_{powder}}, \quad [mg/g]$$

Here K is the value of adsorption [4,5].

If the SAA is added into the magnetic suspension then the polymer and that agent concurrently adsorb on the magnetic powder. Such observation was obtained during the adsorption of the thermoplastic polyurethane, PU, and of the SAA onto the magnetic powder γ-Fe_2O_3.

The data of the thermogravimetrical study of the linking polyurethane and the composition PU+γ-Fe_2O_3 are represented in Fig. 1.

The thermal decomposition of the polymer goes in two stages, see curves 1 and 2 (the first obtained by the thermogravimetrical analysis and the second by the DTA). Regarding the pure linking we see that about 80% of the polymer disappears in the first step with the maximal speed at 593K and the rest – in the range of 723-823K. Two close maximums emerge in the second step on the curves of TG and DTA, see curves 2 and 3.

The thermooxidizing decomposition of the polymer in the mix PU+γ-Fe_2O_3 starts one hundred kelvins earlier due to the catalyzing effect of the filler. The decomposition also possesses two phases with maximums at 488K and 623K, see curves $1'$ and $2'$.The decomposition stops at 673K that is much earlier than the decomposition of the pure linking PU in the same conditions.

The DTA allows us to find interesting changes in the values of the heat effect of the destruction in the presence of the γ-Fe_2O_3 determined by area under the peaks on the curves.

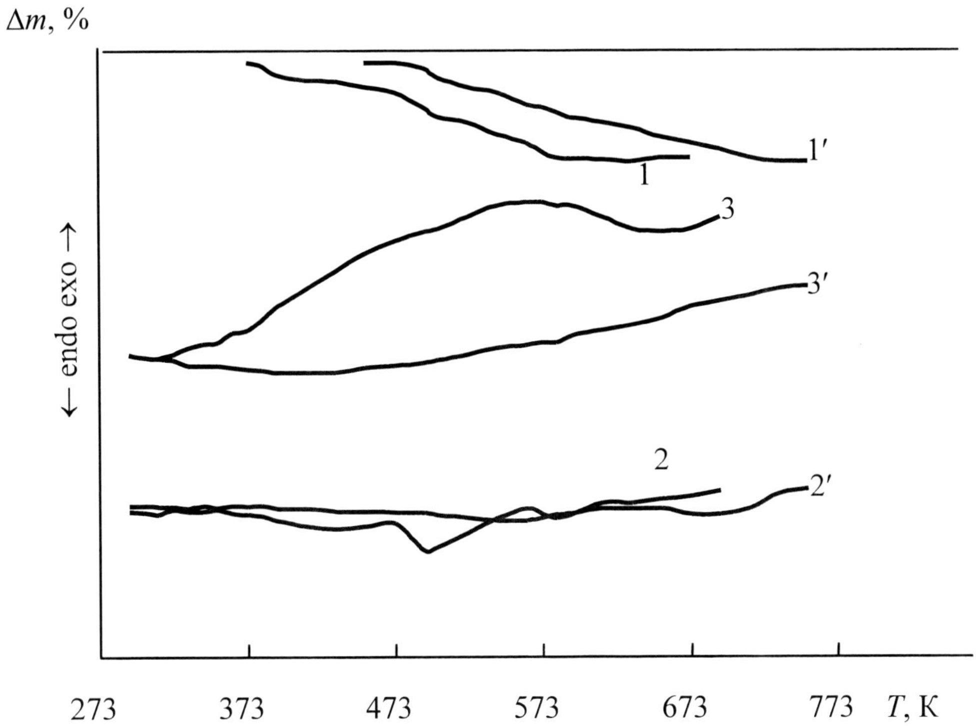

In the paper:"The adsorption of the polyurethane on the surface of the iron oxide particles" by Ligidov M.Kh., Khashirova S.Yu. and Pakhomov S.I.

Figure 1. The thermogravimetrical curves for polyurethane (curves 1,2,3) and for polyurethane doped with gamma iron oxide (curves 1/,2/,3/)

The thermal decomposition is usually accompanied by the exothermal effect. The heat effect following the destruction of the aliphatic groupings of the aromatic heterochain polymers (thermoplastic PU being one of them) on the first stage of the decomposition is generally negligible.

The considerable emission of the heat is observed in such polymers on the second stage of their decomposition, when starts the intense oxidizing of the condensed aromatic fragments.

The thermal decomposition of the polymer in the mix PU+γ-Fe_2O_3 is accompanied by the noticeable increase in the heat released from the very beginning of the process. The area under the peak of the DTA curve for the composition PU+γ-Fe_2O_3 is several times more than that for pure linking agent.

To explain this phenomenon we offer the following mechanism. The heat effect emerging on the DTA curve is the sum of two effects. The first, endothermal one, is caused by the breaking of the bonds and the second, exothermal one, is the consequence of the oxidation. The presence of the γ-Fe_2O_3 makes the first process easier thus lowering the endothermal contribution and so the total amount of the released energy increases.

So the presence of the magnetic filler catalyzes the process of destruction and increases the heat effect during the thermooxidizing decomposition of the linking polymer. Thus we suppose that γ-Fe_2O_3 is the active filler in relevance to the polymer linking agent. The interaction between them results in the change of the decomposition scheme of the polymer.

Our results agree with the data on high surface polarity of the γ-Fe_2O_3 and its ability to enter the chemical reactions with formation of durable chemisorptional layer.

The studies of the adsorption onto the magnetic powder and of the heat effect during the decomposition of the adsorbed polymer in the dependence of the cyclohexane content have been performed. We have determined the ratio of the peak area on the DTA curve to the mass of polymer decomposed (*A*, in relative units) for estimating the value of the heat effect.

The results show that adsorption decreases with the increasing content of the polymer in solution (Fig. 2a, curve 1). This probably is associated with the bigger associates formed by the molecular balls. The surface of the magnetic powder now adsorbs not only separate macromolecules but whole associates which have more problems with orienting themselves to form the chemical bridges to the surface of the iron oxide. Besides, huge associates can easily be removed by washing in the active solvent. This proposition agrees with the increase of the heat effect of the destruction per unit mass of the polymer (Fig. 2, curve 1). The number of contacts with the surface of the iron oxide per one macromolecule adsorbed will apparently increase with decreasing adsorption. Consequently, the catalyzing role of the γ-Fe_2O_3 will be more pronounced, thus increasing the value of the parameter *A*.

The third component, namely the SAA L-α – lecithin, was introduced into the system PU+γ-Fe_2O_3 for more thorough destruction of the aggregates of the magnetic powder and for increasing the total area of the surface of the particles. The presence of the SAA in the solution may produce a secondary effect: it may prevent the formation of the associates thus breaking the intermolecular bonds. Moreover, the molecules of the SAA might adsorb on the surface of the iron oxide particles interacting with the polar centers and decreasing the adsorption of the polymeric macromolecules.

Such complicated behavior of the SAA is well demonstrated on the curves of the dependence of the parameters *K* and *A* versus the solution content (Fig. 2). The value of the adsorption from 1% solution doped with SAA decreases, what can be explained by the predominant adsorption of the SAA on the magnetic particles. The SAA molecules possess light molecular mass and consequently they reduce the loss of weight during the decomposition. The adsorption reaches its maximum value at 3% concentration of the solution. This obviously happens after the formation of the associates in the presence of the SAA molecules (Fig. 2, curve 2). It is worth noticing that the heat effect of the decomposition process is practically independent from the value of adsorption.

We explain this phenomenon from the standpoint that the SAA molecules are adsorbed on the surface of the powder thus preventing the catalytic influence of the γ-Fe_2O_3 on the decomposition of the polymer. Besides, judging on the evident shift of the exothermal peak corresponding to the first stage of the decomposition to the lower temperatures and on the increase in the height of that peak we conclude that the destruction of the system starts from the layer of the SAA being in contact with the magnetic powder. Because the SAA is the low – molecular chain, its thermal decomposition results in the large number of light – volatile products, initiating the process of the oxidizing the polymer.

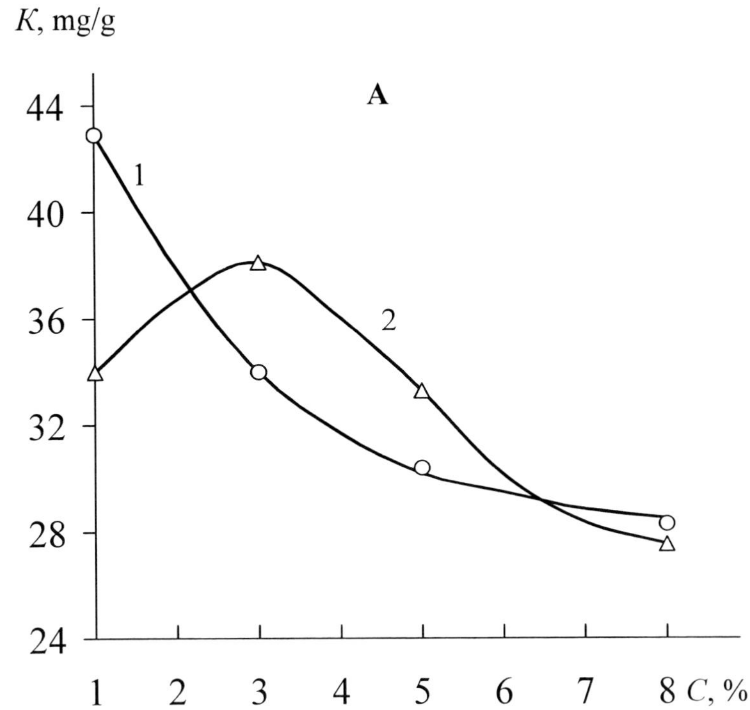

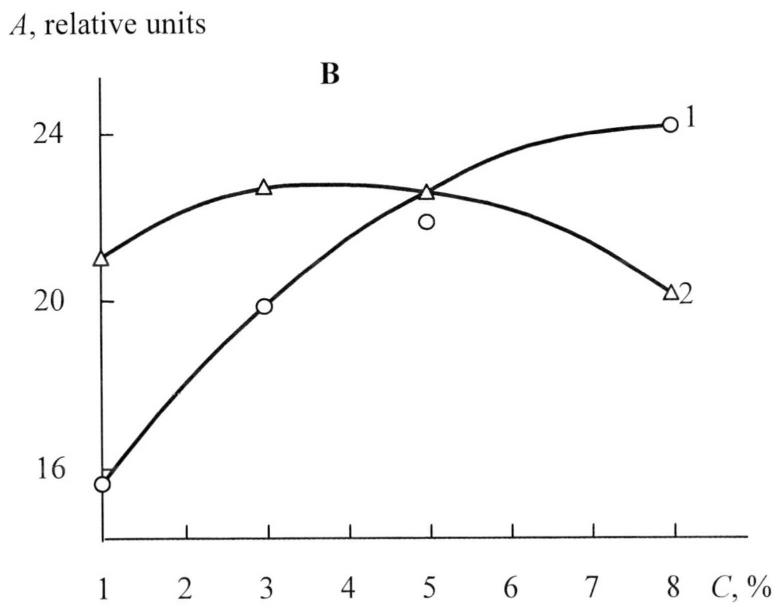

In the paper: "The adsorption of the polyurethane on the surface of the iron oxide particles" by Ligidov M.Kh., Khashirova S.Yu. and Pakhomov S.I.

Figure 2. The dependencies of the polyurethane (curves 1) and of the mix polyurethane + surface – active agent (curves 2) adsorbed on the gamma iron oxide (plot A) and of the heat effect of the destruction (plot B) versus the concentration of the polymeric solution.

The presented above investigations allowed us to establish some features of the adsorption of the thermoplastic polymer, namely polyurethane, on the surface of the magnetic powder and to find out how the surface – active agent influences on the given process.

ACKNOWLEDGEMENTS

This research was supported by the grant R.N.P.2.1.2.25 of Russian Education Agency, the joint tender 2007-3-1.3-24-02-003 of Russian Education Agency and Ministry of Education of Russian Federation and the grant No 06-03-96641 of Russian Foundation for Basic Researches.

REFERENCES

[1] Braginsky G.I. and Timofeev E.N. The technology of the magnetic films (Leningrad, *Khimia,* 1987) 328 p. [in Russian]

[2] Sibadzaki I. // *Settiaku,* 21(9) 330-339

[3] Guchinov V.A., Murzahanova I.I., Bazheva R.Ch. et al. // *Lakokrasochnie materialy i ih primenenie, 12* (1996) 12-14 [in Russian]

[4] Guchinov V.A., Gukepshokova E.B. and Ligidov M.Kh. // *Vestnik KBGU. Ser. Himicheskie nauki,* 1 (1996) 56-60 [in Russian]

[5] Ligidov M.Kh., Ulimbasheva Kh.L. and Guchinov V.A. / in Abstracts of the IV All-union *Conference on the Chemistry and Physico-Chemistry of the oligomers* (Nalchik, 1990) 103 p. [in Russian]

In: Progress in Monomers, Oligomers, Polymers, Composites … ISBN 978-1-60456-878-3
Editors: R. A. Pethrick, G.E. Zaikov et al.

Chapter 21

CARBON NANOTUBES: MANUFACTURE, PROPERTIES, APPLICATIONS

E.G. Rakov
D.I. Mendeleev Russian Chemical Technology University, Moscow, Russian Federation

Nanomaterials find wider application every day. The most promising ones are the carbon nanotubes (CNT) and nanofibres (CNF) and composites filled with them. The increased attention to these materials is caused by both the extra properties of ideal mono- or bi-layer CNTs (mechanical strength, heat and electroconductivity, density of the passing current) and the diversity in the morphology and properties of such nanoobjects.

There are two main groups of obtaining the CNT: the Sublimation/desublimation of the graphite and the Pyrolysis of the hydrocarbons. (The thermal decomposition of the carbon monoxide belongs to the second group). The former ways of manufacture are realized with help of electroarc and laser – thermal plants while the latter method is usually processed on the plants with either volatile catalyst or catalyst on the carrier. Because the pyrolitic methods have certain advantages (number of the versions, softer conditions, scaling opportunities and others) in comparison to the sublimation, they are more popular on the experimental and industrial plants abroad.

If one introduces the CNT or CNF into the polymeric matrix then the composite becomes electroconductive, its mechanical properties improve, thermal stability increases. The inserting of the specified chemical groupings or the adsorbing of the molecules of the linear polymers modifies the filler thus making the bond with the matrix stronger.

The areas of the application of CNTs and CNFs and composites include the state defense, air – space and atomic technologies, electronics, optoelectronics, energetics and informatics, biology and medicine.

D.I. Mendeleev Russian Chemical Technology University embraces the efforts of many scientists working on the developing the technology of the pyrolitic synthesis of common and thin multilayer CNTs and CNFs, as well as of modified ones; producing their stable dispersions in water and organic solvents; designing novel polymer – based composites. Also the pilot plant for manufacturing the carbonic nanomaterials is being built.

In: Progress in Monomers, Oligomers, Polymers, Composites … ISBN 978-1-60456-878-3
Editors: R. A. Pethrick, G.E. Zaikov et al.

Chapter 22

OZONE APPLICATIONS

***S. Rakovsky*[1] *and G.E. Zaikov*[2,*]**

[1]Institute of Catalysis, Bulgarian Academy of Sciences, Sofia 1113, Bulgaria
[2]N.M.Emanuel Institute of Biochemical Physics Russian Academy of Sciences, Moscow, Russian Federation

INTRODUCTION

The ozone application is based on its powerful oxidative action. The ozonolysis of chemical and biochemical compounds, as a rule, takes place with high rates at low temperatures and activation energies [1]. This provides a basis for the development of novel and improved technologies, which find wide application in ecology [2], chemical, pharmaceutical and perfume industries, cosmetics [3, 4], cellulose, paper and sugar industries, flotation, microelectronics, veterinary and human medicine, agriculture, foodstuff industry and many others [5-8]. At present it is very difficult to imagine a number of high technologies without using ozone. Here should be mentioned the technologies for purification of waste gases, water and soils, manufacture of organic, polymer and inorganic materials, disinfecting and cleaning of drinking and process water, disinfecting of plant and animal products, sterilization of medical rooms and instruments, resolving cosmetics problems, sterilization and therapy in veterinary and human medicine, surface cleaning and functionalization in the manufacture of polymer and microelectronics articles, flotation, deodorization and decolourization of gases, liquids and solid substances, oxidation in the intermediate stages of various technologies, etc.

* Correspondence to: N.M.Emanuel Institute of Biochemical Physics Russian Academy of Sciences, Moscow 119334, Russian Federation

1.1. ECOLOGY

1.1.1. Waste Gases

More than 100 contaminants of the atmospheric air have been identified. Among them SO_2, CO_2, nitrogen oxides, various hydrocarbons and dust constitute 85 %.

The main sources of harmful substances emissions, i.e. dust, SO_2 and CO_2 in the air, are the processing plants for coke, briquettes, coals, the thermal power stations, air, water, and road transport [9].The exhaust gases contain also, CO, organic and inorganic compounds, etc.[10]. Dust, sulfur dioxide, carbon dioxide, nitrogen oxides and organic compounds, etc. are the main pollutants released in the environment from metallurgy, the manufacture of fertilizers and petrochemistry. The most typical air pollutants emitted by some chemical productions are listed in Table 1.1.

Table 1.1. Main air contaminants emitted by some chemical productions

No.	Production	Pollutants	No.	Production	Pollutants
1.	Nitric acid	NO, NO_x, NH_3	12.	Ammonium nitrate	CO NH_3, HNO_2, NH_4NO_3 - dust
2.	Sulfuric acid: a) Nitroso b) contact	NO, NO_x, SO_x, H_2SO_4, Fe_2O_3, dust	13	Superphosphate	H_2SO_4, HF, dust
3.	Hydrochloric acid	HCl, Cl_2	14	Ammonia	NH_3, CO
4.	Ocsalic acid	NO, NO_x, $C_2H_2O_2$, dust	15	Calcium chloride	HCl, H_2SO_4, dust
5.	Sulfamidic acid	NH_3, H_2SO_4, $NH(SO_3NH_4)_2$,	16	Chlor	HCl, Cl_2, Hg
6.	Phosphor	P_2O_3, H_3PO_4, HF, $Ca_5F_4(PO_4)_2$, dust	17	Caprolactam	NO, NO_2, SO_2, H_2S, CO
7.	Phosphoric acid	P_2O_3, H_3PO_4, HF, $Ca_5F_4(PO_4)_2$, dust	18	PVC	Hg, $HgCl_2$, NH_3
8.	Acetic acid	CH_3CHO, CH_3OH	19.	Artificial fibers	H_2S, CS_2
9.	Nitrogen fertilizers	NO_2, NO, NH_3, HF, H_2SO_4, HNO_3	20.	Mineral pigments	Fe_2O_3, $FeSO_4$
10.	Carbamide	NH_3, CO, $(NH_2)_2CO$, dust	21.	Electrolysis of NaCl	Cl_2, NaOH

The technologies with ozone participation are very promising for SO_2 utilization as sulfate, CO as carbonate, nitrogen oxides as nitrates.

Scrubbers used previously for removing only acids, for example HCl and HF, etc., now can be used for SO_2 separation via the addition of an oxidizing agent into the water intended for gas treatment. Ozone as compared with the conventional oxidizers such as hydroperoxide, chlorine, sodium hypochloride, perchlorate , shows appreciably higher oxidizing efficiency [11].

The rate constant of ozone reaction with NO is about 1010 $cm^3/(mol.s)$, and with NO_2 it is ~ 107 $cm^3/(mol.s)$ whereby the oxidation of NO by ozone in the liquid phase is characterized by a higher absorption rate and rise in the concentration of the obtained HNO_3. Moreover the oxidation of NO can be accomplished in the exhaust gases or in the course of absorption [12].

The CO oxidation by ozone is carried out in the presence of Fe, Ni, Co and Mn oxides. In most cases ozonation is more appropriate than the conventional methods and sometimes it appears the only possible way for its preparation [13].

The purification of exhaust gases from burners working on liquid and gas fuel is accomplished by using oxidation catalysts of Perrovski type oxides - ABO_3 combined with ozonation. Ozone is injected prior the waste gases flow. A may be La, Pr or other alkali earth element; B may be Co, Mn or other transition metal. A may be also partially substituted by Sr, Ca or any other alkali earth element. A catalyst deposited on ceramic support of honeycomb type may contain $SrCo_{0.3}Mn_{0.7}O_3$. It is oxidized 24% CO in the absence of ozone and 80% in ozone atmosphere [14].

The removal of nitrogen oxides (NO_x) from waste gases is carried out through ozone oxidation in charged vertical column. The lower part of the column is loaded by 5-15% $KMnO_4$, and ozone is blown through the lower and upper part of the column and the waste gases are fed to the center of the column charge The inlet concentration of 1000 ppm NO_x in the waste gas is reduced to 50 ppm in the outlet gas flow. Another method for cleaning of exhaust gases from NO_x involves the application of plasma generator and ozonator [15] which practically leads to the complete removal of nitrogen oxides.

Sulfur containing compounds are removed from gas mixtures by silicon oil scrubbing and ozone oxidation of absorbed pollutants. A model system providing that the gas mixture contains methandiol and deimethylsulphide is fed at a rate of 9-120 m^3/h into silicon oil charged scrubber followed by ozone oxidation of the absorbate. The oxidates are extracted with water and the silicon oil is regenerated. Thus the purified gas mixtures practically do not contain any sulfur [16].

A special apparatus is designed for decomposing acyclic halogenated hydrocarbons in gases. It includes a chamber for mixing of the waste gases with ozone coupled with UV-radiation, ozonator, and inlet and outlet units. This method is very appropriate for application in semiconducting industry whereby acyclic and halogenated hydrocarbons are used as cleaning agents [17].

The mechanisms of ozonolysis of volatile organic compounds such as alkenes and dienes are discussed and the products output is determined by matrix isolation FTIR spectroscopy [18].

The synthesis of a material from zeolite via pulverization, granulation and drying at 500-700°C, cooling to 100-200°C , electromagnetic radiation or ozone treatment is described in [19]. The material thus obtained is suitable for air deodorization, drying and sterilization.

The purification of gases containing condensable organic pollutants can be carried out by gas treatment with finely dispersed carbon, TiO_2, Al_2O_3, Fe_2O_3, SiO_2 and H_2O_2 and ozone as oxidizers [20].

A wet-scrubbing process for removing total reduced sulfur compounds such as H_2S and mercaptans, as well as the accompanying paramagnetic particles from industrial waste gases is proposed [21]. For this purpose a water-absorbing clay containing MnO_2 is used. The collected clay is regenerated by ozone oxidation.

The removal of mercury from waste gases is carried out by catalytic ozonation [22]. The used gases mixed with ozone are blown over a zeolite supported Ni/NiO catalyst. This procedure results in 87% conversion of mercury into mercuric oxide which is isolated by filtration. Pt and CuO_x/HgO system is proposed as another catalyst for this process [22].

1.1.2. Waste Water

In contrast to the cleaning of natural water by ozonation which is experimentally confirmed as the most appropriate, the purification of waste water by ozone is still an area of intensive future research. This could be explained by the great diversity of pollutants in the used water and the necessity of specific approach for each definite case.

The recycling of water from cyanide waste water (copper cyanide contaminant) is accomplished by ozone oxidation combined with UV radiation and ion exchange method [23]. Further, the oxidate is passed through two consecutive columns charged with cationite and anionite. The ozonolysis priority over the conventional chlorinating method is demonstrated by the fact that ozonation fails to yield chlorides whose removal requires additional treatment.

The ozonolysis of CN-ions leading to CNO^--ion formation combined with UV-radiation proceeds at 3 fold higher rate as compared with that without radiation employment. Upon exchange in NaOCN, the Na^+ is substituted by H^+ followed by its decomposition to CO_2 and NH^{4+} via hydrolysis in acid medium. The ammonium ion is absorbed by the cation-exchange resin. The solution electroconductivity on the anion-exchange resin becomes lower that 10 muS/cm and cyanides, cyanates or copper ions have not been monitored. On the basis of these experimental results, a recycling method and apparatus for detoxification of solutions containing cyanides is proposed. The recycled water from the cyanide waste water may be re-used as deionization water in gold - plating. The process ion-exchange resins are regenerated after conventional methods. Actually, this method is not accompanied by the formation of any solid pollutants [23].

The waste water from the electrostatic and galvanic coatings containing CN^--ions and heavy metals is treated consecutively by ozone and CO_2. The cyanides and carbonates being accumulated in the precipitate are already biodegradable [24].

Waste water containing Cr(III) is oxidized by ozone combined with UV-radiation [25]. The radiation reduces the oxidation time about three-fold. The oxidate is consecutively passed through columns charged with cationite and anionite resulting in the obtaining of waste water with electroconductivity below 20 mu.S/cm. This deionized water can be successfully used for washing of coated articles. The Cr(IV) concentrated solution from the anionite regeneration can be further treated by appropriate ion-exchange method yielding a high purity Cr(IV) solution [25].

Purification of waste water containing $ClCH_2COOH$ and PhOH is carried out by ozone treatment and UV-radiation in the presence of immobilized photocatalyst. Thus the rate of decomposition is higher than the rate obtained only in the presence of ozone and UV-radiation [26].

Used water containing small amounts of propyleneglycol nitrate or nitrotoluenes is a subject to combined oxidation with ozone and H_2O_2, under pressure and heating to the supercritical point. Thus a nominal conversion of the contaminants higher than 96.7% is achived [27].

In laboratory experiments for reduction in residual COD in biologically treated paper mill effluents it is subject to ozonation or combined ozonation and UV radiation at various temperatures and pressures [28]. At a ratio of the absorbed ozone to COD less that 2.5:1 g/g, the elimination level with respect to COD and DOC is up to 82% and 64%, respectively. The ozone consumption is essentially higher in the case of the UV combined ozonation at pH>9 and elevated temperature to 40°C.

The decolourization and destruction of waste water containing surface active substances is performed by ozone treatment at ozone concentration of about 80 mg/l. It has been found that for waste water which has undergone partial biological treatment by fluent filters or anaerobe stages up to COD concentration > 500 mg/l and ratio 30D5/COD>0.2 ozonolysis with ozone concentrations up to 1.8 g/l does not change CBR, while after the complete biological treatment at COD<500 mg/l ozonation results in rise of the ratio 30D5/COD from <0.05 to 0.37. This evidences the increase of CBR [28].

The purification of waste water from paper manufacture characterized by enhanced biodegradability, as well as the removal of COD and halogenated compounds is carried out by a method based on ozonation and biofiltration [29].

Waste water from the paper pulp production are clean up by treatment with ozone and activated acid tar. The high values of pH favor the lignin decomposition and carboxylic acids formation since in this case the •OH - radicals and not ozone are the oxidizing agent. The application of activated acid tar has a positive effect on the dynamics of the microbiological growth, substrate consumption and CBR of organic acids. For example, maleic and oxalic acids are decomposed completely at ozonation of waste water in the presence of activated acid tar. For immobilization of biological culture the use of polyurethane foams appear to be very suitable [30].

The used water in collectors of a plant for paper manufacture are ozonized in a foamy barbotating contactor. The results indicate that ozone is very efficient in the oxidation of coloring and halogenated compounds. Its activity is proportional to the amount of absorbed ozone regardless of the variations in the gas flow rate, the inlet ozone concentration and contactor design. The amount of consumed ozone depends on the operating conditions and waste water characteristics. The absorption rate rises in the presence of stages including ozone decomposition, particularly at higher rates [31].

In biological granular activated carbon (GAC) columns the effect of pretreated ozonation on the biodegradability of atrazine is investigated. The metabolism of isopropyl-^{14}C-atrazine gives higher amount of $^{14}CO_2$ than the ring UL-^{14}C - atrazine which shows higher rate of dealkylation that the process of ring cleavage. The pretreatment with ozone increases the mineralization of the ring UL-^{14}C - atrazine and consequently rises the GAC-columns capacity. 62% of the inlet atrazine is transformed into $^{14}CO_2$ in columns charged by ozonized atrazine and water. However, in columns with nonozonized atrazine and water only 50% of the inlet atrazine is converted into $^{14}CO_2$ and in columns supplied only with nonozonized atrazine only 38% of it are converted to $^{14}CO_2$ [32].

The waste water from petrochemical industry contains substantial amounts of phenol. The phenols mixtures appear to be more toxic that phenol itself and possess a synergetic effect. Their decomposition to nontoxic products is important and modern problem. The oxidation of phenols such as *p*-cresol, pyrocatechin, rezorzine and hydroquinone by O_3 in aqueous medium appears to be a promising method for their degradation. At concentrations of 100 mg/l and pH = 11.5 their complete oxidation is performed for 30 min.

The phenol content in used water is reduced from 145-706 mg/l to 2.5 mg/l at ozone consumption of 1.1 - 2.6 g/l. In real conditions the ozone consumption varies from 5 to 10 g/l per g phenol.

The results from the ozonolysis of waste water from coke processing are shown in Fig.1.2.

Table 1.2. Results from the ozonation of waste water from coke processing. [O_3] = 5 mg/l, flow rate 6 l/min and treatment time - 4 h

Pollutant, mg/l	Before ozonation	After ozonation
Monophenols	710	0.8
Polyphenols	380	198
Cyanides	3	29
Thiocyanides	384	0
Thiosulfates	538	0
Sulfides	43	2
H_2O_2	0	12
Bases	114	27

The treatment of concentrated aqueous solutions of phenol (1.0g/l) with ozone causes the appearance of a yellowish coloring after absorption of 1 mol O_3 per 1 mol phenol which gradually fades away. In this case the higher phenols undergo slight oxidation and H_2O_2 is identified in the water after the treatment.

The tests demonstrate that upon varying content of phenols in the used water the ozonation is one of the most promising method for their removal. The ozonation was found to be applicable for the decomposition of rhodanide in neutral and weak acid medium in the temperature range of 9-25^oC. The ozone consumption in this case is 2 mg/mg. The ozone uptake for the oxidation of the cyanide ion (CN^-) is 1.8 mg/mg.

The complex Zn cyanides are oxidized similarly to the simple soluble cyanides. Regardless of the higher resistance of Cu cyanide complexes, the ozone treatment reduces the CN^- content substantially (traces). Copper carbonate is precipitated as a reaction product.

The ozonation can be successfully employed as a method for cleaning the waste water from the manufacture of ammonium nitrate containing CN^-, S^{-2}, CSN^- ions. The purification level is 83, 98 and 95% respectively at O_3 consumption of 0.7 g/l/g water for 10 min contact.

The ozonation of waste water from sulfate-cellulose production results in 60% purified used water which is returned in the process cycle.

The cleaning of waste water from the petrochemical plants by ozone is efficient under wide range of pH (5.8 - 8.70 and temperatures (5 - 50^oC). Thus the level of petrochemicals in the waste water at pH 5.8, ozone consumption of 0.52 mg/mg and 10 min contact is reduced from 19-33 mg/l to 2 mg/l. In Table 1.3 is shown the efficiency of various treatment procedures for deodorizarion of petrochemicals contaminated water.

Table 1.3. Result from deodorization of petrochemicals contaiminated water (grades/dilution)

Petroleum products	Before treatment	Coagulation and filtration	Charcoal and filtration.	Chlorination	Ozonoation	Coagulation, filtration and ozonation
Petrol	5/80	5/60	5/50	5/80	2÷3/10	1/0
Kerosin	5/100	5/80	5/70	5/100	4÷5/25	1/0
Petroleum	5/100	5/60	5/60	5/100	5/20	2/5
Oil	5/10	1/0	1/0	5/10	1÷2/2	1÷0/0

Ozonation is very economical and highly efficient method for decomposing cancerogenic substances such as 3,4-benzopyrene, etc., particularly after the biological purification of waste water.

The application of ozone is also advisable for mercury oxidation in chlorine producing, for decontamination of the used water from the poisonous tetraethyllead, for destroying pesticides traces, etc.

The highest efficiency of cleaning of petrochemicals contaminated water is achieved at the following sequence: primary cleaning (mechanical), secondary (biological) and complete purification (ozonation).

Ozone is also successfully applied for the complete purification of waste water form dyes manufacture. Their color decreases about 10 000 times at exposure to 0.22 g/l ozone and the biological activity of the reaction products is reduced almost to zero. The semi-industrial experiments in CP "Kistenetz" were carried out on the dyes manufactured in the plant:

1. Direct Congo red - $C_{32}H_{26}O_6N_6S_2Na_2$, Mw=698

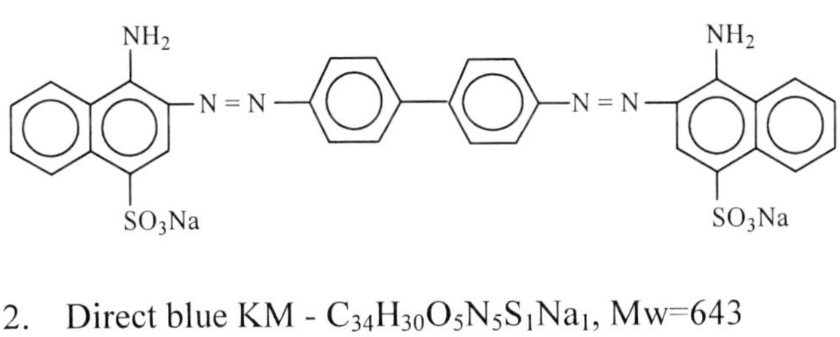

2. Direct blue KM - $C_{34}H_{30}O_5N_5S_1Na_1$, Mw=643

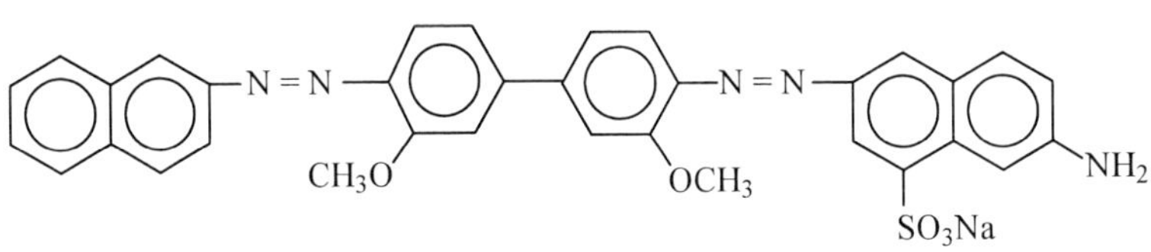

3. Acid blue ZK - $C_{26}H_{20}O_{10}N_3S_3Na_3$, Mw=699

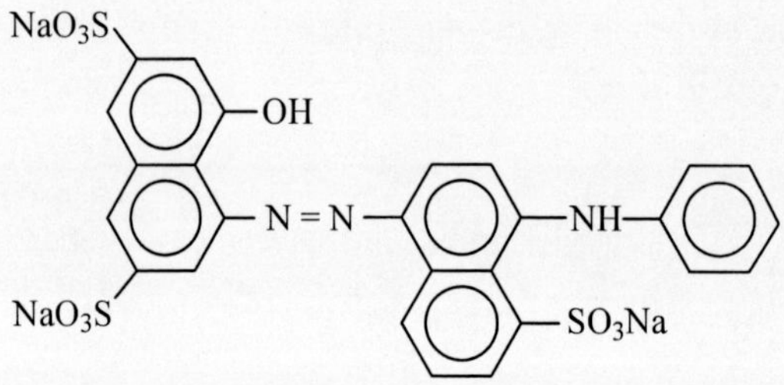

4. Acid chromium green - $C_{16}H_{11}O_{10}N_4S_2Na_2$, Mw=529

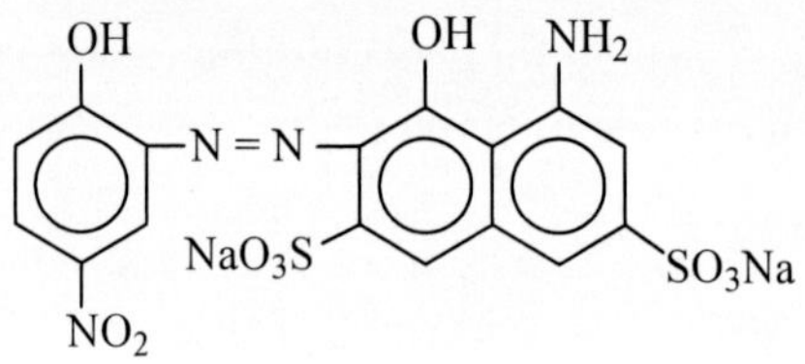

5. Acid black ATT - $C_{22}H_{16}O_9N_6S_2Na_2$, Mw=618

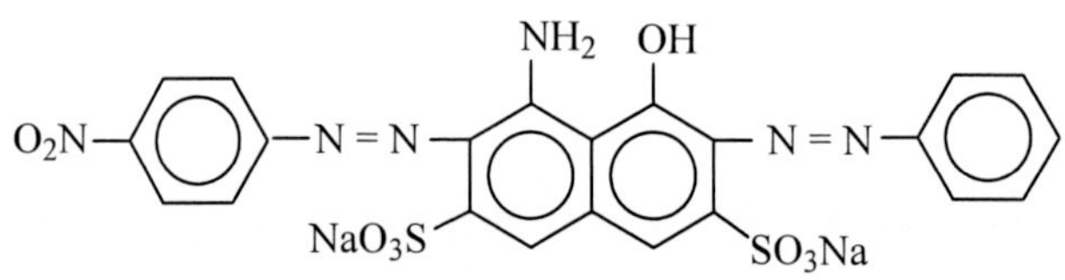

6. Acid sulfonic blue - $C_{32}H_{25}O_6N_5S_2Na_2$, Mw=685

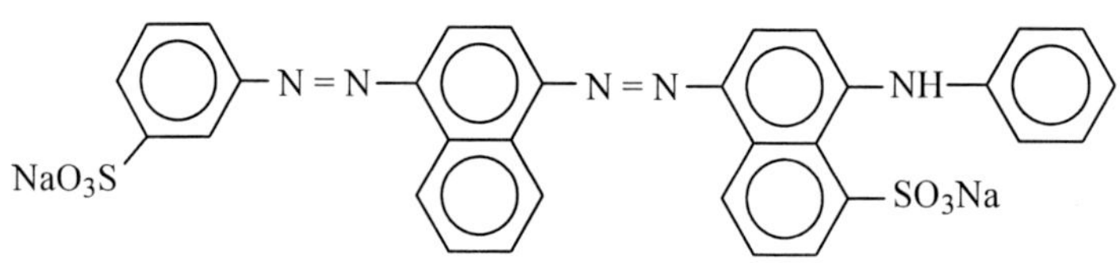

7. Acid chromium black C - $C_{22}H_{19}O_5N_6S_1Na_1$, Mw=502 è $C_{16}H_{13}O_7N_4S_1Na_1$, Mw=428

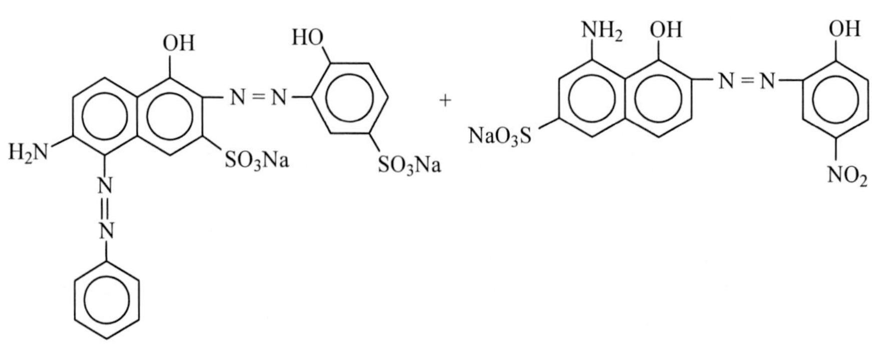

8. Direct black Z - $C_{34}H_{26}O_7N_8S_2Na_2$, Mw=768

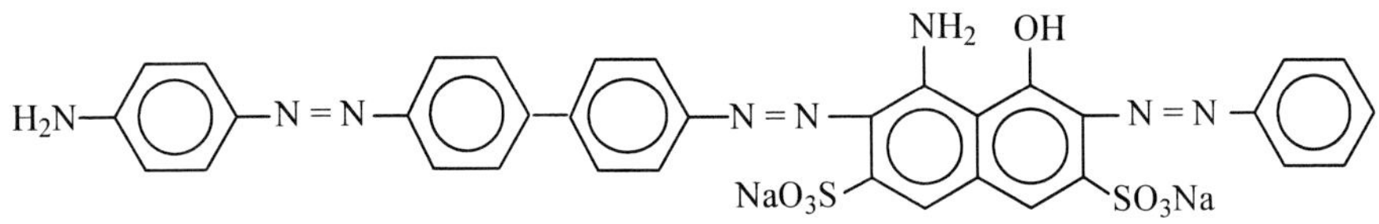

9. Acid chromium yelow R - $C_{17}H_{14}O_3N_3S_1Na_1$, Mw=367

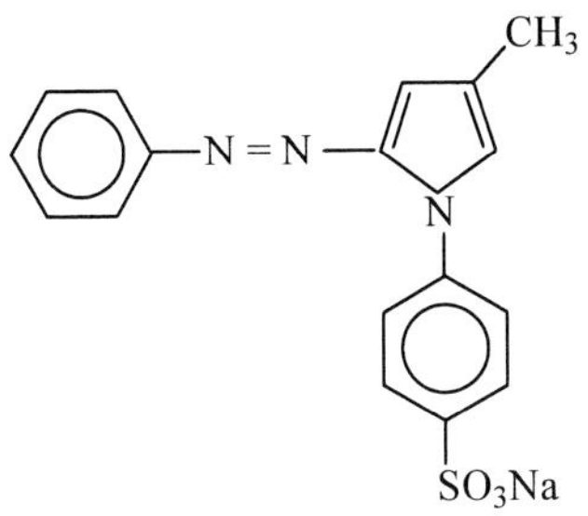

In the process of developing a technology for water purification from various organic contaminants, we have proposed a method and apparatus for quantitative determination of chemical compounds, separated by means of liquid chromatography. Special attention is paid to those compounds susceptible to ozone attack such as olefin hydrocarbons and their derivatives, phenols, amines, thiocarbamates, inorganic compounds, etc. The characterization of the separated compounds is carried out by spectral, massspectral, refractometer, electrochemical analyses, but the detectors in most of these equipments are rather expensive and complex units without sufficient identification capabilities, particularly for the compounds pointed above. In this sense the apparatus designed by us allows to overcome the disantvantages of the known methods and equipments. The scheme of the ozone detector applicable for identification and liquid chromatography separation of chemical compounds is presented in Fig.1.1.

The ocean power stations operate on the basis of the temperature difference between the upper and lower water layers. Taking advantage of the one-step solution of ozone in depth, the COD is decreased which prevents the befouling in the piping systems found in the upper layers. The ozone-enriched water from the lower layers is supplied to a cyclic filter and is directly injected into upper seawater for control of the sea microorganisms growth [33].

The removal of benzofurenes from ash, clays, soils, water and oils by combustion, ozonolysis of supercritical water, cracking-processes of petroleum, as well as thermal, photochemical and biological decomposition, has been the subject of many investigations [34]. The direct blowing of ozone through water pipers results in slimes removal from the inner pipers walls [35].

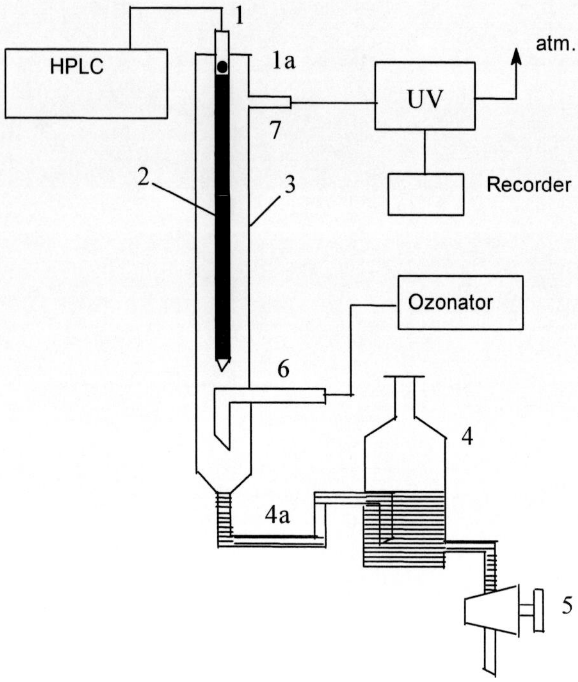

Figure 1.1. Set-up of ozone detector for identification of separated by liquid chromatography chemical compounds. 1 - eluent access; 1a - entry for pouring out the elate as thin film on 2 - solid glass stick; 3 - outer glass jacket; 4 - vessel for eluent accumulation; 4a - hydraulic détente; 5 - tap for pouring out the eluate; 6 - ozone inlet; 7 - ozone outlet.

The ozone blowing at a rate of 2.87 mg/min through aqueous solutions of 15 pesticides with 10 μg/l concentration leads to their transformation and their concentrations are reduced by about 20% for 60 min. The identified intermediates from the rearrangement of EPN, phenitrothione, malathion, diazinon, izoxathion and chloropiriphos are the corresponding oxones, while for phenthione they are the corresponding sulfoxide, sulfone and sulfooxide oxone analogues, and for disulfothione the intermediate is its sulfo analogue. 2-Amino-4-ethylamino-6-chloro-1,3,5-thriazine is identified as symazin and its 4-isopropylammonium derivative is atrazine intermediate. The isopropylthiolane results in obtaining of its 1-oxo-derivative. Upon replacement of ozone by air changes are actually not registered [36].

At the purification of process water in a soil decontamination plant containing pesticides the removal of organophosphorous compounds such as Thiometon and Disulphoton is of particular importance. In this connection four methods have been applied: 1) ozonolysis at pH = 2.5; 2) ozonolysis at pH = 8; 3) UV/H_2O_2 treatment and 4) oxidation by Fenton reagent [37]. The laboratory experiments were carried out with pure compounds soluted in buffered deoionized water and in process water which is extracted by solid-liquid extraction of the contaminated soil. The use of ozone in an acid medium turns to be ineffective since the Thimeton reaction practically stops after the formation of PO-derivatives. In all cases when the oxidation is carried out by HO - radicals, a sufficient removal of the pesticides and their metabolites is achieved. Oxadixyl, a cyclic nitrogen containing compound which is present in high concentrations in the process water is the most stable one, thus being the main soil contaminant. The investigated experimental conditions and the results obtained show that these methods are quite acceptable for universal application [37].

The water-soluble agrochemicals ASULAM and MECOPROP are decomposed for 30 min at ozone blowing (flow rate - 5.1 mg/min and concentration of 10 ppm)) through their aqueous solution (200 ml). The decrease in the flow rate of ozone retards the decomposition process. Similar treatments for 5h give in ACEPHATE and DICAMBA decomposition to 20%, the rate of decomposition being promoted with H_2O_2 addition [38].

The primary ozonation products of organophosphorous pesticides in water such as diazinone, phenthione (MPP) and ediphenphos (EDDP) [39], are identified by means of GC-MS analysis. The massspectra of the ozonation products of 17 organophosphorous pesticides point oxones as the primary reaction products. This fact is also confirmed by the SO_4^{2-} generation resulting from the ozonation of their thiophosphoryl bonds. Oxones are relatively stable towards ozone attack but they are further hydrolyzed to trialkylphosphate and other hydrolysis products. However, with MPP the thiomethyl radicals are first oxidized to thiophosphoryl bonds giving MPP-sulfoxide,. MPP-sulfone, MPP-sulfoxide-oxone are also obtained. Two main oxidation products have been identified at the oxidation of bis-dithio-type ethiones. Phosphate type EDDP is stable towards ozonolysis but its oxidation products are identified after hydrolysis [39].

Pesticides and their degradation products which are not mineralized to CO_2, NH_3, H_2O and inorganic salts can damage and contaminate the water piping. The treatment by photolysis or ozonation may substantially increase the mineralization rate. The photodestructive products of S-triazoles, chloroacetanylide and parakaute are declorinated and/or oxidized. The ozonation of these herbicides results in formation of products whose side alkyl chains are oxidized or removed, and the aromatic ring is oxidized or soluted but dechlorination does not occur. In some cases the medium may cause microbiological effect [40].

A deodorization method for treatment of ill- smelling air containing ammonia, hydrogen sulfide or amines from refrigerator chambers, lavatories, cattle-sheds, etc. involves the ozone contact with air in the presence of porous catalyst of 120 m^2/g specific surface. The used gas passes through a layer of carbon black for adsorption of the residual ozone and the nondegradated smelling substances. The active component of the catalyst may be: transition metal or its oxides supported on porous honeycomb type supports [41].

The contaminated smelling gases in drain and fecal water are dried to 60-30 % humidity, followed by ozone treatment on an oxidizing catalyst. For example, contaminated gas containing 30 ppm methylsulfide is dried to 30% humidity and then is treated with ozone in the presence of honeycomb type catalyst containing 83:12.5 = TiO_2:SiO_2:MnO_2 with a volume rate of 50 000 h^{-1}. In this case the deodorization efficiency is 99 % while at 100 % humidity it drops to 80% [42].

1.1.3. Natural (Drinking) Water

The basic sources of drinking water usually contain various organic admixtures in mg/l: carbon - 30, nitrogen - 0.8, fatty acids - 30, highmolecular organic acids - 0.2. naphthenolic acids - 1.5, phenols - 1.2, luminescent substances: neutral resins - 17%, oils, humus - 55%, acid resins - 19%, hydrocarbons, naphthenolic acids - 4-7%.

The coloring of natural water varying from pale yellow to brown is due to the presence of humus substances, Their decolorization is carried out by adsorption on coagulated $Al(OH)_3$ or $Fe(OH)_3$. The increasing demand for fresh water requires the use of more colored water which however is pre-chlorinated and then subject to adsorption up to their complete decolorization.

The ozone treatment removes the bad taste and odor of water, the resistant phytoplankton being oxidized to 20%. The ozone consumption for decolorization of natural water varies in broad limits - from 1 to 18 mg/l and depends on the humus composition. The process of humus compounds oxidation by ozone is presented in Scheme 7.1.

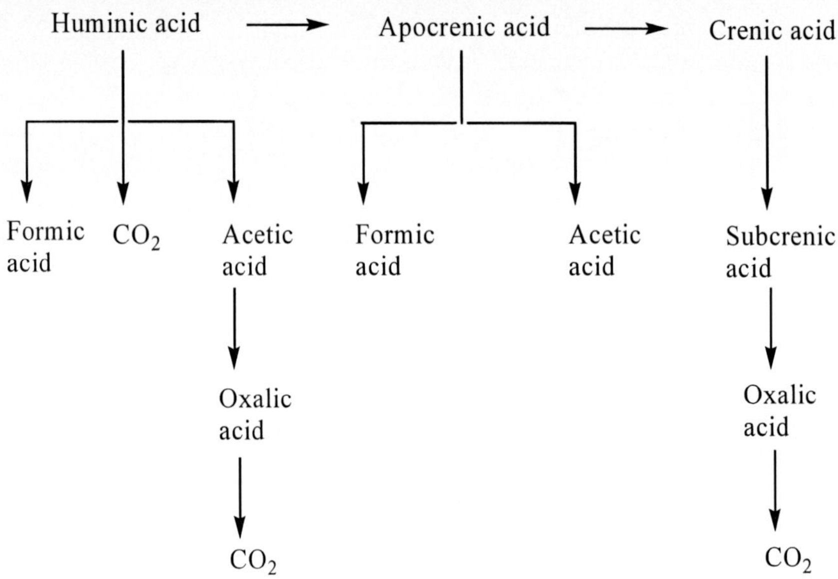

Scheme 7.1.

The various biological processes lead to formation of substances causing the bad taste and smell of water - mercaptanes, sulfides, alcohols, carbonyl compounds, acids, terpenes, amines, etc.

The deodorization effect of ozone is preserved in wide range of pH values, temperatures and ion composition. Due to its high oxidation potential, ozone can also degrade the resistant toxic substances. It is known that during the blooming in the water reservoirs are accumulated biologically active substances among which algotoxine is particularly harmful for the human health. Its complete degradation is achieved by using ozone concentration of 14 mg/l and 50 min contact time.

At ozone concentration of 8 mg/l in water, only 6% of *Scendezmus* algae remain alive after 10-15 min. However is to be noted that the ozone action on the various algae is quite different and specific. Thus *Astetionella* type algae are very ozone resistant.

The oxidative, disinfective, decolorizative and deodorizative properties of ozone have found wide application for preparation of drinking water. Many world corporations such as the French Taileygas, Degramont, Ozonia, the German Fisher, AEG, Siemens, the Swiss Braun & Bovery, the American General Electrics, General Motors etc., are among the biggest producers of drinking water. Twenty of the biggest world ozonator stations for drinking water treatment are listed below (Table 1.5).

Table 1.5. Some of the world biggest ozonation stations for drinking water treatment

City	State	Water, m^3/day	Ozone, kg/h
Moscow	Russia	1 200 000	200
Montreal	Canada	1 200 000	150
Ashford Common	GB	-	146
Coppermills	GB	-	144
Neuilly-sur-Marne	France	600 000	120
Choisy-le-Roi	France	900 000	120
Masan	Korea	-	112.5
Kiev	Ukrain	400 000	80
Helsinki	Finland	495 000	63
Manchester	GB	480 000	50
Orly	France	320 000	50
Singapore	Singapore	230 000	48
Minsk	Belaruss	200 000	47
Mery-sur-Oise	France	300 000	45
Lodz	Poland	190 000	40
		290 000	
Chiba	Japan	270 000	38
Nizhni Novgord	Russia	350 000	30
Bodenzee	Germany	480 000	30
Nant	France	380 000	30
Amsterdam	Holland	125 000	25
Brussels	Belgium	250 000	24
Quebec	Canada	220 000	19
Wrotzlaw	Poland	180 000	17

Ozonation stations for drinking water preparation work in Portugal, Greece, Bulgaria, USA, etc. all over the world.

Recently water treatment stations in Pancharevo and Bistritsa in Sofia, Bulgaria, of capacity 33 and 66 kg O_3/h, respectively are built.

The technology of drinking water ozonolysis leading to its deodorization, disinfecting and decolorization includes gas units for dry air producing, ozonator, barbators or other mixing units, apparatus for HO•-radicals generation, promoting the ozone formation and filter-catalyst for decomposition of the residual ozone. The preferable materials for barbators are zeolites or ceramics on the base of aluminum and/or silicon with pores diameter of 5-100Å, and the catalysts should contain one or more of the following components: Pt, MnO_2, CuO, Ni_2O_3, Zn, YiO_2 and SiO_2 [43].

For desinfection of drinking water, the ozone-containing air is mixed with water in the water pipe, afterwhich the bubbled water is supplied into a special designed chamber filled by porous ceramics charge for intensification of ozone dissolution into water. Some problems concerning the optimization of the UV stimulated oxidation of organic contaminants in soil water with ozone and hydroperoxide is discussed [44].

The disinfection of drinking water by ozonation can be also carried out at the outlet of various water supplying units before and after consummation, such as taps, fountains, water heaters, dental units [45], etc. In this connection are used small dimensional and low capacity personal ozonators coupled or installed additionally to these equipments.

The attempts for treatment of drinking water with ionization radiation appear to be ineffective mainly because of nitrates and H_2O_2 formation. These problems, however, do not occur when the radiation is performed in the presence of sufficient ozone amount [46]. The addition of ozone before and in the course of radiation converts the radiation into a purely oxidation process, a modification of the so called - Advanced Oxidation Process (AOP). The combined ozone-ionization radiation method is unique since two processes for HO-radicals generation are simultaneously induced. This method yields higher concentrations of HO-radicals as compared with other AOP. This in its turn results in precipitation of smaller amounts at equal ozone concentrations or reduces the ozone consumption at one and the same amount of precipitates. This makes the process particularly suitable for remediation of weakly contaminated soil water. The ozone concentrations at this process are higher than those used for drinking water disinfection. Thus, the drinking water subject to this treatment is disinfected more rapidly. The purification of waste water using this method is more efficient than the UV-radiation treatment only. This combination reduces COD without causing the rise of BOD. In this sense this method appears to be more attractive than the conventional two-stage processes of ozonation and biological treatment. The injection of gaseous ozone into the radiation chamber improves the water flow turbulence and substantially increases the efficiency of contaminants degradation.

The European organization "European regulation" has allowed the use of ozone for iron and manganese removal from mineral water, this initiating studies on secondary products formation, amount and composition , particularly during ozonation of bromine water. This water, usually contain BrO_3^- ions and various halogenated organic compounds (HOC). Some projects regulate a standard for PDK??? of HOC of 25 μg/l BrO_3^--ions, while after other it is from 10 to 0.5 μg/l. The present standard for LD of HOC is 100 μg/l [47]. Upon ozonation of bicarbonate mineral water - one sodium-rich water, and the other, calcium rich, it has been established that: 1) the process depends on the initial Fe and Mn concentrations and the Na and Ca concentrations; 2) HOC and bromate ions are secondary products and 3) the specific components characterizing a definite mineral water affect the ozonolysis. This requires definition of the conditions for Fe and Mn removal for each concrete type of mineral water. The removal of iron and manganese is efficient when the amount of ozone is three times greater than the stoichiometric one (0.43 mg O_3/mg Fe and 0.87 mg O_3/mg mn).

The disinfection of swimming pool water is carried out by controlled addition of compounds- bromine and iodine donors (NaBr or NaI) and water oxidation by hypochloride, ozone or potassium peroxymonosulfate [49]. We have designed a compact module system (stationary and mobile) which includes module for air preparation, ozonator, absorption column and unit for the complete decomposition of the residual ozone. Its capacity is 12-15 g/h and it is very suitable for sanitation of drinking water and used water from swimming pool (3.5 l/s capacity).

1.1.4. Soils

The decontamination of dredging wastes containing spilled petroleum is carried out by using a system for the direct injection of ozone (100-1000 ppm) containing air into pipes at the bottom; means for pumping out the spilled petroleum to a tank for its storage; units for its mixing with ozone for decomposition of organic matter in the tank; units for returning of the processed petroleum for the complete degradation of the residual organic matter by means of aerobic bacteria. The efficiency of organic matter decomposition is by about 20% higher than that by using the conventional methods including only treatment by ozone [49].

Another technology for treatment of contaminated soils [50] involves the ozonolysis of the water slurry of the contaminated soil layer, in the presence of H_2O_2 or in its absence and consecutive biological decomposition. The priority of the proposed technology over the conventional ones is that the removal of organic contaminants is by *on-site* treatment of the soils.

The *on-site* recyclation of petroleum contaminated soils by ozonation requires the good knowledge of the geological setting and the recycling system. Usually the recycling stages include: 1) system for air evacuation from the soil; 2) means for injection of ozone combined with air evacuation from the soil; 3) development of means for ozone generation; 4) screening of soils contaminants; 5) use of systems for safety control; 6) organization of the cleaning units - pumping and injecting drills; 7) decision for air injection into the soil; 8) methodology for measurement and regulation of ozone concentration; 9) hydraulic purification of the highly contaminated soil areas, etc. [51].

The treatment by ozone can be successfully applied for decomposing mineral oils, polycyclic aromatic compounds (PACs), phenols and some pesticides to biodegradable, nontoxic compounds. It should be noted that although ozone decomposes a great number of microorganisms, the soil microflora can be easily remediated after this treatment [52]. The latter can be carried out on- and out of the contaminated area. A number of problems related to the ozonolysis of organic compounds, the PACs degradation products, the effect of ozone on the soil microflora, the laboratory and pilot results, etc. are discussed.

A method and apparatus for decontamination of wastes such as soil, porous granulated sludge, are suggested. The polluted materials are placed in a special vessel and ozone is bubbled through. The waste gases obtained from the ozone oxidation are further treated with ozone [53].

A method for ozone treatment both *in-situ* and *on-site* has appeared to be applicable for the degradation of organic substances adsorbed on the solid surface [54].

Wet-scrubber method (based on MnO_2) for removing total reduced S-containing compounds such as H_2S, mercaptanes and non-magnetic particles from industrial waste gases using ozone, is suggested [21].

1.2. INDUSTRY

1.2.1. Manufacture of Organic Compounds

We will focus our attention to some of the numerous ozone applications in organic polymer and inorganic productions, passivation, cleanup and preparation of surfaces for electronics, superconductors, etc.

The reaction of ozone with olefins is the main reaction for ozonides preparation (1,2,4-trioxalanes) which can be converted into a mixture of carbonyl compounds, the exact composition of which depends on the olefin structure. This reaction is used for synthetic purposes as with small exceptions the C=C bond is cleaved quantitatively under very mild conditions. Usually the ozonolysis is carried out at low temperatures (as a rule at $-70 \div -30^{o}C$) as demonstrated in Chapters 2 and 3. Mostly, ozonolysis of olefins, except in solid phase, is carried out in various solvents such as paraffins, halogenated hydrocarbons, oxygen, sulfur and nitrogen containing hydrocarbons like ahcohols, glycols, aldehydes, ketones, acids, ethers, esters, amines, amides, nitriles, sulfoxides, mineral acids, bases and mixtures from them, etc. In most cases the reaction products are not separated from the reaction mixture except for the purposes of ozonide obtaining, and the oxidates are subject to hydrolysis, reduction or oxidation.

The hydrolysis is applied for preparation of aldehydes [55] and ketones [56]. The reductive decomposition is carried out in the presence of Zn/AcOH, SO_3^{-2}, HSO_3^{-1}, $LiAlH_4$, $SnCl_2$, ArP_3, Ph_3P, SO_2, $(CH_3)_2S$, H_2/catalyst. The reduction by metal hydrides leads to alcohol formation in the decomposition products mixture while the use of other reductors yields carbonyl compounds. Actually there are no clear requirements for selection of the most appropriate reductor, but it turned out that dimethylsulfoxide is probably [57] the most preferable reductor in methanol solution. The ozonolysis in the presence of tetracyanoethylene yields directly carbonyl moieties omitting the reduction step [58]. A novel modification of the conventional methods of ozonolysis is exemplified by the ozonolysis of silica gel adsorbed alkenes (acetylenes) [59] and selective ozonolysis of polyalkenes, controlled with the help of dyes introduced in the reaction mixture [60]. Some examples of ozonolysis using various reductors for ozonides decomposition are given in Table 1.4.

Ozonolysis is a powerful chemical method for economical and ecologically pure preparation of various carbonyl compounds [69].

The ozonolysis of cyclic alkenes in protic solvents followed by reductive decomposition of the hydroperoxides formed is a classical method for dialdehydes synthesis [70]. This method is applied for the preparation of 3-ethoxycarbonylglutaric dialdehyde by ozonolysis of ethyl 3-cyclopentenecarboxylate. The dialdehyde is an intermediate in the synthesis of dolazetrone mesitilate which is an active medical substance in ANZEMET anti-emetik. Zn/AcOH, phosphines, amines and sulfides such as 3, 3'-thiodipropionic acid and its salts are the most suitable reductors for this system. Their efficiency is comparable to that of methyl sulphide but without its shortcomings. The polymer immobilized 3,3'-thiodipropionic acid is also very active in this reduction reaction [70].

Table 1.4. Preparation of aldehydes, ketones and alcohols during alkenes ozonolyis with consecutive reduction

Nî.	Substrate	Conditions	Product	Yield, %
1.	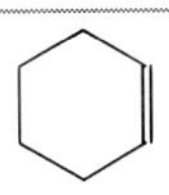	1. EtOAc, -20^0- -30^{0}C 2. $Pd/CaCO_3/H_2$	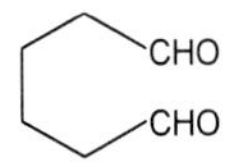	61
2.	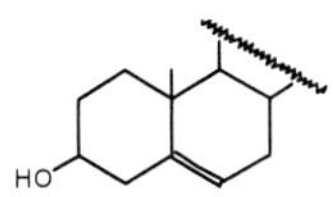	1. n-C_6H_{14} 2. Zn/HOAc	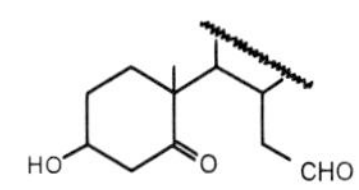	59
3.	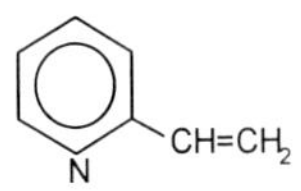	1. MeOH/-40^{0}C 2. Na_2SO_3/H_2O	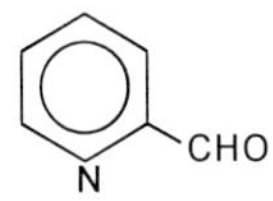	65
4.	$Me(CH_2)_5CH=CH_2$	1. MeOH/-30^0÷ -60^{0}C 2. Me_2S	$Me(CH_2)_5CHO$	75
5.	n-Bu-OOCCH=CHCOO-Bu-n	1. MeOH/-65^0÷ -70^{0}C 2. $(MeO)_3P$	n-Bu-OOCCHO	78
6.	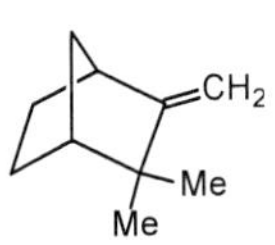	1. MeOH/-35^0÷ -75^{0}C 2. NaI/MeOH/HOAc	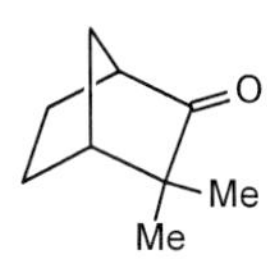	75
7.	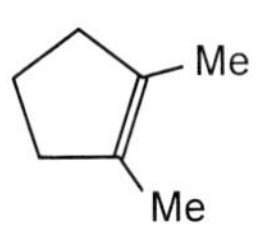	EtOAc/$(CN)_2C=C(CN)_2$, -70^{0}C	$MeCO(CH_2)_3COMe$ + 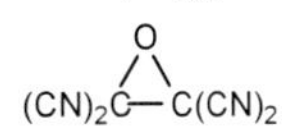	61
8.	$Me(CH_2)_5CH=CH_2$	1. n-Pentane/-38^0 - -42^{0}C 2. $LiAlH_4/Et_2O$	$Me(CH_2)_5CH_2OH$	93
9.	$H_2C=CH(CH_2)_8COOH$	1. MeOH/0^{0}C 2. $NaBH_4$/NaOH/ EtOH/H_2O	$HOCH_2(CH_2)_8CO$ OH	91

The efficient conversion of oximes into the corresponding carbonyl compounds can be also carried out through ozone oxidation [71].

The application of ozone is a new and convenient way for the preparation of cyanoacetylaldehyde (3-oxopropylonitrile) (1) and its stable dimethyl acetal (3,3-dimethoxypropylonitrile) (2) used as valuable intermediates for organic synthesis [72]. For this purpose the ozonolysis of (E)-1,4-dicyano-2-butene or 3-butylonitrile is carried out. Then the oxidates are treated by DMS yielding 1. Further, compound 1 can be used either directly in the next reactions or is transformed into 2. The 2 output amounts to from 67 to 71%. The acetal can be again hydrolyzed to 1 by treatment with ion-exchange resin Amberlyst-15 [72].

The easily available allylphenyl ethers are ozonized at -40°C, treated with DMS giving solutions of the corresponding phenoxyacetaldehydes which are purified by column chromatography. Their reaction with 1-methyl-1-phenylhydrazine yields the corresponding hydrazones [73].

Quinolinealdehyde derivatives can be synthesized by ozonolysis of the respective quinolineolefins at -60 - -72°C in methanol or ethanol solution with consecutive reduction with DMS at the same temperatures. Then the reaction mixture is heated to room temperature for 1 h. The yield of the products amounts to 29%. [74].

6-Chloro-2-hexanone is prepared from 1-methylcyclopentane via three-step scheme: a) ozonolysis of 1-methylcyclopentane in carboyilic acid solution to 1-methylcyclopentanol; b) conversion of 1-methylcyclopentanol into 1-methylcyclopentyl hypochloride using NaClO; c) cyclization of 1-methylcyclopentyl hypochloride in 6-chloro-2-hexanone [75].

Ozonation is a stage in the synthesis of optically pure (R)-(+)-4-methyl-2 cyclohexen-1-one from (R)-(+)-pulegone and hydroxy ketone from (-)-*cis*-pulegol [76].

Polycarbonyl compounds and aldehyde-acids are the ozonolysis products of (+)-4a-[1-(triethylsiloxy)-ethenyl]-2-carene [77].

Carbonyl oxides obtained from vinyl ethers ozonolysis undergo rapid [3+2] cycloaddition with imines giving the corresponding 1,2,4-dioxazolidines in yields of 14-97% [78].

Carbonyl oxides can be also used for the synthesis of 3-vinyl-1,2,4-trioxalanes (α-vinylozonides) by a [3+2] cycloaddition with α,β-unsaturated aldehydes. Howeve, α,β-unsaturated ketones are practically inactive in this reaction. The reaction of 3-vinyl-1,2,4-trioxalanes with ozone leads to the formation of the corresponding dizonides [79].

The carbonyl oxides prepared from the ozonolysis of enol ethers (for example, 1-metoxy-4,8-dimethylnone-1,7 diene) undergo stereoselective intramolecular cycloaddition with inactivated alkenes yielding bicyclic dioxalanes. The latter are easily converted into β-hydroxycarbonyl species and 1,3 diols by catalytic hydrogenation thus providing a new approach to the synthesis of 1,3-oxygenated products [80].

The ozonolysis of vinyl ethers is discussed in detail in Ref. 81.

The synthesis of 1,2-dioxalanes is carried out by ozonolysis of 1,1-disubstituted nonactivated olefins. Thus the ozonolysis of cyclopropyl-1,1-disubsituted olefins does not produce carbonyl oxides, but formaldehyde oxide. The latter can react with the initial olefin leading to the formation of 2-dioxalanes with 10% yield. At the addition of "foreign" olefins or aldehydes other dioxalanes and normal ozonides can be obtained [82].

The cycloalkenes ozonolysis in the presence of methyl pyruvate results in tri-substituted ozonides formation. The latter contain three reaction centers (peroxides, proton on the ozonides cycle and methoxy-carbonyl group) accessible for various functionalization. The cyclohexene ozonolysis in the presence of methylpyrovate gives ozonide whose treatment with PPh_3 or Et_3N yields $CHO(CH_3)_4CHO$ and $CHO(CH_3)_4COOMe$ (after esterification), respectively. This method proved to be very convenient and practical way for the synthesis of linear compounds containing various terminal groups from symmetrical cycloalkenes, in two steps and with good yields [83].

The direct conversion of olefins into esters is accomplished during mono-, di- and thi-substituted olefins ozonolysis in 2.5 M methanol sodium hydroxide-dichloromethane solution. The methyl esters are obtained in high yields. Thus, 3-benzyloxy-1-nonene (5b) is transformed into 2-benzyloxyoctanoate (7b) in 78% yield [84].

The ozonolysis of acyclic alkenes including terpenes is reviewed in Ref. 85.

The role of ozonolysis as ecological process for the selective and specialized oxidation of petrochemical olefins and cyclic alkenes, for the manufacture of biologically active substances and normal organic compounds is thoroughly discussed in Ref.86.

In Ref. 87 are summarized the data on ozonolysis of acyclic and cyclic mono-, di- and trienes in the various stages of the synthesis of insect pheromones and juvenoids.

Here should be also mentioned the importance of fulerene (C-60) ozonolysis [70] resulting in the obtaining of mixture containing various oxidation products such as ketone, ester and epoxides species. The intermediate in this reaction can act as oxygen atom carrier thus yielding phenols [88].

N-acylated esters of (cyclohexa-1,4-dienyl)-L-alanine are ozonized aiming at the synthesis of novel unnatural amino acids. The combined reduction and ozonolysis followed by condensation with a suitable nucleophile results in transformation of the aromatic ring of L-Phe to isooxasolyl, N-phenylpyrazolyl and to bicyclic pyrazolo[1,5-a]pyrimidine groups. The preparation of heterocyclic alanine derivatives is reported [89].

Thymidine diphospho-6-deoxy-a-D-ribo-3-hexulose synthesized by D-glucose ozonolysis of methyl-glucophosphate. is used as a central intermediate in the biosynthesis of di- and tridioxy sugars [90].

The ozonolysis of vinyl halides followed by reductive regio- and stereo-controlled intramolecular cyclocondensation is a key stage in the synthesis of amino sugars [91].

The ozonolysis of pyrols, oxazoles, imidazoles, and isooxazoles demonstrates another application of this reaction to the organic synthesis. Pyrols are efficient protective groups of the amino functions in the synthesis of α-aminoalcohols, α-aminoketones, α-aminoaldehydes and some peptides. The oxazole ring is also known as a protective group in the peptide synthesis [92].

The 1-substituted imidazoles ozonolysis leads to the formation of the corresponding N-acylamides, which are important amine or acyl derivatives [94].

Some novel tetraacetal oxa-cages and complex tetraquinone oxa-cages are synthesized from alkylfuranes by three stage reaction. Oxo rings in the tetraacetals are obtained by the ozonolysis of *cis*-endo-1,4-diones (norbornene derivatives) in dichloromethane solution at -78°C and consecutive reduction with DMS. The tetraquinone oxa-cages are obtained in the cis-endo-1,4-diones ozonolysis in dichloromethane solution at -78°C and TEA treatment [94].

A method for cleavage and oxidation of C_{8-30} olefins to compounds containing terminal carboxylic acid groups is reported in Ref. 95. The process is promoted by oxidation catalysts, such as Cr, Mn, Fe, Co, etc.

The conversion of ethane to methanol and ethanol by ozone sensitized partial oxidation at near atmospheric pressure has been quantitatively studied. The effect of temperature, oxygen concentration in the inlet gas, contact time in the reactor and ozone concentration in oxygen has been evaluated. The selectivity in regard to ethanol, methanol, as well as the combined selectivity towards formaldehyde-acetaldehyde-methanol-ethanol is also discussed [96].

Upon studying the ozone-induced oxidative conversion of methane to methanol and ethane to ethanol it has been established that these reactions do not occur in the absence of ozone which clearly suggests that the partial oxidation is initiated by the oxygen atoms generated from the ozone decomposition [75, 97].

Ortho-selective nitration of acetanilides with nitrogen dioxide in the presence of ozone, at low temperatures results in ortho-nitro derivatives formation in high yields [98].

The ozone mediated reaction of nonactive arenes with nitrogen oxides in the presence of suitable catalysts is reported as a new method for the synthesis of the corresponding nitro derivatives with high yields [99].

The ozone induced reaction of polychloro benzenes and some related halogeno compounds with nitrogen dioxide is a novel non-acid methodology for the selective mono nitration of moderately deactivated aromatic systems [100]. In the presence of ozone and preferably with methanesulfonic acid as a catalyst, the polychloro benzenes undergo mononitration with nitrogen dioxide at low temperatures giving polychloronitrobenzenes in nearly quantitative yields [100].

The ozone mediated reaction of aromatic acetals and acylal with nitrogen dioxide is suggested as a novel methodology for the nuclear nitration of acid sensitive aromatic compounds under neutral conditions [101].

Mineral and acid-free nitro compounds are prepared from CH_2Cl_2 soluted pyridine or its derivatives treated with $NO_2/O_2/O_3$ - mixture for 8 h [102].

The nitration of aromatic compounds with nitrogen oxides in the presence of ozone is a catalytic process [103].

The stereoselective synthesis of vinyl ethers is accomplished by N - (arylidene (or alkylidene) amino) - 2-azetidinones reaction with ozone and $NaBH_4$ treatment resulting in di- and trisubstituted olefins derivatives [78].

The steroeselective synthesis of (2s, 3s) norstatine derivatives is carried out through aldehydes ozonolysis in the presence of lithium methoxyallene [82].

An interesting method of succinic acid preparation from butadiene rubber ozonolysis is suggested [104].

The preparation of α-phenyl ketone, ω-carboxylate-ended telehelic methyl methacrylate oligomers by the ozonolysis of regioregular methyl methacrylate-phenylacetylene copolymers is described in Ref. 105. The oligomers have a number molecular mass varying from 1600 to 4500 (with respect to the initial copolymers) and polydespersity less than 2.

Ozone is also very convenient agent in the manufacture of organic ceramics [106].

The ability of ozone to destroy the double C=C bonds in organic compounds is the reason for its wide application in the preparation of bifunctional compounds. This is the principle that lies in the organization of the manufacture of dodecanedicarboxylic acid (1) and azelaic acid (2). The initial substrate for the preparation of (1) is cyclododecane which is obtained from the trimerization of butadiene and partial hydrogenation and (2) is prepared on the base of oleic acid:

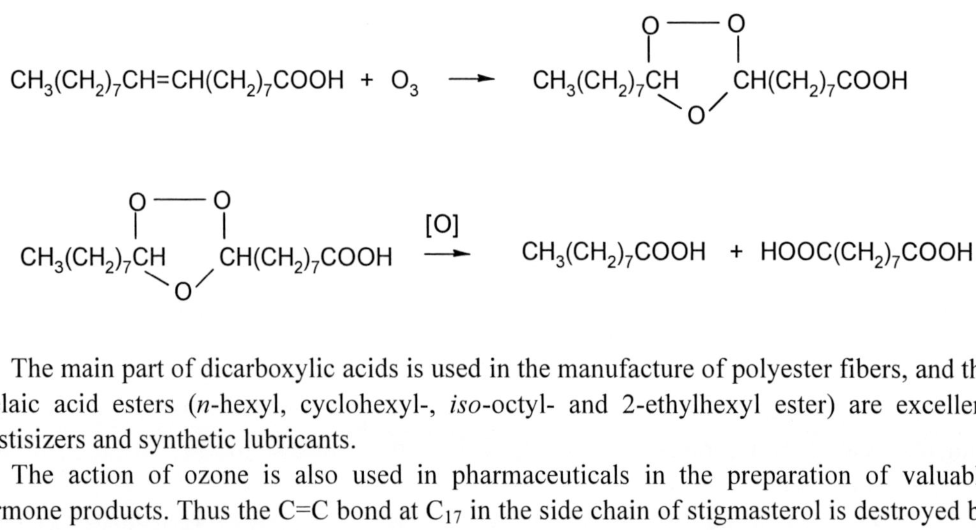

The main part of dicarboxylic acids is used in the manufacture of polyester fibers, and the azelaic acid esters (*n*-hexyl, cyclohexyl-, *iso*-octyl- and 2-ethylhexyl ester) are excellent plastisizers and synthetic lubricants.

The action of ozone is also used in pharmaceuticals in the preparation of valuable hormone products. Thus the C=C bond at C_{17} in the side chain of stigmasterol is destroyed by ozone yielding progesterone - an initial source for many hormones such as cortisone, male and female sex hormones, etc.

The method of selective decolorization of fabrics containing cellulose materials, such as cotton and oxidizing dyes [107], includes the following steps: application of oxidation blocking agent to the fibers; contact of the fibers with the oxidizing reagent in gas or evaporated state in the presence of moisture till the oxidation and decolorization of the dyer is carried out; interruption of this contact before the beginning of a substantial destruction of the fiber. The oxidizing agents are selected among ozone, chlorine and nitrogen oxides and sulfur dioxide flow. The application of ozone for decolorizing indigo-painted cotton jeans after this procedures results in jeans material which does not turn yellow for 6 months while the untreated goes yellow much more rapidly [107].

The changes in the composition and chemistry of UV/ozone modified wool fiber surfaces [108] is investigated by means of photoelectronic spectroscopy (XPS). The oxidation of the disulfide sulfur to sulfone groups ($^{-}SO_3H$) containing S^{6+} approaches almost 90% conversion. This is much more that the conversion levels by using oxygen plasma. Ozonolysis results in rise of C-O groups content, particularly of the carbonyl ones [108].

The process for producing cellulose fibers and moldings such as fibers, filaments, threads (yarn), films, membranes in the form of flat, pipe and empty fibrous membranes, etc. is carried out using ozone [109]. They are produced by extrusion of cellulose solutions in tertiary amine aminooxides and in some cases in water (particularly N-methylformaline N-oxide and water); regenerating bath and water for washing by treatment with hydrogen peroxide, peracetic acid, ozone or chlorine dioxide for the regeneration of tertiary amines oxides. The introduced compounds can be enzymatically or catalytically destroyed before solvent and water regeneration.

Pitch-based carbon or graphite fibers of high tensile strength are manufactured by primary treatment of the pitch-based fibers with high concentrated ozone for a short period; then the treatment by ozone-free gases follows; in the next step they are subject to carbonization of graphitization to give elliptical fibers of long, wool similar structure. Upon treatment in the absence of ozone the fibers stitched during graphitization [110].

The EPR analysis of thermal decomposition of peroxides in ozonized polypropylene fiber for grafting shows that the decomposition of the peroxide groups begins at about 70°C. The generation of several peroxides radicals is registered; the access of the fiber to the spin sample is enhanced through oxidation. Small amounts of RO_2 - radicals with lifetime of couple of weeks have been identified. The integral intensity of the EPR-signal rises with time and ozonation temperature [111].

Ozonolysis modifies the diffusion pattern of liquid monomers in polypropylene matrix as their addition occurs in the amorphous phase. The appearance of intra-morphological structure and peroxides localization upon ozonolysis of polypropylene (granules and fibers) is monitored using electron spin resonance and transmission electron microscopy [112].

Upon ozonolysis, UV-radiation and plasma treatment polymer peroxides are generated on the surface of films from polypropylene, polyurethanes and polyester fibers [113]. Their thermal and reductive-oxidative decomposition have been studied by means of functional analysis using peroxidase and iodide. However, their disposition in the polymer specimen is quite various thus impeding their analysis and depends on the treatment agent. For example, the treatment with plasma generates easily accessible peroxides in polyurethane films while the UV-radiation and ozonation leads to formation of peroxides uncapable of reacting with aqueous solution of peroxidase. The redox decomposition of the peroxide groups by ferro-ions at 25°C has shown that less than 50% of the peroxides may react with ferro-ions at rate constants similar to those of hydrogen peroxide in aqueous solution. The thermal degradation of peroxides does not follow first order kinetics, most probably because of the generation of various peroxide species, characterized by different rates of decomposition. The lowest rate constant of decomposition observed at 62°C is 3.10^{-3} /min which does not depend on the polymer nature and the method of peroxides generation [113].

Ozone-induced graft polymerization onto polymer surface is an important and convenient method for polymer modification [114].

Ozone-induced graft copolymerization of polyethylene glycol monomethyl ether methacrylate onto poly(etherurethane) improves the hydrophility and water absorption. The autoaccelerated effect in ozone-induced polymerization has been also discussed [115].

The modification of the surface properties of polypropylene and block copolymer is carried out by ozone treatment. Thus, the wettability of the polymers is improved - the contact angle of water becomes $\leq 67^0$. The break of the polymer chain and carboxyl groups formation are accelerated by using high ozone concentrations. The cut-off fragments resulting from the ozone treatment are removed from the surface by ultrasound and organic solvents [116].

The low temperature nonelectrolytic nickel plating onto three types of polypropylene is carried out by substrate pretreatment with ozone. The latter modifies the polymer surface for galvanization while the combination of polar with anchor effect as a result of the ozone etching enhances the adhesive properties of the polymer surface. The washing of the material after ozonolysis is obligatory for ensuring a good adhesion of the material [117].

The surface properties of polypropylene (I), 100 μm films from (I), ethylene-propylene block polymer (II), or ethylene-propylene polymer prepared by the random method (III) can be improved by ozonation at concentrations of 1.38, 0.64 and 0.41 mol %, respectively. The adhesion of dyes, coloring agents, dye layers, on (I) is substantially improved and depends on the substrates in the following order: (II) > (III) > (I). It has been found that for each sample there are optimum conditions of ozonation. Reactive dyes such as epoxy and acrylo urethane resins impart a better adhesion force that the conventional nonreactive acrylic or vinylchloride resin [118].

Freshly extruded, 50 μm C_2H_4-ethyl acrylate-maleic anhydride copolymer film is treated by 500 ml/m^2 oxygen containing 10 g/m^2 ozone and is calendered with 200 μm monolayer C_2ClF_3 - polymer film at 15°C to produce moisture-proof packaging material with intra-layer adhesion of 800 g/15 mm and thermosealing force of 3.6 kg/15 mm against 250 and 1.3, respectively in the absence of ozone treatment [119].

Methylmethacrylate polymers with good thermal degradation resistance are prepared by ozonolysis [120] whereby the end unsaturated groups are converted into carbonyl or carboxyl ones. Ozone/air mixture is blown through a 10 g Acrypet VH solution in CH_2Cl_2 at -78°C followed by exposure only to air for 60 min and after solvent removal the solid residue is dried in vacuum at 297°C for 8 h. The polymer thus obtained is characterized by an initial temperature of thermal degradation of 315°C against 297°C for the polymer without ozone treatment [120].

The surfaces of propene polymer moldings are treated with ozone for improving their hydrophility. The ozone treatment of the surfaces for 8 h reduces the contact angle of water from 100 to 81^0 [121].

Ozonation of PVC latex is also carried out for removal of vinyl chloride residues. The aqueous dispersions of saturated polymers are treated with ozone and the vinyl monomers are removed [122].

Graft polymerization of vinyl monomers onto Nylon 6 fiber is carried out after ozone oxidation of the fibers or films from Nylon 6 with vinyl monomers such as acrylamide, methylmethacrylate and vinyl acetate. The molecular mass of Nylon 6 decreases slightly upon ozonation. For acrylamide graft polymerization system the preliminary treatment in air or vacuum by γ-rays radiation prior ozonation results in higher graft percent. For methyl methacrylate the apparent graft percent does not rise with the ozonation time. However the apparent graft percent for vinyl acetate is increased with the ozone time treatment [123].

The adhesion of PVC-, fluropolymer, or polyester-coated steel panels is improved by treatment with 5-50% solution of H_2O_2 or ozone [124].

Synthesis of water-soluble telehelic methyl-ketone-ended oligo- N,N-dimethylacrylamides by the ozonolysis of poly(N,N-dimethylacrylamide-stat-2,3-dimethylbutadiene)s is reported in Ref. 125.

The synthesis of telehelic methylmethacrylate and styrene oligomers with fruorophenyl ketone end groups is accomplished by the ozonolysis of copolymers containing 4-flurophenyl butadiene units [126].

Graft polymerization of acrylic acid is carried out into preliminary ozonized siloxane matrixes [127].

The manufacture of base discs for laser recording material is realized by covering of the plastic substrates surface with solid polymer layer possessing directing channels and/or signal holes. The surface of the plastic substrate is preliminary cleaned by UV/ozone exposure before the formation of the hardened polymer layer [128].

The controlled degradation of polymers containing ozonides in the main chain takes place in the presence of: 1) periodate suported on Amberlyst A26 (SPIR); 2) diphenylphosphine deposited on polystyrene (SPR); boron hydride supported on Amberlyst A26 (BER) [129]. Poly (butyl) methacrylate-copolymers are prepared by emulsion polymerization. These materials and the homopolymer poly (butyl) methacrylate are ozonized at various temperatures and treated by any of the reagents described above, thus giving telechelic oligomers in 99% yield. Molecular mass variations with ozonation temperature change has been observed. The end aldehyde groups are registered using ^{1}H and ^{13}C -NMR spectroscopy; the end hydroxyl groups are observed by means of FTIR and ^{13}C-NMR; the presence of hydroxyl groups is confirmed by tosylate formation; the carboxyl groups are identified by FTIR and ^{13}C-NMR and quantitatively determined by titration; in the SPR-generated oligomers the content of aldehyde groups constitutes about 80% of the end functional groups. The oligomers obtained in BER and SPIR contain 99% hydroxyl and carboxyl groups [129].

Polyethylene fibers are subject to ozone treatment for modifying their surface [130]. The analysis of the surface is carried out by means of X-ray photoelectron (ESCA) and IR (FTIR) spectroscopy. Carbon (C) and oxygen (O) were the main atoms monitored with ESCA (C-1s, O-1s areas) on the treated fibers. The analysis of C-1s peaks (C_1, C_2 and C_3) reveals that the oxidation level depends of the ozonation time. The components of 1s peak (O_o, O_1, O_2) are very useful for carrying out the surface analysis. They demonstrate the presence of carbonyl groups (1740-1700 cm^{-1}) even onto untreated fibers whose intensity rises with treatment time. Ozonolysis is directed from the surface to the fiber bulk. This is confirmed by the great enhancement of the carbonyl bond band after 3 h ozonation. The thermal analysis suggests structural and morphological changes of the fiber when ozonolysi time treatment exceed 2 h [130].

The processing of fiber-reinforced plastics is performed by blowing an ozone-oxygen mixture (flow rate of 0.4 l/min) for 5 h through a CH_2CL_2 solution containing glass reinforced fiber particles (1mm diameter) filled with $CaCO_3$ and unsaturated polyester. The fibers emerge on the surface till a fine powder is precipitated [131].

The manufacture of laminates through heat-sealing method involves the application of electric crown, UV-radiation or ozone exposure [132].

The manufacture of pour point depressants for oils , particularly useful for diesel oils, is carried out by oxidative degradation of waste polyethylene and/or polypropylene (I) with ozone at 30-150°C. 1000 g waste (I) is exposed to ozone action at 150°C for 5 h giving the pour point depressant. The addition of 1% depressant to diesel oil reduces the solidification of the oil from -20°C to -35°C and the temperature of filter plugging from -9 to -20°C [133].

1.2.2. Inorganic Productions

Stainless steel parts are treated by ozone for surface passivation. The parts are heated in oxidative or inert atmosphere at a temperature of condensation lower or equal to -10°C. The unreacted ozone is re-used [134].

The main passivating agent is oxygen in combination at least with ozone. The system is particularly appropriate for passivation of metal (for example, stainless steel, Ti, etc.) equipment, used in chemical plants and exposed to strong corrosion action at high temperatures and pressures [135].

For increasing the corrosion resistance of metals and alloys they were exposed from 1 s to 10 min in cold plasma under pressure 1-103 Pa and 100-5000 V in atmosphere containing O_2, O_3, N_2, H_2, air, CO_2, CO, N-oxides, H_2O (gas), combustible gas and/or inert gas. Thus, 17% C0-ferrite stainless steel is subject to plasma treatment for 4 min at 103 Pa, 100 mA and 250 V in nitrogen-oxygen mixture with 20% oxygen. The corrosion resistance was evaluated by treatment of the sample with a solution containing 17 ml 28% $FeCl_3$, 2.5 ml HCl, 188.5 ml H_2O and 5 g NaCl. The sample shows relatively good resistance as compared with the untreated one [136].

Electrochemical tests reveal the influence of dissolved ozone on the corrosion behavior of Cu-30 Ni and 304L stainless steel in 0.5N sodium chloride solutions [137]. These experiments include: measurements of the corrosion potential as a function of the time and ozone concentration; cyclic polarization experiments; isopotential measurements of the current density and study of the film components. The results of these experiments show that for Cu-30 Ni and 304L -stainless steel the corrosion potential is shifted to the more noble values (300 mV) at $[O_3] < 0.2\text{-}0.3$ mg/l. At higher concentrations it remains unchanged. The dissolved ozone reduces the corrosion level for Cu-30 Ni - alloys which is evaluated by the substantial decrease in the current density at constant applied potential. The improvement of the corrosion resistance should be related to the decrease of the thickness of the corrosion products film and to the higher oxygen fraction as compared with the chloride in the same film. For stainless steel the differences in the passivating films in ozonized and nonozonised solutions are negligible as it is shown by spectroscopy [137].

Laboratory experiments have been carried out to study the ozone application for acid oxidizing leaching of chalcopyrite with 0.5 M H_2SO_4. For evaluating the reaction mechanism the effect of particles size distribution , stirring and acid concentration, the dissolution reaction and reaction kinetics on the leaching have been investigated. The reaction rate is governed by ozone diffusion to the reaction mixture. Ozone is an efficient oxidizer and the process is most effective at 20°C [138].

The rate of acid leaching of chalcopyrite depends on the use of ozone as an oxidizing agent in sulfuric acid solutions. The leaching of chalcolyrite follows a parabolic law [139]. The use of ozone as an oxidizer provides conditions for the formation of elemental sulfur on the leached surfaces. The rate of leaching is reduced with temperature as the ozone solubility decreases with temperature. The results show that ozone is the best oxidizing agent for acid oxidizing leaching of chalcopyrite and may be applied in pilot plants for regular manufacture.

We have studied the possibility of using ozone for improvement Ag extraction form polymers deposits in a flotation plant in the town of Rudozem. It has been found that upon blowing of ozone (1% vol concentration, flow rate - 300 l/h) through the empty shaft of the flotation machine stirrer (5 m^3 volume) the degree of Ag extraction is increased by 1-2% [139].

The redox leaching of precious metals from manganese-containing ores carried out by other authors show also positive results [140].

Molybdenite flotation from copper/molybdenum concentrates by ozone conditioning results in relatively pure copper-free molybdenum. The process including multi-step ozone flotation proves to be a technical and economical profitable method [141].

The manufacture of potassium permanganate is accomplished by melting of Mn-containing compounds with KOH, dissolution of the melt and the solution oxidation. For reducing the energy consumption $Mn(NO_3)_2$ is used as a Mn-containing compound. It can be easily alloyed with KOH in a 1:5 - 1: 10 ratio at 250-300°C . Then the product is dissolved in 20-25% KOH solution, the solid residues is removed, soluted in 3-5% KOH and the solution is oxidized by ozone-air mixture [142].

The manufacture of silicon carbide ceramics is performed as melted organosilicon polymer is oxidized with 0.001% vol ozone. Polycarboxylstyrene is the prefered polymer. The ceramics obtained is characterized by high thermal resistance and acceptable physical properties [106].

Mixtures containing In- and Sn-compounds are molded and sintered in a furnace in air atmosphere containing ≥ 1000 ppm ozone. The ITO ceramics thus obtained are characterized by high density at lowtemperature sintering for a short time [143, 144].

A method for oxidizing carbonaceous material, and especially for bleaching gray kaolin for subsequent use as coating or filler for paper in the presence of ozone is reported in Ref. 145..

The manufacture of mercury(I)chloride includes the reaction of Hg with hydrochloric acid in the presence of water, subsequent removal and drying at 95-105°C. Ozone (0.1-0.1 g/g product) is bubbled through the reacting mass to increase the product yield and quality [146].

Arsenic acid is prepared from $(As_2Cl_6)_2$- -ions by ozonolysis [147].

The removal of color and organic matter in industrial phosphoric acid by ozone and the influence on activated carbon treatment is described in Ref.148. Industrial phosphonic acid containing 42-45% P_2O_5 and 220-300 mg/l organic matter (OM) is subject to combined treatment with ozone and activated carbon. The independent ozonation results in removal of the initial dark color of the acid and the organic matter. It is only through absorption on activated carbon that the level of OM could be reduced to 80% per 25 g/kg P_2O_5. The ozonation prior adsorption enhances the efficiency of the activated carbon effect and decreases its specific consumption [148].

The fabrication of high-Tc superconductors is carried out using ozone-assisted molecular beam epitaxy (MBE). It includes the simultaneous evaporation of the elementary components and application of ozone as a reactive oxygen source. The ozone priority over the other oxygen forms is that it is rather stable and could be produced and supplied to the substrate in a very pure state using simplified apparatus ensuring a well defined flow of oxidizing gas. Upon application of ozone the untempering through a thermal treatment for removing the inner stresses could be ignored. In order to prepare films with high temperatures of superconductivity the growth should be carried out at relatively low stresses in the system. In addition, the surface during film growth can be analyzed by reflectance high energy electron diffraction. The most recent improvements in ozone-assisted MBE for family of $YBa_2Cu_3O_7$-delta films are described. The results show that this technique is very appropriate for growth of high quality superconductive films and could be ideal for the manufacture of such structures [149].

The growth of superconductive oxides under vacuum conditions compatible with MBE requires the use of activated oxygen. The atomic oxygen or ozone appear to be such species [150]. The characteristics of a radio-frequency plasma source for molecular beam epitaxial growth of high-Tc superconductor films (200 Å) from $DyBa_2Co_3O_7$ on $SrTiO_3$ are described and discussed in Ref. 150.

An apparatus for the preparation of pure ozone vapor for use *in-situ* growth of superconducting oxide thin films is designed [151]. Pure condensed ozone is produced from distillation of diluted ozone-oxygen mixture at 77K. The condensed ozone is heated until the pressure of its vapors approaches the necessary for an adequate flow of ozone- gas to the chamber of thin films growth. The thin films from $YBa_2Cu_3O_7$ with zero resistance at temperatures of about 85K grow at ozone pressure in the chamber of 2.10^{-3} Torr even without subsequent untempering. It should be noted that in contrast to other highly reactive oxygen species ozone can be prepared and stored in very pure form which makes it very convenient for studies on the kinetics of growth and oxidation of films with a well defined gas [151].

A patent for preparation of oxides superconductors M-M^1-M^2-M^3 includes ozone oxidation where: M = elements of III B groups such as: Y, Sc, La, Yb, Er, Ho or Dy; M^1 = elements from II A group, such as: B, Sr, Ba or Ca; M^2 - Cu and one or more elements from I B group such as Ag or Au; M^3 = O and on or more elements belonging to VI A group like S or Se and/or elements from VII A group as F, Cl or Br. The superconductors thus prepared have high critical temperature and high critical current densities [152].

The manufacture of bismuth-, copper- alkaline earth oxide high-temperature superconductors is carried out through calcination and/or sintering in ozone-containing atmosphere with $Bi(OH)_2$, $Ca(NO_3)_2$ and $Sr(NO_3)_2$ and subsequent treatment by $CuCO_3(OH)_2$; the components are dried, calcinated in ozone containing air, molded and dried in air atmosphere. Their temperature of superconductivity is 107K and the critical current density is 405 A/cm^2 [153].

A lowtemperature method for preparation of superconductive ceramic oxides is described in Ref.154. It includes the treatment of the substrate heated surface by ozone to the complete evaporation of the other components thus forming a superconductive ceramic oxide.

The manufacture of rare earth barium copper oxide high-temperature superconductor ceramics is carried out by sintering at 930 - 1000°C in ozone atmosphere containing oxygen. Y_2O_3, $BaCO_3$ and CuO are mixed, calcinated, pulverized, pressed and sintered at 950°C in the presence of ozone-oxygen mixture after which they are gradually cooled to the critical superconductive temperature 94K [155].

A review devoted to the growth of co-evaporated superconducting yttrium barium copper oxide (Y Ba_2CuO_7) thin films oxidized by pure ozone is presented in [156].

The removal of organic pollutants from the surface of supports for microelectronics purposes is conducted by UV/ozone treatment [157].

Low temperature silicon surface cleaning is carried out by fluoric acid (HF) etching, washing , cleaning with deionized water, N--gas blowing, /UV-ozone treatment. This is a treatment preceding the process of silicon epitaxy [158 -160].

The method for total room temperature wet cleaning of silicon surface comprises the use of HF, H_2O_2 and ozonized water and is with 5% more economical than that the standard procedure of wet cleaning [161].

The removal of resist films supported on semiconductive substances is accomplished by using ozone [162].

The application of ozone for reducing the temperature and energy in the process of very large scale integration (VLSI) is described in [163, 164]. The main problems under discussion are as follows: cleaning of Si-support with ozone by two methods: dry process combined with UV and wet one - with ozonized water; stimulation of Si thermal oxidation with ozone; deposition of SiO_2 under atmospheric pressure and low temperature via chemicals evaporation using tetraethylorthosilicate and ozone; 4) deposition of Ta_2O_3 at atmospheric pressure and low temperature applying $TaCl_5$ and ozone; 5) tempering with ozone and UV-radiation of Ta_2O_5 films used for dielectric for memory units , this treatment reduces the current permeability into the film; 6) etching of photoresistant materials by $O_2/O_3/CF_4$ and the effect of ozonator charge and injection of CF_4 [163, 164].

The cleaning of synchrotonic radiation optics with photogenerated reactants has a number of priorities over the methods of discharge cleaning [165]. Upon discharge cleaning, the discharge particles should react with the surface contaminants until its sheilding by the rough discharge elements which may pollute or destroy the surface. Contrary, if the particles can be photon generated near the surface, the problem with the protection drops off and in some cases the cleaning can be more efficient. An estimation of the various methods for cleaning was made comparing the rates of polymethylmethacrylate films removal. A number of various light and geometry sources have been tested. The highest rate of cleaning was achieved upon using UV/O_3 cleaning method at atmospheric pressure. This method has been widely applied for cleaning semiconductive surfaces from hydrocarbon contaminants. It is noted that it is also effective in removing graphite-like pollutants from synchrotonic radiation optics. It proves to be more simple, economical and selective method as compared with other cleaning method. Although it requires the drilling of a hole in the vacuum chamber, the cleaning of the optics can be carried out without dissemble which saves much time. This method is successfully applied for cleaning grates and reflectors in several beams [165].

Native oxide growth and organic impurity removal on silicon surface is carried out by ozone-injected ultrapure water [166]. In order to manufacture high-efficient and reliable ULSI-units the further integration and minimization is in progress. The cleaning with $H_2SO_4/H_2O_2/H_2O$ is accompanied by serious problems: a great amount of chemical wastes is obtained which must be suitably treated. The cleaning technology with ultrapure water includes ozone injection in concentrations of 1-2 ppm. This method is very efficient in removing the organic impurities from the surface for a short time and at room temperature. The process wastes can be treated and recycled. The ozonized water components can be easily controlled [166].

The reaction mechanism of chemical vapor deposition using tetraethylorthosilicate and ozone at atmospheric pressure is reported and discussed in Ref.167.

Covering of semiconductor devices with silica films is carried out via CVD using $Si(OEt)_4$ and ozone. The deposition process is followed by heating in oxygen atmosphere with simultaneous UV radiation. The SiO_2-film thus formed can improve the water resistance of semiconductive devices [168].

The fabrication of nondoped silicate glass film with flat structure by O_3-TEOS deposition includes three stages: 1) substrate treatment; 2) formation of Al conducting layer on it using oxygen plasma through heating; 3) formation of SiO_2-film on the substrate by plasma using ozone and TEOS [169].

In Ref. 170 are reviewed the future trends for interlayer dielectric films production and their formation technologies in ULSI multilevel interconnections. The properties of the interlayer dielectric films and their preparation technologies should satisfy the following three requirements: 1) available for disposition (setting) on large surface; 2) low dielectric constant; 3) low deposition temperature. Two techniques have been developed for the selective deposition of SiO_2-films which is the best way for achieving a complete planirization of the interlayer surface of dielectric films: low temperature liquid phase deposition using a saturated aqueous solution of hydrofruorosilicon acid H_2SiF_6; 2) half-selective SiO_2-films deposition at 390°C with $Si(OC_2H_5)_3$ and ozone and preliminary treatment with tetraflurocarbonic (CF_4) plasma on TiW or TiN surfaces [170].

Lead zirconate and titanate thin films are successfully prepared by reactive evaporation. The elements Pb, Zr and Ti are evaporated in ozone-oxygen mixture. The films obtained have equal thickness and composition per a large surface (in the range of ±2% from the 4-inch support) [171].

Ceramic coatings on substrates are formed in the presence of ozone [172]; the substrates (silicon checks) are covered (dipped) in a solution of one or more (partially) hydrolyzed pre-ceramic silicon alholates with general forrmula $R_xSi(OR)_{4-x}$ (R = C_{1-20} alkyl, aryl, alkenyl or alkinyl; x = 0-2). Further, the solvent is evaporated to form a coating which is subsequently heated in the presence of ozone up to 40-100°C thus converting into a ceramic coating. It in its turn may be covered also by a protective layer containing Si, or Si and C, or Si and N, or Si, C and N, or SO_2 and oxide. These coatings which are abrasive-, corrosive and thermo-resistant have also a small number of defects and suppress the diffusion of ionic contaminants such as Na and Cl ions and are particularly convenient for electronic units [172].

Method of forming zinc oxide light-shielding film for liquid crystals shields includes the injection of vapors of alkyl zinc compound with ozone or atomic oxygen in the activation chamber after which they pass through a chamber for ZnO film deposition at low temperature heating at about 200oC. The method is applicable for large scale production of these films [173].

An evidence for a new passivating indium rich phosphate prepared by UV/O_3 oxidation of indium phosphide, InP, is provided in Ref.174. The phosphate does not exist as crystal compound and its composition is $InP_{0.5}O_{2.75}$. The passivating ability of the latter with respect to InP surface is discussed.

For improving the light-absorption characteristics of oxide optical crystals they are heated in to ozone-oxygen atmosphere [175, 176]. Thus the light absorption from the optical crystal in a wave range different from that with which the crystal affects its own absorption is reduced to the most possible level. Devices using such oxide optical crystals with improved absorption characteristics work with high efficiency as optical amplifiers, optical isolators, optical recording medium, optical generators.

The manufacture of solar cell modules with transparent conducting film covered by amorphous Si layer and electrode on the backside linked to the transparent isolator layer is described in Ref.177. The preparation includes the application of a laser beam for electrode molding and exposure in oxidizing atmosphere containing 0.5-5% O_3 or ≥10% O [177].

A method for strong oxidation in ozone atmosphere is proposed for surface activation of photoconductive PbS films [178].

Adhesion-producing materials for electroless plating and printing circuits contain particles of thermostable material slightly soluble in the oxidizing agent; the particles are dispersed in the thermostable resin which becomes almost insoluble in the oxidizing agent at hardening. These materials find application for printing circuits and are treated by an oxidizing agents (for example, chromic acid, chromate, permanganate or ozone) to create a concave material surface [179].

1.3. MEDICINE

Upon exposure of living organisms to ozone, depending on doses, the different vital systems can be suppressed or stimulated to a certain extent.

For the first time already in 1898 Binz [180] showed interest in the effect of ozone on the blood and plasma and found that ozone exhibits a somnolent effect. Brinkmson and Lambert [181] proved that ozone inactivates the cell enzymes and decreases the oxygen exchange between blood and tissues.

A thorough investigation of the effect of ozone on the human organism should allow to extend the range of its application in medicine. The operation mechanism of ozone at small (therapeutic) concentrations, up to 100 $\mu g/m^3$, in the human cell has been already studied. Ozone enhances the cell membrane resistance to damaging effects. Human blood, treated with ozone, is able to keep erythrocyte membrane intact for 5 to 15 days. A method for preparation of nonhemolyzed plasma has been developed which facilitates laboratory diagnostic investigations and preserves the blood without inducing hemolysis [182]. *In vivo* and *in vitro* experiments have shown that ozone stimulates erythrocytes gas-transportation function which provides a continuous optimum tissue oxygenation. Besides, ozone strengthens leukocytes antibacterial function, the high activity of the mieloperoxidase enzyme system being preserved for 15 days. Also, ozone could find application in disinfection of biological materials, blood in particular. Encouraging results of HIV inactivation by means of ozonolysis have been reported in Refs 183-189. Blood, plasma, infected with retrovirus, whose transcriptase (RT) activity was $1.4x10^5$ cmp/ml (enzyme for retrovirus replication) was separated into three fractions - A, B and C. Fraction A and fraction B (after cryoprecipation) were exposed to ozone, 12 μg/m^3, at 0°C for 30 min. Fraction C was not treated with ozone. Subsequently, after 13 days of incubation, the following changes in RT activity were observed: fraction A = 130 000; B = 0 and C = 175 000.

In acute massive blood losses the adaptive potentialities of the human organism are increased. Besides, the operating strength of the intracellular mechanism of hemoglobin antioxidative resistance is enhanced, as deduced from superoxide dismutase activation in erythrocytes. Phrophylaxis in hemorrhages by means of ozone inhalation is due to the activation of biological processes which provide erythrocytes gas-transportation function. The application of ozone in medicine opens prospects to invent means for correcting the respiratory systems by changing the molecular regulation mechanism of hemoglobin affinity to oxygen in various hypoxic conditions. Methods can be developed to preserve blood cell vitality and dynamic activity during conservation and transfusion. There is growing interest in looking for techniques for ozone use in cardiovascular diseases. Combined effects and hyperbaric oxygenation in conditions related to hypoxia have also been studied.

The ability of ozone to oxidize toxic products formed under various pathological conditions, has attracted clinicians attention. Ozone either directly oxidizes double bonds (in unsaturated compounds) or, by activating free radical processes, initiates a danger of enhancing the spontaneous process of lipid peroxide oxidation. Kontorshticova et al. [190] studied the lipid and protein spectra of blood plasma exposed to ozone. Blood, taken from a dog 15 min after suffering from hemorrhagic shock, was exposed for 5 min to oxygen-ozone mixture at a rate of 1.0 l/min and ozone concentration of 0.048 mg/l/ The total protein content and separate protein fractions were analyzed. The fatty acids content was determined by gaschromatography. The protein and lipid spectra reflect processes which take place in hypoxia and subsequent oxygenation in the organism of test animals. These are: a) elimination of lipid depots of nonesterified fatty acids; b) nonoxidized or partially oxidized products of lipid and protein nature; c) blood saturation with oxygen and activation of oxygenation processes. An increase (by 20 %) in α_2-globulin content zeruloplasmin and haptoglobin, which possess antioxidative properties, is observed in the protein spectrum. The content of fatty acids is decreased by about two fold. The total protein and β_2-globulin concentration also decrease. These changes are consistent with a direct effect of ozone on peptide bonds in proteins and on double bonds in unsaturated fatty acids.

We have studied the effect of ozone [191] on POL of rabbit brain and liver. In this connection the rabbits were exposed to ozone action in a special chamber for 2, 3, 10 and 15 h. The ozone concentration in the gas phase was varied in the range 50-250 ppm. We have found a substantial enhancement of the content of endogenous POL? fluorescence products (FP) and TBA-reacting products (TBA -rp) after 2 h exposure to 250 ppm ozone concentration. However, after 4 h exposure to the same ozone concentration the rabbits die. When the concentration of exposed ozone was 50 ppm the increase in FP content in the rabbits brain is increased by about 1.8 times after 10 h, and in the liver - by about fold 2 compared with that of the control test. The level of TBA-rp was 3.1 and 3 fold higher, respectively. It has been observed that the amount of FP in the brain between the 10th and 15th h does not change, but the FP content in the liver microsomes rises 2.6 times and that of TBA-rp - 3.6 fold. A series of test animals, which were preliminary injected with 56 mg/kg potassium salt of DL-α-tocopherol 10 h before the test, were also studied under the same conditions. The results from FP determination after 10 and 15 h of exposure reveal that their amount in the brain has increased by 1.09 and 1.3 times, respectively and in the liver microsomes - 1.5 and 1.9 fold. Simultaneously there is no marked difference in the DL-α-tocopherol content before and after exposure of the test animals as demonstrated by the HPLC data. These facts indicate that the brain is either more stable or more protected in regard to ozone action since DL-α-tocopherol protects first of all the biomembranes.

The effect of tocopherol and more than 20 its homologues and derivatives was investigated in model reaction of cumene oxidation and in enzymatic and enzyme-free POL in rabbits brain and liver microsomes [192-196]. It has been shown that the inhibitor efficiency depends rather on their diffusion coefficients through the membrane biolayer as well as their mobility than on the chemical nature of the substituents. A correlation between the rate of POL inhibition and the suppression of the luminol-dependent chemiluminescence in biomembranes has been established. However, no relationship was found between the action of 2,6-di-*tert*-butyl-phenol inhibitor and its derivatives and the chemiluminescence produced during the interaction between superoxide anion and luminol.

Fairly high doses of ozone can be employed to saturate blood with oxygen and deactivate nonoxidized and partially oxidized products. Therapeutic doses of ozone, 0.048 mg/l, are applied to correct pulmonary insufficiency in postreamination periods.

High concentrations of ozone, 0.2 mg/l, can be used to model pathological conditions which allow to study morphological changes in pulmonary tissue. The effect of ozone upon varying the mode of administration in organism has also been investigated [197].

An isotonic NaCl solution saturated with ozone was injected intracutaneously or a saturated aqueous solution was administered orally to rabbits. In animals, injected with ozonized water, the red blood cell level remained stable in the course of experiment (10 days). An increase in leucocyte level (2 fold) was observed at the expense of monocytes. Changes of the granulocytes were not pronounced. The functional activity of neutrophilic cells, evaluated by phagocytosis activity is increased 1.5 time. On intracutaneous administration of ozonized isotonic NaCl solution, the leucocyte level was diminished after the first hour, enhanced after the third hour and remained high by the end of experiment with respect to the initial level. By comparing the two methods, it was established that the toxic effect was more pronounced with the intracutaneous administration than the oral one. Thus, for example, the activity of the process of lipid peroxide oxidation, evaluated through malonic dialdehyde determination in erythrocytes, increased by 3-fold upon intracutaneous injection and did not change after oral administration. On exposure of blood from healthy donors aged 25-35 to 3 mg/l ozone the activity of the aspargataminotranspherase was increased whereas that of lactic dehydrogenase attained a maximum concentration and then fell down to the initial one.

Ozone therapy, being an effective means to treat a number of diseases, attracts the attention of practicing clinicians, dermatologists in particular [198]. Attempts have been made to treat various dermatoses; psoriasis, herpes, microbial eczemas, etc. [199]. The obtained results were positive. In this connection a joint project with the dermatology department at the Sofia Medical Academy is developed. Its object is to study the effect of ozone on dermatologically injured skin and development of ozone generating units.

The effect of ozone has focused scientist's attention for more than two decades. Of particular interest is the oxidative effect of ozone on cells, especially tumor cells [200-202]. Cells from alveolar adenocarcinomas, breast adenocarcinomas and uterine carcinosarcoma and normal cells from lung diploid fibroplast as a reference were incubated in a special chamber. They were exposed to ozone (0.3 to 0.7 ppm) for 8 days. The number of cells was determined every 48 h. Ozone levels from 0.3 to 0.5 ppm inhibited the accumulation of tumor cells by 40-60% and did not affect the normal cells. The inhibition was 90% at a concentration of 0.8 ppm.

Many viruses are sensitive to ozone, for example *vesicular stomatitis*, *encefalomyocarditis*, *poliovirus type 2 and 3*, *coxsacknevirus*, *echovirus* and *adenovirus* [203-205]. The authors have studied the inactivation of rotaviruses HRV type 2 from humans and SA-11 from apes for various exposures of ozone (0.05, 0.1, 0.2 mg/l) at 4°C. HRV are considerably more sensitive to ozone than SA-11. In this connection ozone might find therapeutic application in oncology.

1.3.1. Therapeutic Application

As a therapeutic means the so-called medical ozone is employed in concentrations from 0.05 to 5%. The types of application and the relevant concentration range of medical ozone/oxygen mixtures are specific for each pathological change (Table 1.6).

Table 1.6. Therapeutic doses

Types of applications and appropriate range of doses of ozone/oxygen mixtures for medical purposes

Colonic insufficiency and external exposure in closed system ↓

Intravenous

Intramuscular injection ↓

Hypodermic

Blood treatment↓ Production of ozonized

Stimulating treatment ↓ water

Circulatory ↓ Cleaning of wounds and disinfection

0 10 20 30 40 50 60 70 80 90 100

Ozone concentration, μg/ml

As seen the ozone concentrations administered by injection reach 40 mg/l. Upon extracorporeal blood treatment, stimulation and circulatory treatment ozone is employed in concentrations up to 30 mg/l whereas for colonic insufflations, external gas application in closed system, cleansing of wounds and in ozonized water production it can be used up to 100 mg/l level.

Ozone therapy is successfully applied in a wide variety of diseases as shown in Table 1.7.

As a rule, the ozone-oxygen mixture is employed in cases of arterial circulatory disturbances. This method was first described by Dr. Lacoste in 1951 and since then has attracted the attention of many scientists. Rokitansky [207] was the first who made a statistical evaluation, according to which four out of ten patients with arterial-metabolic

Table 1.7. Some examples of diseases and medicine fields of ozone application

Disease	Medical field
Anal fissure	-Dermatology / proctology
Acne	-Dermatology
Abcess	-Surgery / dermatology
Allergies	-General medicine, dermatology, -allergology
Virus infections	-AIDS, various special fields
Arterial circulatory disturbances	-Surgery, vascular surgery
Fistulare	-Proctology, drmatology, urology, gynecology
Mucous colitis	-Proctology, gastroenterology
Cystitis	-Urology
Cerebral sclerosis	-Gerontology, internal medicine and neurology
Decubitus (ulcers)	-Surgery, gerontology and dermatology
Furunculosis	-Dermatology
Gangrene	-Surgery
Ulcus cruris	-Dermatology
Hepatitis	-General medicine, internal medicine
Hypercholesterolaemia	-Internal medicine
Herpes (AIDS)	-Dermatology, internal medicine
Climacterium/menopause	-Gynecology
Cirrhosis of the liver	-Internal medicine
Raynaud's disease	-Surgery, vascular surgery
Constipation	-Internal medicine
Radiation scars (following irradiation/treatment)	-Radiology, dermatology
Osteomyelitis	-Surgery
Oncological additives	-Oncology and relevant fields
Parkinson's disease	-Neurology
Mycosis/fungus infections	-General medicine, dermatology
Polyarthritis	-Internal medicine, neurology
Spondylitis	-Orthopedics, surgery
Stomatitis	-Odontology, dentistry
Sudeck's disease	-Surgery, orthopedics
Thromboflebitis	-Internal medicine, dermatology
Wound heling disturbances	-Surgery

circulatory disturbances treated with ozone did not need operation The results of other authors confirm his conclusions [208, 209].

Rectal application method (called ozone enema) was reported in 1935 and its successful application has been proved in cases of *mucous colitis* and *fistulae*. It was established that the oxygen content in the blood was increased [210, 211].

Ozone has shown an antalgic effect, this property being used through injections in case of pains of the spinal column [21]. Care should be taken of sensitive patients, since the initial pain on the injection is considerable. Usually 30-50 cm^3 ozone are injected intracutaneously in the lumber and sciatic region to relieve pain which is observed after 1-2 h and may continue for a couple of days.

Wolff [213[reported that as early as 1915 ozone was provided for local gas treatment in cases of *fistulae*, *dicubitus* (*ulcers*), *ulcus cruries*, *osteomyelitis* as well as badly healing wounds. For that purpose plastic bags filled with gas mixtures were employed. This therapy became particularly useful for inaccessible ulcers, fistulae and damage from X-radiation.

Payr [214] applied subcutaneous injection of ozone gas in cases of venous circulatory disturbances and varicosis. For the same cases, he also used intravenous injections. He also reported that Dr Fisch had tested this method on himself without side effects.

The intraarterial application of ozone therapy is promising for orthopedic practice and for accident surgery. Intramuscular injection of oxygen-ozone mixture is mainly applied in some cases with obstinate carcinoma patients or as a supportive therapy together with the so-called 'minor" autohemotherapy. This is a successful therapy: painless, harmless and widely applied in inflammatory processes.

Dorstewitz [215] employed autohemotherapy to treat hepatitis. 50 - 100 ml blood from the patient are treated with ozone and then reinjected. 20 patients with herpes simplex and herpes zoster were treated in the same way for 5-10 days. A complete healing of skin lesions was observed. In one of the cases of extensive herpes in the trigeminal region with keratitis, healing was achieved within 10 days, although a persistent opacity of the keratic tissue remained. In six cases of herpes simplex, upon double treatment with ozone, a positive effect was observed after 28 h. It is desirable to start therapy at an early stage of illness.

Moreover, the so-called hematological ozone therapy (HOT) has found wide application [216]. The therapy has been employed since 1947. In a special hyperbaric chamber the air is enriched with ozone by means of UV-Cs lamp. It has been suggested that singlet oxygen is formed under these conditions. 60-70 ml blood from patients are placed in the HOT apparatus and enriched with oxygen-ozone. The blood thus treated is reinjected intravenously or intramuscularly into the patients. HOT therapy is applied in the following affections: chronic ulcers and gastritis, prophylaxis in recoveries from myocarditis; lipid metabolism disturbances hypercholesterinae; chronic liver trouble and recoveries from infectious hepatitis; prophylaxis in acute hepatitis, donor blood preliminary treated with ozone-mixture migraine, chronic nephritis with increased creatine and urea levels; acute artery retinae occlusions, chronic polyarthritis; scab treatment after burning; acute bleeding of the eye. Most likely, HOT therapy enhances prostoglandine synthesis. Following HOT therapy, the cholesterol level of 32 patients was examined (322 determination tests). It was found that the cholesterol level dropped at an average of 20% with respect to the initial values. It was also demonstrated that this therapy caused no toxic effect on the marrow by blocking the enzyme system.

Major autohemotherapy is applied if the artery in question is not accessible to palpation. However, in cases of milder circulatory disturbances, intramuscular and/or subcutaneous injections may be sufficient.

Rokitansky [217] widely recommended an external ozone gas therapy in the advanced stages of gangrene which prevents from further inflammation. He has indicated that as a result of intravascular therapy applied in gangrene of the toes of the foot, the average rate of amputations at the upper thigh is reduced from 15 to 8 % in stage III. In some cases during the postoperational period, an extensive formation of necrosis with purulent secretion might start to develop. However, by making use of intraarterial ozone injection and local ozone-oxygen gas treatment, it became possible to improve circulation and oxygenation of the tissue. Ozone therapy was also used to treat gas gangrene. Guinea-pigs were treated every day with 450 μg ozone. The death-rate decreased from 97 to 72 % [218]. Peripheral artery circulatory disturbances are treated with intraarterial ozone injections at a maximum concentration of 33 μg/l. In autohemotherapy the concentrations should not exceed 40 μg/l because of risks to develop hemolysis [219].

In cases where the microorganisms come into direct contact with ozone in a sufficient quantity, killing of the microbes most certainly take place via the oxidative decomposition of the capsid/protein shell of the virus, thus exposing their DNMA and RNA components to open attack; the presence of DNA and A catabolism products could be demonstrated empirically [220].

In the treatment of virally produced diseases such as herpes or hepatitis, however, the circumstances are quite different. In this case, a completely different method of application must be applied, which is safer for the patient, the choice here generally being in favor of "major autochaemotherapy" whereby the patients own blood receives an admixture of ozone upon withdrawal from the body and is immediately put back into system whereas during ozone therapy session lasting for 5 min a wound receives almost 800 mg of ozone, as acute case of virus infection needs only 10 mg of ozone in autohaemotherapy. It is for that reason that, as an active mechanism, a killing off of the virus does not come into question, such as is the case for wound treatment: instead there are two processes to be discussed here which probably take place simultaneously: 1) virus inactivation through ozone or ozone peroxides and 2) a peroxide intolerance of those body cells infected by viruses. Therefore, this means that in the ozone/oxygen therapy, not only the reaction of ozone itself, but also those of its reaction products must come under discussion. Ozone reacts with unsaturated fatty acids in the cell forming ozonides and peroxides. The intervention of ozone and/or its peroxides ought to take place at the acceptors of the free virion, whereby the virus-to-cell contact, and thus the entire reproductive cycle is interrupted. The free pair of electrons on the nitrogen of the N-acetylglucosamine probably represents the point of attack for the electrophilic ozone molecule or its peroxides, so that the virus is blocked for further reaction by oxidation, making it effectively inactive.

The infected cell, however, produces hydrogen peroxide as a defensive function. Apart form this, the electrophilic reaction of ozone can also take place with unsaturated fatty acids, as a membrane constituent of the virally infected body cell, and injected peroxides into the cell. The enhanced quantity of peroxide in the cell affect the cell processes through two ways: 1) the peroxides destroy the membranes of the human cell before reproduction cycle of the infected viruses has been completed or 2) produce a synergetic effect, destroying the microorganisms which have penetrated the cell. Intravascular ozone/oxygen injection using small quantities of ozone constitutes a direct intervention in the metabolic processes, particularly in that of erythrocytes, participating in the normal glucose and pentose phosphate cycle. The final product is 1,3-diphosphoglycerate which is of decisive importance for the deoxygenation of hemoglobin.

Under the effect of peroxides, obtained as a result of ozone reaction with unsaturated fatty acids in the erythrocytes membranes, the detoxifying mechanism of these substances is immediately put into action via the glutathione system. In order to maintain the glutathione redox system, the pentose/phosphate path must be accelerated in order to cover the ADPH debit as a reducer of GSSG. This means, simultaneously, an increased breaking down of sugar. The decisive product of the accelerated glucose metabolism is the presence of 2,3-diphosphoglycerate (2,3DPG) which represents a key compound in the curative effect of ozone:

$$HbO_2 + 2{,}3\ DPG \Leftrightarrow Hb^{\bullet}2{,}3\ DPG + O_2$$

Every increase in 2,3-DPG levels facilitates the release of oxygen through a shifting of the HbO/Hb balance in favor of deoxygenated hemoglobin. As hemoglobin is very stable, particularly in the case of diabetics, ozone therapy is particularly indicated for this decrease and brings about an improvement in the peripheral oxygen supply. It is possible, by measurement of the arterial and venous partial pressure of O_2, to observe *in vivo* the change in the oxygen situation during the course of treatment. However, not only the increase in arterial pO_2 but primarily the reduction in venous pO_2 is decisive for an increased supply of oxygen, which is equal to an increase in the arterial/venous pO_2 difference, even up to normal values of $pO_2 = 60$ mm Hg.

The immunoglobulins are proteins produced from lymphoid cell, which are unified according to similar general principles of their structural organization. The functional integrity of various classes of immunoglobulins is reduced to their participation in homeostasis, the most important factor of immunologic defense systems. Immunologic processes are associated with the presence of antibodies. Five classes of immunoglobulins are known: IgG, IgA, IgM, IgD and IgE. Changes of normal levels cause certain troubles, for example, increased levels of IgG lead to chronic liver deseases. The effect of ozone on the immune system was investigated by monitoring changes of immunoglobulins levels in the blood [221].

Nineteen patients with cervix carcinoma and 20 patients with ovarial carcinoma were entered into a study. They were injected intravenously with ozone-oxygen mixture (540 μg ozone) for 5 days. However, the therapeutic effect was not confirmed.

34 patients with chronic rheumatic troubles underwent autohemotherapy through intramuscular injections with 5 ml blood from patient enriched with 300-350 μg ozone.

1000 μg ozone were administered intravenously into 47 patients every week for 6 months (50-60 ml blood from patient enriched with ozone). Statistical data processing indicated insignificant changes of the IgA, IgG, IgC and IgM levels. Therefore, no immunodeficient effects were observed. Certain changes of IgG levels can be evaluated as immunostimulating.

The inhibiting effect of ozone on tumor tissue has been investigated by Sweet [222]. Zanker confirmed this effect, his attention being attracted to the effect of ozone on tumor metabolism.

Steiner et al. [223] studied the changes in the level of: DPG, DPN, F-6-PK, GAPDH, HBDH, ICDH, LDH and NAD in carcinomatous and healthy ovarial tissues (*in vivo*) under the influence of ozone. Cytostatics like Adriblastin and Xoloxan were also used for reference purposes. The lysolecithin levels decreased in the tumor tissue but did not change in the healthy one, which indicates that ozone has no influence on the tumor metabolism. However, the ozone therapy has a positive effect in employing ozone in tumor tissue.

Brinkman and Lamberts [224] were the first who reported on the effect of ozone on erythrocyte(RBC's) behavior. These authors exposed volunteer patients to the influence of ozone (1 ppm) for 10 min and then measured the oxygen evolved from hemoglobin. Upon prolonged ozone exposure oxygen is not bound with hemoglobin which causes blood-vessel occlusions in fingers and toes. Other authors [225] have observed changes in red blood cell biochemistry, the RBC's being taken from mice and apes. Brikley et al. [226] have investigated the morphological and biochemical changes in human red blood cells for ozone concentration of 0.5 ppm. This study indicated a 20% increase of erythrocyte shortness, a 20% drop in acetylcholine esterase activity, a 15% fall in glutathione levels (GSH) and a comparatively slight increase in the activity of lactic dehydrogenase (LDH). Some authors [227, 228, 230] have suggested that, due to inhaled ozone, the amount of oxygen free radicals is increased. As a result of this, hydrogen peroxide, which might be a notable oxidative metabolite, also increases its levels. The catalase activity in red blood cells is diminished by 50% and the intracellular hydrogen peroxide is formed upon ozone exposure 6-6 ppm.

Changes in erythrocyte membranes stability have been observed [229]. Ozone concentrations of 0.25 ppm from 1 to 6 h change the erythrocyte spherical shape which may change the normal capillary permeability. Schulz et al. [230] have found that exposures of 0.7 ppm ozone for 2 h increase the methemoglobin levels and decrease the total number of red blood cells. In some mice and Guinea-pigs the hemoglobin and hematocrit levels were increased at high concentrations of ozone as compensating response against the decreasing number of erythrocytes.

Many authors [231-233] have investigated the ability of vitamin E to protect erythrocytes and lung tissue from the influence of ozone because, being a biological antioxidant, ozone interferes with the metabolism and protects other vitamins, hormones and enzymes from oxidation. Chow and Kanes [231] found that under vitamin E deficiency in mice blood during a 7 day period with 0.8 ppm ozone, the activity of glutathione peroxidase, pyruvate kinase and LDH in the erythrocytes was enhanced while the total glutathione level was decreased. The latter maintains the normal functional state of erythrocytes, low redox potential of the cell constituents which preserves the thiol groups of a number of enzymatic proteins in reduced state. The principal function of glutathione is to eliminate hydrogen peroxide formed during direct oxidation of certain drugs. The toxic effect of hydrogen peroxide is due to its ability to oxidize hemoglobin to methemoglobin and to form hydroperoxides with unsaturated fatty acids which lead to changes in the lipids of the cell membranes.

The effect of ozone on erythrocyte membranes has been investigated with respect to the SH-groups in some enzymes. At an ozone concentration of 40 mmol/min over 5 ml membrane suspension, the activity of ATPase is totally lost. The latter can be regenerated by means of phosphatidilserine [234]. Peterson [235] observed changes in T-lymphocyte and B-lymphocyte levels for an ozone concentration of 784 $\mu g/m^3$ for 4 h.

Canada and Airriess [236] studied changes in the concentration of vitamin E, vitamin A and vitamin C (antioxidants) in the human serum after exposure to ozone concentration of 0.37 ppm for 2 h. These changes were insignificant. The alkali phosphotase in the serum was increased at ozone concentrations of 10 ppm for 1 h. This was explained by the fact that the blood tissues let the enzymes pass because of a visible danger to lungs at these concentrations. Other authors [57, 237, 58, 238] have followed the concentration levels of prostoglandines PGFa and PGEa during exposure of mice to ozone at concentrations of 4 ppm for 8 h. They observed an increase by 186 and 200% respectively, which was attributed to arachidonate oxidative induction by plasma membranes. Veninga and Wagenaar [239] established that creatine kinase (CPK) levels were increased after exposure to 0.02 ppm ozone for 2 h but, unexpectedly there were no changes at higher ozone concentrations. Probably, this effect is associated with some adaptation of the organism. There was response in serum lipids [240] at 1.1 ppm ozone for 24 h; the activity of lecithincholesterol transferase, free cholesterol, and lipoproteins was increased, however, there was a descending tendency in triglyceride levels. The bacterial effect of ozonized blood serum on gram negative bacteria (*Klebsiella*, *Pseudomonas* and *Salmonella*) has also been studied. This effect is dependent on the serum ozonation time, type of bacteria and bacteria membrane structure [241]. Prolonged exposures to ozone (0.4 to 1 ppm, 6 h daily) caused a drop in serum albumen levels and increased the α-globulin and β-globulin levels. The total protein level was not changed [242].

It has been established that ozone affects the central nervous system (CNS) [243]. Exposures to > 0.6 ppm ozone concentrations cause a headache and lethargic states. Changes in the visual sharpness (0.2 ppm) and eye muscle may occur [244]. A fall in the cerebral levels of catechol-*o*-methyl tranferase on exposure of dogs to 1 ppm of ozone (8, 12 or 18 h) was observed [245]. The reduction of enzyme activity was in a good correlation with a decrease in catechol amounts in the brain.

Takanashi and Miura have reported a considerable fall in the activity of metabolite enzymes in the liver: benzopyrene hydroxylase, 7-ethoxycumarine-*o*-diethylase and aniline hydroxylase. No effects were observed on *p*-nitroanisole-N-dimethylase on exposure to ozone of 0.8 ppm for 7 h. A 20% drop in cytochrome P-450, NADPH cytochrome, P-450 reductase, and cytochrome B5, included in mitochondria and/or ergastoplasma membranes, was established. Fluorescent pigments in the liver, being an end products in lipid oxidation, were increased upon exposure to 1.5 ppm ozone for 30-60 min. The effect of ozone concentration 1.0, 0.5 and $0.2.10^{-5}$ M has been investigated on mice for exposure time 2, 3 and 15 h [249, 250]. The quantity of cytochrome P-450 in liver microsomes at 1.10^{-5} M and 2 h exposure is 73.4 % with respect to the control level and B5 - 93.3 %, respectively. After 3 h under the same conditions these parameters are already 75 and 69.6 %, respectively and after 4 h the mice die. The cytochrome level at ozone concentration of $0.2.10^{-5}$ m and exposure time of 15 h is 91.3 and 93.2%, respectively. Simultaneously 10 h before the exposure of mice to ozone atmosphere the potassium salt of α-tocopherol is administered intraperitonitously - 56 mg/kg. This leads to increase in P-450 level to 158.9 % and that of B5 to 136.6%. In the control mice this leads to increase of their levels with 135 and 116%, respectively.. These data allow to assume that tocopherol forms adducts with cytochromes and ozone stimulates the adduct formation.

Zelac and Cromroy [251] have established that ozone is a better mutant than radiation. Humans exposed to 0.5 ppm ozone for 8 to 10 h manifested chromatid and chromosomal aberrations. No increase in chromosomal aberration was observed at 0.4 ppm ozone for 4 h.

In tracheotomy with dogs, with acquired interrupted respiration accompanied by hypertension and bradicardia during exposure to ozone (50 ppm, 4 min), the symptoms decreased due to stimulated *parasympathetic vagus nerves* [252, 253]. Exposure of rabbits to ozone (0.2 ppm, 5 h) gave rise to pathological changes without myocardial damage but cell membranes did not remain intact.

Ozone causes a multiplex effect on the endocrine system [254]. The authors report an inhibiting effect on ^{131}I(ozone 4 ppm, 5 h) which was observed even 12 days after exposure. Plasma thyrotropin was reduced by 50%. The thyroid gland increased its size. The authors suggest that decreased thyroid hormone levels appear to be prophylactic against ozone damage.

Gordon et al. [255] have reported that upon inhalation of ozone (0.8 ppm, 1 h) the activity of choline esterase was diminished not only in the blood, but also in the diaphragms of Guinea-pigs by 14% and lung tissue (16%). The investigations]256] of the effect of ozone on the immune system show that exposures of 0.6 ppm for 2 h did not affect the number of T-lymphocytes. However, an exaggerated spleen was observed during ozone exposure of 0.31 ppm for 103 h. Kidneys manifested dilated tubules.

1.3.2. Toxic Action

Mechanisms, according to which ozone damages biological tissue, are not clearly defined. However, several processes should be considered which lead to disturbances in normal cell functioning. Ozone oxidizes polyunsaturated fatty acids (PUFA) in the cell membranes according to Criege's mechanism [257]. Also, it contributes to the increase of malonic dialdehyde content and interacts with natural antioxidants (vitamin C, vitamin A and vitamin E). On the other hand, ozone reacts directly with aminoacids in tissues, proteins, and small peptides [240]. It inactivates the following enzymes: lysozymes, betaglucoronidase and acid phophatase, while urease is resistant to ozone.

Very likely, the cellular regulation of the electrolytes becomes worse, and in consequence of this, the enzymes become inactivated and the mitochondrial metabolism is changed. Normal cell functioning is disturbed which might lead to cell destruction [258].

Lung tissue, being vulnerable to ozone attack, has been the object of many investigations. Biochemical studies indicated changes of enzyme levels, the enzymes being an integral part of the intracellular protection from disturbances. These enzymes bind with free radicals, lipid peroxides, and hydrogen peroxide which are factors for changes of tissues. Chow and Hussain [259] have reported enhanced activity of glutathione peroxidase in mice inhaled with 0.8 ppm ozone for 3 days. Other natural antioxidants, such as superoxide dismutase, catalase, vitamin E, and vitamin C, destroy the free radicals, thus contributing to reduce the risk of permanent tissue damage due to ozone. For example, bronchia constriction, caused by ozone, fades away or disappears after treatment with vitamin C or with combination between vitamin E and vitamin C (synergistic effect). Complementary biochemical investigations of ozone toxicity involve measurements of the enhanced synthesis of collagen which induces fibrosal changes in lung tissue [260, 261].

Exposure of mice to small doses of ozone [262] gave rise to changes in two enzyme systems: 1) a decrease in the hepatic ascorbic acid (HAA) concentrations and 2) a change in the creatine phosphokinase (CPK). HAA was considerably changed for 30 and 120 min after cessation of the ozone exposure whereas CPK demonstrated changes only 15 min after that. Compared to reference animal, no significant changes were observed after 24 h for HAA and 30 min for CPK. Plasma histamine lactodehydrogenase was not affected with various doses of ozone. It is assumed that this is due to the ability of organisms to adapt themselves to the harmful agent, which, on the other hand, stimulates metabolite processes and enzyme activity. At higher doses of ozone the enzyme activity gradually falls down, probably due to some compensatory balanced conditions. At ozone concentrations of 0.8 ppm (8 h) the activity of the total lactate dehydrogenase (LDH) in the lungs was enhanced but not in the plasma and erythrocytes. The LDH-5 fractions from LDH isoenzymes considerably decreased in terms of their distribution in the lung and plasma, while fraction LDH-4 was increased. No change of isoenzyme activity in the lungs, plasma and erythrocytes was observed for ozone concentration of 0.5 ppm [263].

Saymor [264] has studied the behavior of mice upon prolonged exposure (120 days) to small doses of ozone. The DNA and RNA levels in the lung were decreased whereas the protein level was increased.

In vivo and *in vitro* high ozone concentrations are prerequisite for suppressed mitochondrial oxygen consumption and oxidative phosphorization and for increased membrane permeability. This can be associated with thiol groups oxidation and lipid oxidation [265].

Ozone exhibits a 100% bacteriological effect at concentrations of 0.023 mg/l without producing spores. Spores are destroyed at 0.61 mg/l [247]. At the same concentration, 87.5% of the hepatic virus B is inactivated. Large amounts of blood can be conserved without clotting by bubbling with oxygen-ozone mixture. Parallel to this, all the microbial materials in blood are oxidized, thus sterilizing it [266].

A detailed evaluation of the short-term influence of ozone on human health has been made in Refs 260-270. The transitory effect of ozone on the respiratory function was investigated by various respiratory measurements.

Miller et al. [271] have made comparative models of ozone operation regions in the respiratory tract, and Criebel and Smith suggested a new pharmacological model of ozone acute influence on human lung and have compared the effect of ozone on the lung (ozone 120 ppb, 2.5 h) in children and adults. They observed a considerable lowering of PFR and FEV values with adults compared to children. They succeeded in measuring the PFR baseline: 3.9 and 2.7% with adults and children, respectively. Adults made more symptomatic complaints than children. Stock and Kotchmar [272] reported that there is almost no risk of asthmatic attack in patients suffering from asthma at ozone concentrations over 120 ppb.

Many authors have investigated the influence of ozone on pulmonary lysosomes and mitochondria. Lysosomes contain quite many enzymes, mainly hydrolases, such as protease, lipases, nucleases, phosphatases, phosphodiesterases, etc. Dillar et al. [273] have studied the changes in lysosome levels upon exposure of mice to ozone (0.7 ppm). The activity of pulmonary lysosomal cathepsin A, cathepsin D, acidic phosphatase, β-N-acetylglucose aminidase, and benzyl arginine β-naphthylamide amidohydrolase was increased. This is due to inflammatory processes caused by ozone. The specific activity of protease and peptidase in the lung is enhanced when related to chronic obstructive genetic lack of α-1-antitrypsin factor.

Mitochondria are contained in every cell. A highly organized and specialized transportation chain, which transfers protons and electrons to molecular oxygen, is located in the internal mitochondrial membrane. It is composed by enzymes, redox systems, which organize proteins and lipids, and is called respiratory chain. The external membrane contains enzymes from the tricarboxylic acid cycle, fatty acid β-oxidation, etc. It has been established that at high ozone levels (2 ppm) the oxygen consumption of pulmonary mitochondria in rodents is decreased whereas this consumption is increased at 0.8 ppm ozone after inhalation for 20 days[274]. Mice, exposed to 0.8 ppm ozone, obtained an acute ozone intoxication due to blocking of the alveolar capillary function, followed by hemolysis and lethal hyloxia [275, 276]. The authors in Ref. 277 have studied in detail the cellular, biochemical, and functional effects of ozone on the lung.

Clearance of foreign materials (particles) from the lung and airways depends on the function of macrophages, ciliated cells and secretory cells, and on the physical and chemical properties of the alveolar cells. All these are affected by ozone exposure. Single acute exposure of animals and humans to ozone concentrations less than 0.6 ppm have been shown to accelerate clearance of particles from the tracheobronchial tree whereas acute exposures to ozone levels greater than 0.6 ppm caused a delay in particles clearance [79, 261]. Repeated exposures (2 h daily for 14 days) to 0.1 ppm ozone gave rise to acceleration in alveolar clearance of latex particles but had no effect on tracheobronchial clearance. These results, related to morphological studies, are consistent with certain adaptation [278-283].

The alveolar macrophage (AM) take part in defending the alveoli and airways against inhaled particles and pathogens. To carry out this important functions, AM maintain active mobility, phagocytic activity, membrane integrity and enzymatic capacity. Ozone exposures of AM *in vivo* has been shown to alter the number and morphology of AM, depending on dose and duration of exposure. A transient decrease in the absolute number of AM has been observed following exposures to high ozone concentrations [284, 285]. AM recovered from rats exposed for 16 h to 0.05, 0.10, 0.20 or 0.40 ppm ozone showed no differences in cell yield or cell viability. An increase in AM numbers has been also observed after *in vitro* ozone exposure [286]. Recruitment and persistence of AM result from chemotaxis and migration of cells into the lung alveoli and from local proliferation. A primary function of AM is the maintenance of sterility in the lung by phagocytosis of foreign particles. Changes in the phagocytic activity of AM has been observed following exposure to ozone. A marked depression of this activity was observed after exposure to 1.2 ppm ozone for 5 h [287] but enhanced phagocytic function has been observed after prolonged exposures for 20 days (3 times daily) to 0.8 ppm ozone. This difference in the phagocytic function of AM for various ozone concentrations and exposure conditions can be explained by changes in the cell populations. Ozone exposures also affect the secretory activity of AM and their ability to release factors which stimulate the migration of neutrophils and other cell types into the lung. A culture monolayer of rabbit AM was exposed for 2 and 6 h to 0.1, 0.3 or 1.2 ppm ozone. Biochemical functions of alveolar macrophage can also be affected. Ozone exposure caused a reduction in intracellular concentrations of acid phosphatase and β-glucoronidase, as well as in lyposomal enzymes [289, 290].

Exposure of the upper airways of mice to ozone at 0.5-0.8 ppm for 1-24 days has induced a two-fold increase in IgA^{+}B-lymphocytes. In the deep lung, exposure to 0.7 ppm ozone for 14 days causes an accumulation of T-lymphocytes [291, 292]. The number of leucocyte cells in controls, $3.8.10^5$, was increased to $7.5.10^5$ cells in exposed mice. Ozone exposure of lymph nodes of euthymic mice indicated that the histopathologic canges were more extensive in the athymic animals [293].

Changes in the immune function depend on the organ, dose and duration of ozone exposure. Concanavalin A (Con A) stimulates the cell-mediated arm of the immune system. Con A reactivity was unaffected at days 4 or 7 of 0.7 ppm ozone exposure (20h/day) but increased by 1.5 and 3 fold after 14 and 28 days, respectively. In contrast, splenic cells of mice manifested decreased reactivity to T-cell mitogens *in vitro* (phytohemagglutinin, PHA, 38% decrease, Con A, 75 5 decrease) but not to a B-cell mitogen (lipopolysaccharide, LPS0 or to alloantigens [294].

In rats, 7 days of exposure to 1 ppm ozone enhanced the reactivity of splenic cells to PHA, Con A and LPS [295]. Ozone inhalation gives also rise to nonspecific changes. Exposure to ozone at concentrations of 0.4 to 0.8 ppm can induce weight losses and/or thymic and splenic atrophy. A similar effect has been observed if steroids are injected into mice, i.e. ozone can produce steroid-mediated effects on lymphocytes.

The immune response of the human organism to a foreign agent involves the following cell quartet: macrophage, B lymphocytes, T helper, and T suppresser. Amoruso et al. [296] have determined the anion peroxide radical (O_2^-), produced by alveolar macrophage from mice exposed to ozone levels of 3.2 to 10.5 ppm. They found decreasing concentration levels. The authors proposed that this decrease in anion peroxide radical amount cannot be due to a direct effect of ozone on cell viability but rather to an infection caused by ozone, i.e. the immune system of experimentally tested animals enhances its activity. Alveolar macrophage from test animals, if stimulated by phorbal myristat acetate, also decreased the concentration of O_2^-. This explanation seems to a some extent confusing since the bactericide action of ozone which is very strong and in case of infection it would rather prevent than stimulate it.

Lung lesions observed after one week of exposure to 0.2, 0.5, 0.8 ppm ozone were more expressed than those observed after 3 months of continuous exposure. Some researchers [297, 298] suggested that the enzyme levels in the lung are increased which may account for the development of adaptation and tolerance to ozone.

Epithelial cels type I and squamous cells appear to be most sensitive to ozone [299]. An exposure to 0.5 ppm ozone increases the volume but reduces the area of epithelial cell types I [300]. These cells are regenerated to the normal condition after a week in fresh air and develop certain tolerance to ozone reexposure [3301, 302]. The authors have studied tracheal cells (volume, density) by electron microscopy. They have found that all changes due to chronic ozone exposure were reversible, the normal values being regained if test animals are exposed to fresh air.

Pulmonary fibroses, defined as abnormal accumulation of collagen, are developed after prolonged exposure to ozone levels greater than 0.5 ppm [303, 304]. Chronic ozone effects at concentrations close to that in atmospheric air are hardly noticeable and cannot be measured. Other scientists [305] have exposed mice to 1 ppm ozone for 40 days and found increased levels of total lung collagen but this effect rapidly faded away after 10 days of exposure to fresh air. Chronic exposure to 0.06 ppm ozone (13h/day, 5 days weekly, 18 months) caused changes in the lung function including a decrease in the general respiratory effect.

1.3.3. Ozone Solutions

Ozone disinfectant and sterilizing properties were used to freshen water in Germany at Wiesbaden yet in 1902. On this basis, in 1934 the dentists started utilizing ozonized water for inflammatory processes. This was reported at the Sixth World Congress on Ozone held in Washington [306]. The following properties of ozone are used in this case: a) disinfectant and sterilizing effect; b) haemostatic effect, especially in cases of running hemorrhages; c) accelerated wound healing, improved oxygen local supply and support to metabolic processes [307, 308]. Ozonized water, in the form of spray or stream, is used for the antrum in the following cases: gingivitis, paradontosis, white-gum, stomatitis. Preference is given to the spray mode, particularly in cases of caries for cleaning and disinfection, as well as for injuries and bleeding in the antrum. It is applied in cleaning nerve canals and pulpa tissue, in case of painful gingivitis and stomatitis. Ozonized water has also found wide application in orthopedy: artificial denture and crown cleaning prior to fixing, stopping of hemorrhages during fixing.

The ozone solutions in olive oils (ozone is present as ozonides) have found wide application in dermatology for treating: fistulae, decubitus and ulcer cruries. It is also successfully applied in mycosis and fungus finger and fingernail infections. Simply, patients has to put their fingers into a plastic sleeve covered by ozone-olive oil. Salzer et al. [309] have investigated the effect of ozonides olive oil on patients suffering from soorkolpitis. An especially positive effect was observed in its application in the vagina. The numbers of examined bacteria decreased very rapidly. These authors suggest that the positive effect of treatment is due to ozonides formed, as well as to tissue enrichment with oxygen. The same authors used autohemotherapy (100 ml blood enriched with oxygen-ozone; 1 ml contains 33 gammas ozone) in cases of chronic placenta insufficiency in pregnant women (20 patients, about thirtieth week). HPLC analysis indicated increased values. This therapy is also useful for weak bleedings during pregnancy.

Wermeister [310] described a technique, utilizing ozone gas of 80-100 μg/ml concentration, for wound cleansing and disinfection. Such an effect has not been observed during enzyme treatment. Often, the bacterially infected spot is washed out by ozone-oxygen mixture which give certain positive effect.

Ozonides that are prepared by means of the reaction between ozone and olefins do not cause inflammatory processes in skin and are not decomposed by catalase. Ozonides participate as an oxidizing agent without liberating peroxide groups which might oxidize cells. About 1000 subcutaneous injections of castor-oil ozonides have been made without tissue response [311]. A 4% solution of the ozonide of sorbic acid in ethyl diglycol (ozonide SV) is used in cosmetics to preserve face creams and as disinfectant. The lethal dose is 10 ml/kg. In wheat-germ oil, containing vitamin E, the ozonide keeps the vitamin E levels unchanged.

1.3.4. Sterilization

A number of methods have been developed, according to which ozone is utilized for sterilizing bottles, cans, containers and glass vessels that find wide application in the pharmaceutical industry and cosmetics [312-314].

Usually sterilization is performed at sufficiently high temperatures which may cause damage on same materials used in food processing industry, pharmaceutical industry and cosmetics. The effect of ozone on bacteria is well known, especially its toxicity and reactivity characteristics, that is why attention should be given to dosage and exposure time.

In [315} is discussed the application of ozone as sterilizing agent in the pharmaceutical industry. Physiological solution (0.9% NaCl), pH = 7, is treated with oxygen-ozone mixture containing 15% ozone for 15 min. The solution is allocated to glass or polyethylene banks which are sealed up. Analysis showed that after 2 h the microbe number was 0 and after 6 h the ozone was completely destroyed. Sterilization for 2 h with 0.5 ppm ozone caused a 100% lost of *Legionella pneumoniae* [278].

An apparatus has been designed to produce oxygen-ozone mixture for sterilization of milk, diary products, wines, butter and liquid medicines. The experimental set-up is described in detail in a patent [279]. The quality of wine, beer and liquor is improved after exposure to ozone. Ozone levels of 0.01 ppm and rates of 1l/min are employed in the pharmaceutical industry and food processing industry. Ozone has been employed fro cleaning dialyzers [316] and contact lens [317].

1.3.5. Ozonators

Portable ozonators have been designed for medicine purposes covering the whole range of terapheutical doses from 0 to 100 μg/l. These units are manufactured in "Shumen" plant Ltd., Shumen, Bulgaria as P-1 trade mark.

REFERENCES

[1] Razumovskii, S.D., S.K. Rakovski, D.M. Shopov, G.E. Zaikov. *Ozone and Its Reactions with Organic Compounds (in Russian)*. Publ. House of Bulgarian Academy of Sciences, Sofia, 1983.

[2] Proskuryakov, VA, LI Shmidt. *Purification of the Sewage Waters from the Chemical Industry*. Khimya, Leningrad, 1977.

[3] Bailey, P.S. "*Ozonattion in Organic Chemistry*", Academic Press, New York, v.I, 1978, v.II, 1982, *Adv. Chem. Ser. - A,* (Ed. HH Wasserman), 39 -I, 39 - II, 1978, 1982.

[4] Kuznetzov, IE, TM Tritzkaya. *Protection of the Air Space Pollute by Dangerous Compounds*. Khimya, Moscow, 1979.

[5] Itogi Nauki i Tekhniki - *Technology of Organic Compounds*, Moscow, VINITI, 7 (1983).

[6] Itogi Nauki i Tekhniki -*Protection and Reproduction of Natural Resources,* Moscow, VINITI, 13 (1983).

[7] Golubovskya, EK. *Biological Bases of Water Purification*. Moscow, Visshaya Shkola, 1978.

[8] Bokris, OM (Ed.). *Environmental Chemistry*. Moscow, Khimya, 1982.

[9] Pergut, EA, DO Gorelik. *Instrumental Methods to Control Air Pollution*. Leningrad, Khimiya, 1981.

[10] Chanlett, ET. *Environmental Protection*., McGraw-Hill, 1979.

[11] Flaekt AB. HCl/SO_2 - scrubbing. *Res. Discl., 326, 453 (1991).*

[12] W Ruck Ozonization of Biologically Purified Sewage Water. *Vom Wasser,* 80, 253-272 (1993).

[13] Mccoustra, MRS, AB Horn. Towards a laboratory strategy for the study of heterogeneous catalysis in stratospheric ozone depletion. *Chemical Society Reviews,* 23: 3, 195-204 (1994).

[14] Tabata, K., I.Matsumoto, Y.Fukuda. Oxidation catalysts for waste gases from burners *Jpn.Kokai Tokkyo Koho JP 01,270,928 [89,270,928] (Cl. B01D52), 30Oct 1989, Appl. 88/98,632, 21 Apt 1988.*

[15] Pollo, I., J. Jarosszynska- Wolinska, J. Malicki, J. Ozonek, W. Wojcik. Removal of nitrogen oxides from waste gases. *Pol. PL 137,426 (Cl. B01D53/14), 31 Jul 1987, Appl. 245,514, 30 Dec 1983, 2pp.*

[16] De Guardia A., A.Bouzaza, G.Martin, A. Laplanche. Aprocess of cleaning gases containing organic sulfur compounds (VOCs) by silicon oil scrubbing and ozone oxidation of absorbed pollutants.*Pollut.Atmos., 152, 82-92, (1996).*

[17] Takeyama, Kenichi, K. Nitta. Apparatus for decomposing acyclic halogenated hydrocarbons in gases.*Jpn.Kokai Tokkyo Koho JP 01,236,925 [89,236,925] (Cl. B01D53/34), 21Sep 1989, Appl. 88/61,038, 15 Mar 1988, 3 pp.*

[18] Moortgat, G.K., .Horie, B.C.Zahn. Ozonolysis of VOC: recent studies by matrix isolation FTIR-spectroscopy.*Pollut.Atmos., 29-44, (1991).*

[19] Tomita, K. Material with adsorptive, oxidative and catalytic properties for air treatment. *Jpn.Kokai Tokkyo Koho JP 02,152,547 [90,152,547] (Cl. B01J20/18), 12 Jun 1990, Appl. 88/304,185, 02 Dec 1988, 4 pp.*

[20] Vicard, J.F., G.Vicard. Method of cleaning gases containing condensable organic pollutants.*PCT Int.Appl. WO 92 19,364 (Cl. B01D53/34), 12 Nov 1992, Fr Appl. 91/5,853, 07 May 1991, 15pp.*

[21] Iannicelli, J. Wet-scrubber process for removing total reduced sulfur compounds from industrial waste gases.*U.S. US 4, 923,688 (Cl. 423-224; B01D53/54), 08 May 1990, Appl. 799,494, 19 Nov 1985, 8pp.*

[22] Buettner, F., P.Koch. Mercury removal from waste gases by catalytic ozonization. *Ger. offen DE 9,931,891 (Cl. B01D53/36), 04 Apr 1991, Appl. 25 Sep 1989, 3pp.*

[23] Wada, H., T. Naoi, Y. Kuroda. Recycling of water from cyanide waste water by ozone oxidation combined with UV radiation and ion exchange method.*Nippon Kagaku Kaishi, 9, 834-840, 1994*

[24] Matt, K., H.G.Guttenberger. Process and apparatus for the detoxification of solutions containing cyanides and heavy metals. *Eur. Pat. Appl. EP 330,028 (Cl. C02F1/78), 30 Aug 1989, DE Appl. 3,805,906, 25 Feb 1988, 8pp.*

[25] Waga, H., T.Naoi, Y.Kuroda. Recycling of water from chromium (VI) waste water by ozone oxidation combined with UV radiation and ion exchange.*Nippon Kagaku Kaishi, 4, 306-313, (1995).*

[26] Tanaka, K. Oxidation treatment promoted by photocatalyst and ozone.*PPM, 27(12), 10-15, (1996).*

[27] Welch, J.F., J.D.Siehwarth. Method for processing of organic compounds. *U.S. US 4,861,497 (Cl. 210-759; C02F1/74), 30 Aug 1989, Appl. 169,851, 18 Mar 1988, 5pp.*

[28] Oeller, HJ, I Daniel, G Weinberger. Reduction in residual COD in biologically treated paper mill effluents by means of combined ozone and ozone/UV reactor stages. *Water Sci.Technol., 35(2-3 Forest Industry Wastewaters V), 269-76, (1997), /Elsevier/.*

[29] MoebiusC.H., M. Cordes-Tolle. Enhanced biodegradability by oxidative and radiative wastewater treatment. *Water Sci.Technol., 35(2-3 Forest Industry Wastewaters V), 245-250, (1997), /Elsevier/.*

[30] Nakamura, Y., T.Sawada, F.Kobayashi, M.odliving. *Microbial treatment of kraft pulp wastewater pretreated with ozone.Water Sci.Technol., 35(2-3 Forest Industry Wastewaters V), 277-282, (1997), /Elsevier/.*

[31] Zhou, H., D.W.Smith. Process parameter development for ozonation of kraft pulp mill effluents. *Water Sci.Technol., 35(2-3 Forest Industry Wastewaters V), 251-259, (1997), /Elsevier/.*

[32] Huang, C.M., M.K.Banks. Effect of ozonation on the biodegradability of atrazine in granular activated carbon (GAC) columns.*J.Environ.Sci.Health, Part B, B31(6), 1253-1266, (1996).*

[33] Sumimoto, H., T.Yamazaki. Injection of ozone into seawatwr piping for biofouling prevention in ocean thermal power plants. *Jpn.Kokai Tokkyo Koho JP 03,111,670 [91,111,670] (Cl. F03G7/05), 13 May 1991, Appl. 89/247,809, 22 Sep 1989, 3pp.*

[34] Kawamoto, K. Techniques for decomposition of dioxins. *Kogaito Taisaku, 27(7), 617-27, (1991).*

[35] Ukita, S. Removing slimes from water pipers by direct blowing of ozone. *Jpn.Kokai Tokkyo Koho JP 03 80,996 [91 80,996] (Cl. C02F1/50), 05 Apr 1991, Appl. 89/218,115, 23 Aug 1989, 4pp.*

[36] Miriguchi, Y., H.Hayashi, T.Umetani. Laboratory experiment on decomposition of agrochemicals by ozonation.*Osaka-shi Suidokyoku Suishtsu Shikensho Chosa Kenkyu narabini Shiken Seiseki, 46, 10-16, (1994) /Pub. 1995/.*

[37] Munz,C., R.Galli, R. Egli. Oxidative treatment of process water in a soil decontamination plant: I. Laboratory studies. *Chem.Oxid., 2, 247-63, (1992), /Pub. 1994/.*

[38] Kojima, H., E.Katsura, H.Ogawa,.H.Kaneshima. Studies on prevention of envirovmental pollution by agricultural chemicals used in golf course I. Decomposition of water-soluble agricultural chemicals by ozone.*Hokkaidoritsu Eisei Kenkyushoho, 41, 71-73, (1991).*

[39] Ohashi, N., Y.Tsuchiya, H.Sasano, A.Hamada. Ozonization products of organophosphorous pesticides in water.*Jpn.J.Toxicol.Environ.Health, 40(2), 185-92, (1994).*

[40] hapeman-somish, c.j. Mineralization of pesticide degradation products. *ASC Symp.Ser., 459 (Pestic. Transform. Prod.: Fate Signif. Environ.), 133-47, (1991).*

[41] Terui, s., K.Sano, K.Nishikawa, A.Inone. Deodorization method.*Jpn.Kokai Tokkyo Koho JP 02 139,017 [90,139,017] (Cl. Bo1D53/36), 29 May 1990, Appl. 88/293,929, 21 Nov 1988, 3pp.*

[42] Terui, s., K.Sano, T.Kanazaki, K.Mitsui, A.Inoue. Deodorization of waste gases using ozone.*Jpn.Kokai Tokkyo Koho JP 01 56,124 [89 56,124] (Cl. B01D53/36), 03 Mar 1989, Appl. 87/209,305, 25 Aug 1987, 4pp.*

[43] Ogawa, K., N.Seki. Apparatus for tap water treatment by ozone.*Jpn.Kokai Tokkyo Koho JP 04,243,597 [92,243,597] (Cl. C02F1/78), 31 Aug 1992, Appl. 91/7,486, 25 Jan 1991, 4pp.*

[44] Hoshima, Y. Apparatus and method for dissolving ozone into water.*Jpn.Kokai Tokkyo Koho JP 02 40,289 [90 40,289] (Cl. C02F1/78), 09 Feb 1990, Appl. 88/190,318, 28 Jul 1988, 4pp.*

[45] Filippi, A., F.Tilkes, E.G.Beck, H.Kirschner. Water disinfection in dental treatment units with ozone. *Dtsch.Zahnaerztl.Z., 46(7), 485-7, (1991).*

[46] Gehringer, P. Radiation- induced oxidation for water remediation. *Oesterr. Forschungszent. Zeitbersdorf, [Ber.] OEFZS 1995, (OEFZS - 4738), 21pp.*

[47] Charles, L., D.Pepin, Ph.Puig. The use of ozone for iron and manganese removal from mineral water: problems and control strategies. *J.Eur.Hydrol., 27(2), 175-191, (1996).*

[48] Thornhill, R.W. Sanitation of swimming pool and other water. *Brit. UK Pat. Appl. GB 2,219,790 (Cl. C02F1/50), 20 Dec 1989, GB Appl. 88/14,222, 15 Jun 1988.*

[49] Nitsuta, Y., Y.Nitta. Decomposition of organic matter in dredging wastes containing spilled petroleum.*Jpn.Kokai Tokkyo Koho JP 04,367,799 [92,367,799] (Cl. C02F11/00), 21 Dec 1992, Appl. 91/165,204, 11 jun 1991, 3pp.*

[50] Gordon, G., G.E.Pacey. Slurry-ozone technology for on-site treatment of contaminated soils. *Chem.Oxid., 2, 230-46, (1992), /Pub. 1994/.*

[51] Bruckner, F. *In situ* remediation of petroleum contaminated soils by ozone treatment. *ECOINFORMA 92, Int.Tag.Ausstellung Umweltinf.Umweltkommun., , 2, 211-22, (2nd 1992), /Pub. 1993/,*

[52] Preusser, M., H.Ruholl, H.Schneidler, E.Wessling, C.Wortmann. Ozone treatment for the cleanup of soil contaminated with organic pollutants.*Wiss.Umwelt, 3, 135-40, (1990).*

[53] Rimpler, M. Method and apparatus for decontamination of wastes such as soil or sludges by oxidation with ozone. *Ger. Offen. DE 19,512,448, (Cl. A62D3/00), 10 Oct 1996, Appl. 19/512,448, 03 Apr 1995, 9pp.*

[54] Wessling, E. Experiences in development and application of the ozone method for the destruction of organic noxious substances at the solid phase- both in-situ and on-site.*Erzmetall,* 44,(4), *196-200, (1991).*

[55] M Mori, H Yamakoshi, M Nojima, S Kusabayashi, KJ Mccullough, K Griesbaum, P Kriegerbeck, IC Jung. Ozonolysis of Cyclopentadiene Derivatives - Competitive Participation of [3+2] and [3+4] Cycloadditions of Carbonyl Oxide Moieties to alpha, beta-Unsaturated Carbonyl Groups. *Journal of the Chemical Society - Perkin Transactions* I : 12, 1335-1343, (1993).

[56] KJ Mccullough, K Teshima, M Nojima. Unprecedented Formation of a Cyclic Tetramer from the Acidolysis of Indene Ozonide - Isolation and Characterisation of a Novel Dodecaoxacycloicosane Derivative. *J.Chem.Soc. - Chem.Commun.*: 11, 931-933, (1993).

[57] T Sugiyama, H Yamakoshi, M Nojima. Ozonolysis of beta-(Alkoxycarbonyl).-Substituted and beta-Acyl-Substituted Vinyl Ethers - Cycloaddition Chemistry of the

Derived alpha-Keto Ester O-Oxides and alpha-Diketone O-Oxides. *J.Org.Chem,* 58: 16, 4212-4218 (1993).

[58] K Ishiguro, T Nojima, Y Sawaki. Novel aspects of carbonyl oxide chemistry. *J.Phys.Org.Chem.,* 10: 11, 787-796 (1997).

[59] Nojima, T., Ishiguro, K., Sawaki, Y. Mechanistic study on the reaction of phenyldiazomethanes with singlet oxygen: Formation and cycloreversion of 1,2,3,4-dioxadiazole intermediates. *J Org Chem,* 62, 20, 6911-6917 (1997).

[60] Ishiguro, K., Nojima, T., Sawaki, Y. Novel aspects of carbonyl oxide chemistry. *J Phys Org Chem,* 10, 11, 787-796 (1997).

[61] R Fukagawa, M Nojima Synthesis and silica gel-catalysed decomposition of 3-, (1-arylcycloalkyl).-substituted 1,2,4-trioxolanes and 1,2,4-dioxazolidine. *Journal of the Chemical Society - Perkin Transactions 1* : 17, 2449-2454 (1994).

[62] M Mori, H Yamakoshi, M Nojima, S Kusabayashi, KJ Mccullough, K Griesbaum, P Kriegerbeck, IC Jung Ozonolysis of Cyclopentadiene Derivatives - Competitive Participation of [3+2] and [3+4] Cycloadditions of Carbonyl Oxide Moieties to alpha, beta-Unsaturated Carbonyl Groups. Journal of the Chemical Society - *Perkin Transactions I*: 12, 1335-1343 (1993).

[63] KJ Mccullough, K Teshima, M Nojima Unprecedented Formation of a Cyclic Tetramer from the Acidolysis of Indene Ozonide - Isolation and Characterisation of a Novel Dodecaoxacycloicosane Derivative. *J.Chem.Soc, - Chem.Commun.* : 11, 931-933 (1993).

[64] T Sugiyama, H Yamakoshi, M Nojima Ozonolysis of beta-(Alkoxycarbonyl).-Substituted and beta-Acyl-Substituted Vinyl Ethers - Cycloaddition Chemistry of the Derived alpha-Keto Ester O-Oxides and alpha-Diketone O-Oxides. *J.Org.Chem,* 58: 16, 4212-4218 (1993).

[65] KJ Mccullough, T Sugimoto, S Tanaka, S Kusabayashi, M Nojima. Cycloadditions Between Carbonyl Oxides and Dicarbonyl Compounds - Isolation and Characterisation of Novel Polycyclic 1,2,4,6-tetroxepane Derivatives. *Journal of the Chemical Society - Perkin Transactions 1* : 6, 643-651 (1994).

[66] S Satake, Y Ushigoe, M Nojima, KJ Mccullough. The synthesis of novel 3,4-dihydro-1,2,5,7,4- tetroxazocine derivatives via extended [3+3+2] cycloaddition reactions between a carbonyl oxide, a nitrone and an aldehyde. *Journal of the Chemical Society - Chemical Communications,,* : 14, 1469-1470 (1995).

[67] K Teshima, SI Kawamura, Y Ushigoe, M Nojima, KJ Mccullough. Ozonolyses of indene and of 1-alkyl- and 1,1-dialkyl- substituted indenes in protic solvents. Remarkable effects of the substituent steric bulk and the solvent nucleophilicity on the direction of cleavage of the primary ozonides and on the mode of capture of the carbonyl oxide intermediates by the solvents. *Journal of Organic Chemistry,* 60: 15, 4755-4763 (1995).

[68] KJ Mccullough, S Tanaka, K Teshima, M Nojima. Polycyclic ozonides and peroxides derived from Hexamethyl(Dewar benzene). *Tetrahedron,* 50: 25, 7625-7634 (1994).

[69] (69) Schobert, B.D. Ozonolysis and reduction in fine chemikals industry. *Chim.Oggi, 13(6), 21-4, (1995), /Pub.1995/.*

[70] Appell, RB, IA Tomlinson, I Hill. New reagents for reductive quenching of ozonolysis reaction. *Syn. Comm., 25 (22), 3589- 3595, (1995)*

[71] (71) Yang, YT, TS Li, YL Li. Ozone oxidation of oximes to carbonyl compounds. *Synthetic Comm., 23(8), 1121- 1124, (1993).*

[72] Jachak, M., M.Mittelbach, U.Kriessmann, H.Junek. Ozonolysis of olefins. 6.Cyanoacetaldehyde by ozonolysis of (E)-1,4-Dicyano-2-Butene or allyl cyanide. *Synthesis- Stuthgart, 3, 275 - 276, (1992).*

[73] Jellen, W, M Mittelbach, H Junek. Ozonolysis of olefins. VIII.Synthesis of phenoxyacet-aldehydes by ozonolysis of allyl phenyl ethers. *Monatsh.Chem., 127(2), 167-72, (1996).*

[74] Matsumoto, H, H Kanda, Y Obada, H Ikeda, T Murakami. Preparation of quinolinaldehyde derivative as intermediate for quinoline type mevalonolactones. *Jpn.Kokai Tokkyo Koho JP 08 03,138 [96 03,138] (Cl. C07D215/14), 09 Jan 1996, JP Appl. 94/28,596, 25 Feb 1994, 10pp.*

[75] Allen, DE, ChE Ticker, ChC Hobbs, R. Chidambaram. Process for preparing 6-chloro-2-hexanone from 1-methylcyclopentane. *U.S. US 5,491,265 (Cl. 568-347; C07C45/51), 13 Feb 1996, Appl. 395,266, 28 Feb 1995, 7pp.*

[76] Lee, H.W., S.K.Ji, I.-Y.Ch.Lee, J.H.Lee. Convenient and practical synthesis of (R)-(+)-4-methyl-2-cyclohexen-1-one. *J.Org.Chem., 61(7), 2542-3, (1996).*

[77] Mikaev, FZ, FZ Galin. Synthesis of (+)-4α-[1-(trimethylsiloxy)ethyl]-2-carene. *Izv. Akad. Nauk SSSR, ser. khim., 10, 1984-7,(1995).*

[78] Mccullough K.J., M.Mori, T.Tabuchi, H.Yamakoshi, S.Kusabayashi, M.Nojima. [3+2] cycloaditions of carbonyl oxides to imines: An alternative approach to the synthesis of 1,2,4-dioxazolidines. *J.Chem.Soc.-Perkin Transactions, 1(1), 41-48, (1995).*

[79] Mori, M., T.Tabuchi, M.Nojima, S.Kusabayashi. Synthesis of 3-Vinyl-1,2,4-trioxolanes by a [3+2] cycloaddition of carbonyl oxides with alpha, beta-unsaturated carbonyl compounds. *J.Org.Chem., 57(6), 1649-1652, (1992).*

[80] (80) Casey,M., A.J.ulshaw. Cycloadditions of carbonyl oxides- a novel stereoselective approach to 1,3-diols. *SYNLETT, 3, 214-216, (1992).*

[81] Kuczkowski, R.L. Ozonolysis of vinyl ethers. *Advances in Oxygenated Process, 3(Series: Advances in Oxygenated Process: A Research Annual 3(1991)).*

[82] Reiser, R., R.-. Seeboth, C.Suling, G.Wagner, J.Wang, G.Schroder. Synthesis of 1,2-dioxolanes by ozonolysis of 1,1-disubstituted nonactivated olefins.*Chem.Ber., 125(1), 191-195, (1992).*

[83] Hon, Y.-S., J.-L. Yan. Synthesis of bifunctional compounds from cycloalkenes via ozonide intermediates. *Tetrahedron, 53(14), 5217-5232, (1997).*

[84] Marshall, JA, AW Garofalo, RC Sedrani. The direct conversion of olefins into esters through ozonolysis. *SYNLETT., 643-645, (1992).*

[85] Odinokov. VN. Â.Í. Controlled ozonolysis of acyclic alkenes in organic synthesis. *Bashk. Khim. Zhurn., 1(3), 29-33, (1994).*

[86] Odinokov. VN. Ozonolysis in specialized transformations of petrochemical olefins. *Bashk. Khim. Zhurn., 1, (1), 11-14, (1994).*

[87] (87) Ishmuratov, GYu, RYa Kkharisov, VN Odinokov, GA tolstikov. Ozonolysis of unsaturated compounds in the synthesis of insect pheromones and juvenoids. *Uspehi Khimii,* 64, (6), 580-608 (1995).

[88] Malhotra, R., S.Kumar, A.Satyam. Ozonolysis of [60] fulerene. *J.Chem.Soc., Chem. Communications, 11, 1339-40, (1994).*

[89] Zwilichovsky, G., V.Gurvvich. Ozonolysis of (cyclohexa-1,4-dienyl)-L-alanine. An approach to the synthesis of new unnatural amino acids. X-ray molecular structure of 2-hydroxy-7-methyl-3-phenyl-pyrazolo[1,5-a] pyrimidine. *J.Chem.Soc., Perkin Trans.1., 19, 2509-15, (1995).*

[90] Mueller, T., R.R.Schmidt. Thymidine diphospho-6-deoxy-a-D-ribo-3-hexulose-synthesis of a central intermediate in the biosynthesis of di- and trideoxy sugars. *Angew.Chem., Int.Ed.Engl., 34(12), 1328-9, (1995).*

[91] Pitzer, K., T.Hudlicky. The synthesis of 4-amino-4-deoxy-D-mannose and glucosamine: a new methodology for the design of homohiral amino sugars from non-carbohydrate precursors. *Synlet., 8, 803-5, (1995).*

[92] Kashima,Ch., T.Maruyama, H.Arao. Ozonolysis of azoles and its appliication to the organic synthesis. *Yuki Gosei Kagaku Kyokaishi, 54(2), 132-8, (1996).*

[93] Kashima, C., K.Harada, A.Hosomi. Ozonolysis of 1-substituted imidazoles.*Heterocycles, 33(1), 385-90, (1992).*

[94] Wu, H.-J., Ch.-Ch. Lin. Novel oxa-cage compounds: synthesis, strukture and the formation mechanism of tetraacetal oxa-cages and complex tetraquinane oxa-cages. *J.Org.Chem., 60(23), 7558-66, (1995).*

[95] McVay,K.R., D.G.Gaige, W.S.Kain. Process and oxidation catalysts for forming two terminal carboxylic acid groups from an ozonide in the presence of fine oxygen bubbles.*U.S. US 5,543,565 (Cl. 562-523; C07C51/16) 06 Aug 1996, Appl. 376,173, 20 Jan 1995,7pp.*

[96] Gesser, H.D., G.Zhu, N.R.Hunter. Conversion of ethane to methanol and ethanol by ozone sensitized partial oxidation at near atmospheric pressure.*Catalysis Today, 24, 321- 325, (1995).*

[97] Gesser, H.D., N.R.Hunter, P.A.Das. The ozone sensitised oxidative conversion of methane to methanol and ethane to ethanol. *Ñatalysis Letters, 16(1-2), 217-221, (1992)*

[98] Hitomi, S., T.Ishibashi, M.Takashi, T.Kenkichi. Ortho-selective nitration of acetanilides with nitrogen dioxide in the presence of ozone.*Tetrahedron Letters, 32(45), 6591-94, (1991)*

[99] Suzuki, H., T.Murashima, K.Shimizu, K.Tsukamoto. The ozone mediated reaction of nonactivated arenes with nitrogen oxides. *Chemistry Letters, 817-18, (1991)*

[100] Suzuki, H., T.Mori, K.Maeda. Ozone mediated reaction of polychloro benzenes and some related halogeno compounds with nitrogen dioxide: A novel non-acid methodology for the selective mono nitration of moderately deactivated aromatic systems. *SYNTHESIS- Stutgart, 8, 841-45, (1994).*

[101] Suzuki, H., S.Yonezawa, T.Mori, K.Maeda. The ozone mediated reaction of aromatic acetals and acylal with nitrogen dioxide: A novel methodology for the nuclear nitration of acid sensitive aromatic compounds under neutral conditions. *J.Chem.Soc.-Perkin Trans., 1(11), 1367-69, (1994).*

[102] Suzuki, H., T.Murashima, K.Tsukamoto. Mineral acid- free preparation of nitro compounds from pyridine or its derivatives. *Jpn.Kokai Tokkyo Koho JP 04,282,368 [92,282,368] (Cl. C07D213/61), 07 Oct 1992, Appl. 91/67,642, 08 Mar 1991, 3pp.*

[103] Suzuki, H., T.Murashima. Nitrogen oxide (NO_x). Nitration of aromatic compounds in presence of ozone. *Kagaku to Kagyo (Tokyo), 46(1), 43-5, (1993*).

[104] Lyakumovich, AG, YD Samuilov, YaM Goldberg, MN Kondrashova. Method of succinic acid preparation. *Russ. RU 2,036,895 (Cl. C07C51/34), 9Jun 1995, Appl. 93,015,890, 25 Mar 1993.*

[105] Ebdon, JR, NJ Flint. Preparation of α-phenyl ketone, ω-carboxylate-ended telehelic methyl methacrylate oligomers by the ozonolysis of regioregular methyl methacrylate-phenylacetylene copolymers. *Macromolecules, 27, 6704, (1994).*

[106] Sawaki, T, S Watanabe, K Shimada. Manufacture of silicon carbide ceramics. *Jpn.Kokai Tokkyo Koho JP 02 26,868 [90,26,868] (Cl. C04B35/56), 29 Jan 1990, Appl. 88/176,938, 18 Jul 1988, 7pp.*

[107] Wasinger, E, D Hall. Method of decolorizing of fabrics. *U.S. US 5 261,925 (Cl. 8-111; D06L3/14), 16 Nov 1993, US Appl. 560,357, 31 Jul 1990, 7pp.*

[108] Bradley, RH, IL Clackson, DE Sykes. UV/Ozone modification of wool fibre surfaces. *Applied Surface, 72(2), 143-47, (1993).*

[109] Connor, HG, D Budgell. Process for producing cellulosic fibers and moldings. *PCT Int. Appl. WO 96 18,761 (Cl. D01F2/00), 20 Jun 1996, DE Appl. 4,444,700, 15 Dec 1994, 21pp.*

[110] Miura, M., T Naito, T Hino, H Kuroda. Pitch-based carbon or graphite fibers with high tensile strenght and their manufacture. *Jpn.Kokai Tokkyo Koho JP 01,250,415 [89,250,415] (Cl. D01F9/14), 05 Oct 1989, Appl. 88/74,471, 30 MAR 1988, 9pp.*

[111] Queslati, R, S Roudesli. EPR analysis of thermal decomposition of peroxides in ozonized polypropylene fiber for grafting. *European Polymer Journal, 27(12), 1383-90, (1991).*

[112] Catoire, B, V Verney, R Hagege, A Michel. Study of polypropylene peroxidation by ozonization using electron spin resonance and transmission electron microscopy. *Polymer, 33(11), 2307-11, (1992).*

[113] Kulik, EA, M Ivanchenko, K Kato, S Sano, Y Ikada. Peroxide generation and decomposition on polymer surfase. *J.Polym.Sci., Part A- Polymer Chemistry, 33(2), 323-30, (1995).*

[114] fujimoto, k., y.takebayashi, h.inoue, y.ikada. Ozone-induced graft polymerization onto polymer surface. *J.Polym.Sci., Part A- Polymer Chemistry, 31(4), 1035-43, (1993).*

[115] Wang, Ch, A Wang, B Che, C Zhou, I Su, S Lin, B Wang. Ozone-induced graft copolymerization of polyethylene glycol monomethyl ether methacrylate onto poly(etherurethane). *Gaofenzi Xuebao, (1), 114-18, (1997).*

[116] Jian, Y., Sh. Shiraishi. Modification of the surface properties of polymers by ozone treatment. I. Modification of the surface properties of polypropylene film by ozone treatment. *Hyomen Gijutsu, 40(11), 1251-5, (1989).*

[117] Jian, Y., SH Shiraishi. Modification of the surface properties of polymers by ozone treatment. II. Low temperature nickel plating onto polypropylene with ozone treatment method. *Hyomen Gijutsu, 40(11), 1256-60, (1989).*

[118] Jian, Y, SH Shiraishi. Surface improvement of polymers by ozone treatment. III. Improvement of the coating properties of polypropylene by ozone treatment. *Hyomen Gijutsu, 41(3), 273-7, (1990).*

[119] Yamamoto, H. Manufacture of moistureproof packaging materials with sealing resin layers. *Jpn.Kokai Tokkyo Koho JP 02,198,819 [90,198,819] (Cl. B29C47/04), 07 Aug 1990, Appl. 89/16,184, 27 Jan 1989, 6pp.*

[120] Yoshito, M. Manufacture of methylmethacrylate polymers with good thermal degradation resistance. *Jpn.Kokai Tokkyo Koho JP 03,45,604 [91,145,604] (Cl. C08F8/06), 27 Feb 1991, Appl. 89/181,558, 13 Jul 1989, 2pp.*

[121] Shiraishi, SH, U Ken, K Suchiro, K Nitsuta. Ozone modification of propene polymer moldings. *Jpn.Kokai Tokkyo Koho JP 03,103,448 [91,103,448] (Cl. C08J7/12), 30 Apr 1991, Appl. 89/239/357, 14 Sep 1989, 5pp.*

[122] Marshall, RA, DK Parker, JW Hershberger. Ozonization of PVC latex for removal of vinyl chloride residues. *U.S. US 5,498,693 (Cl. 528-483; C08F6/24), 12 Mar 1996, Appl. 313,504, 26 Sep 1994, 5pp.*

[123] Iwasaki, T. Graft polymerization of vinyl monomers onto Nylon 6 fiber by ozone oxidation method. *Kobunshi Ronbunshu, 50(2), 115-20, (1993).*

[124] Hosoda, Y, K Ikishima, A Yanai. Adhesion improvement of PVC-, fluoropolymer-, or polyester- coated steel panels. *Jpn.Kokai Tokkyo Koho JP 04,197,472 [92,197,472] (Cl. B05D3/10), 17 Jul 1992, Appl. 90/331,521, 29 Nov 1990, 4pp.*

[125] Ebdon, JR, NJ Flint. Synthesis of water-soluble telehelic methyl-ketone-ended oligo-N,N- dimethylacrylamides by the ozonolysis of poly(N,N- dimethylacrylamide- stat-2,3- dimetylbutadiene)s. *J.Polym.Sci., Part A- Polymer Chemistry, 33(3), 593-97, (1995).*

[126] Rl, L, JR Ebdon, P Hodge. Synthesis of telehelic methyl-methacrylate and styrene oligomers witth fluorophenyl ketone end groups by the ozonolysis of copolymers containing 4-fluorophenyl butadiene units. *Polymer, 34(2), 406, (1993).*

[127] Eftushenko, AM, IP Chikhachova, GV timofeeva, ST Stavrova, VP zubov. Graft polymerization of acrylic acid into ozonated siloxane matrixes. *Visoko molek. soed., ser. A i B, 35(6), Â312-15, (1993).*

[128] Tashibana, SH, R Yokoyama. Manufacture of base disk for laser recording material. *Jpn.Kokai Tokkyo Koho JP 01,248,339 [89,248,339] (Cl. G11B7/26), 03 Octl 1989, Appl. 88/75,079, 28 Mar 1988, 4pp.*

[129] Rimmer, S, JR Ebdon, JM Sherpherd. Polymer-supported reagents as aids in controlled degradation of polymers containing ozonides in the main chain to give telechelic oligomers. *Reactive and Functional Polymers, 26, 145-55, (1995).*

[130] Chtourou, H, B Riedl, BV Kokta, A Adnot, S Kaliguine. Synthetic pulp fiber ozonation: an ESCA and FTIR study. *J.Appl.Polym.Sci., 49, 361-73, (1993).*

[131] Vallet, A, M Delmas, TH Fargere, G Sacher. Processing fiber-reinforced plastics and ceramics for recycling. *PCT Int.Appl. WO 96 18,484 (Cl. B29B17/02), 20 Junl 1996, FR Appl. 94/15,174, 16 Dec 1994, 16pp.*

[132] Kawabe, K, Y Hatakeyama. Heat-sealing method for laminates. *Jpn.Kokai Tokkyo Koho JP 09 01,665 [91 01,665] (Cl. B29C65/02), 07 Janl 1997, Appl. 95/172,699, 15 Jun 1995, 8pp.*

[133] Czerwiec, W, A Kwaitek. Manufacture of pour point depressants from waste polyethylene and/or polypropyllene. *Eur. Pat.Appl. EP 383,969 (Cl. C08F8/50), 29 Aug 1990, Appl. 89/103,098, 22 Feb 1989, 3pp.*

[134] Shimizu, SH, N Iwata. Surface treatment of stainless steel parts la. *Jpn.Kokai Tokkyo Koho JP 08 337,867 [96 337,867] (Cl. C23C8/18), 24 Dec 1996, Appl. 95/141,448, 08 Jun 1995, 6pp.*

[135] Lagana, V. System for passivation of metal surfaces. *EUR Pat. Appl. EP 504,621 (Cl. C23C8/12), 23 Dec 1992, IT Appl. 91/MI715, 18 Mar 1991, 7pp.*

[136] Berneron, R, P de Gelis. Increasing corrosion resistance of metals and alloys. *EUR Pat. Appl. EP 340,077 (Cl. C23C8/36), 02 Nov 1989, FR Appl. 88/5,091 18 Apr. 1988, 10pp.*

[137] Lu, HH, DJ Diguette. Effect of dissolved ozone on the corrosion behaviour of Cu-30 Ni and 304L stainless steel in 0.5N sodium chloride solutions. *Report 1989, Order No.AD-A213804, 37pp.*

[138] Havlik, T, M Skrobian. Acid oxidizing leaching of chalcopyrite with ozone. *Rudy, 37(10), 295-301, (1989).*

[139] Havlik, T, M Skrobian. Acid leaching of chalcopyrite in the presence of ozone. *Can. Metall.Q., 29(2), 133-9, (1990).*

[140] Clough, T, JW Siebert, AC Riese. Redox leaching of precious metals from manganese-containing ores. *Pat. Specif.(Aust.) AU 607,523 (Cl. C22B11/04), 07Mar 1991, US Appl. 213,884, 30 Jun 1988, 40pp.*

[141] Ye, Y, WH Jang, MR Yalamanchili, JD Miller. Molybdenite flotation from copper/molybdenum concentrates by ozone conditioning. *Trans.Soc.Min., Metall., Explor., 288, 173-9, (1990), /Pub. 1991/.*

[142] Chkoniya, TK. Manufacture of potassium permanganate. *U.S.S.R. SU 1,541,190 (Cl. C01G45/12), 07 Feb 1990, Appl. 4,246,700, 24 Mar 1987.*

[143] Saski, T, H Sasaki, K Iida. Manufacture of ITO ceramics by firing in air atmosphere containing ozone. *Jpn.Kokai Tokkyo Koho JP 08 169,769 [96 169,769] (Cl. C04B35/64), 02 Jul 1996, Appl. 94/316,464, 20 Dec 1994, 4pp.*

[144] Saski, T, H .Sasaki. Manufacture of ITO sinters. *Jpn.Kokai Tokkyo Koho JP 07 277,821 [95 277,821] (Cl. C04B35/495), 24 Oct 1995, Appl. 94/66,030, 04 Apr 1994, 4pp.*

[145] Caropreso, FE, HM Castrantas, JM Byne. Method for oxidizing carbonaceous material, and especially for bleaching gray kaolin for subsequent use as coating or filler for paper. *U.S. US 4,935,391 (Cl. 501-146; C04B20/06), 19Jun 1990, Appl. 331,423, 31 Mar 1989, 6pp.*

[146] Emel'yanova, OV, BV Emel'yanov. Manufacture of mercury (I)chloride. *U.S.S.R. SU 1,567,520 (Cl. C01G13/04), 30 May 1990, Appl. 4,342,566, 14 Dec 1987.*

[147] Ruhlandtsende, AD, AD Bacher, U Muller. Formation of arsenic acid from $(As_2Cl_8)^{2-}$ -ions and ozone crystal structure of $PPh_4Cl.H_3AsO_4$. . *Zeitschrift fur Naturforschung, Section B, 47(12), 1677-80,*

[148] Belfadhel, H, A Ratel, A Ouederni, RS Bes, JC Mora. Removal of color and organic matter in industrial phosphoric acid by ozone: effect on activated carbon treatment. *Ozone: Sci.Eng., 17(6), 637-45, (1995).*

[149] Achutharaman, VS, KM Beauchamp, N Chandrasekhar, GC Spalding, BR Johnson, AM Goldman. Fabrication of high-Tc superconductors ysing ozone-assisted molecular beam epitaxy. *Thin Solid Films, 216(1), 14-20, (1992).*

[150] Locquet, JP. Characterization of a ratio frequency plasma sourse for molecular beam epitaxial growth of high-Tc superconductor films. *J.Vacuum Sci.&Tech., A.Vacuum surfaces and films, 10(5), 3100-3103, (1992).*

[151] Berkley, DD, AM Goldman, BR Johnson, J Morton, T Wang. Techniques for the drowth of superconducting oxide thin films using pure ozone vapor. *Rev.Sci.Instrum., 60(12), 3769-74, (1989).*

[152] Usui, T, Y Osanai. Ozone oxidation in preparation of oxide superconductors. *Jpn.Kokai Tokkyo Koho JP 01 122,921 [89 122,921] (Cl. C01G3/00), 16 May 1989, Appl. 87/278,772, 04 Nov 1987, 5pp.*

[153] Odan, K, H Miura, Y Bando. Manufacture of bismuth copper alkaline earth oxide superconductors by using ozone. *Jpn.Kokai Tokkyo Koho JP 01 224,258 [89 224,258] (Cl. C04B25/00), 07 Sep 1989, Appl. 88/45,774, 01 Mar 1988, 3pp*

[154] Goldman, AM, DD Berkley, BR Johnson. Preparation of superconductive ceramic oxides using ozone. *PCT Int. Appl. WO 90 02,215 (Cl. C23C14/26), 08 Mar 1990, US Appl. 234,421, 19 Aug 1988, 29pp.*

[155] Nakamori, T. Manufacture of rare earth barium copper oxide high-temperature supercondctor ceramics. *Jpn.Kokai Tokkyo Koho JP 01 179,752 [89 179,752] (Cl. C04B35/00), 17 Jul 1989, Appl. 88/648, 07 Jan 1988, 3pp*

[156] Johnson, BR, KM Beauchamp, DD Berkley, JX Liu, T Wang, AM Goldman. Growth of co-evaporated superconducting yttrium barium copper oxide ($YBa_2Cu_3O_{7-x}$) thin films oxidized by pure ozone. *Proc. SPIE-Int.Soc.Opt.Eng., 1187(Process.Films High Tc Supercond.Electron), 27-36, (1990).*

[157] Vig, JR. UV/ozone cleaning of surfaces. *Proc. Electrochem. Soc., 90-9 (Semicond.Clean.Tachnol. 1989), 105-13, (1990).*

[158] Suemitsu, M, T Kaneko, N Nobuo. Low temperature silicon surface cleaning by fluoric acid (HF) etching/UV-ozone cleaning (HF/UVOC) method. (1).Optimization of the HF treatment. *Jpn.J.Appl.Phys., Part 1, 28(12), 2421-4, (1989).*

[159] Kaneko, T, M Suemitsu, N Miyamoto. Low temperature silicon surface cleaning by hydrofluoric acid etching/UV-ozone cleaning (HF/UVOC) method. (II).*In situ* UVOC. *Jpn.J.Appl.Phys., Part 1, 28(12), 2425-9, (1989).*

[160] Inada, A, K Kawasumi. Removal of organic contaminants from semiconductor substrates by ozone. *Jpn.Kokai Tokkyo Koho JP 01 135,023 [89 135,023] (Cl. H01L21/30), 26 May 1989, Appl. 87/292, 016 20 JNov 1987, 2pp.*

[161] Ohmi, T. Total room temperature wet cleaning of silscon surface. *Semicond. Int.,* 19, (8), 323-4, 326, 328, 330, 332, 334, 336, 338, (1996).

[162] Kamikawa, Y, K Matsumura. Removal of resist films using liquid ozone. *Jpn.Kokai Tokkyo Koho JP 02 106,040 [90 106,040] (Cl. H01L21/304), 18 Apr 1990, Appl. 88/259,204, 20 Oct 1988, 3pp.*

[163] Maeda, K. Ozone applied in VLSI (very large scale integration) manufacturing. *Kurin Tekunoroji, 4(2), 62-9, (1994).*

[164] Choi, H, K Kyoo, SD Jung, H Jeon. Effects of UV/O3 and SC-1 step in the HF last silicon wafer cleaning on the properties of gate oxide. *Han'guk Chaehyo Hakhoechi, 6(4), 395-400, (1996).*

[165] Hansen, RWC, J Welske, D Wallace, M Bissen. Cleaning of optical surfaces with photogenerated reactants. *Nucl.Instrum.Methods Phys.Res., Sect.A, 347(1-3), 249-53, (1994).*

[166] Ohmi, T, T Sagawa, M Kogure, T Imaoka. Native oxide growth and organic impurity removal on silicon surface with ozone-injected ultrapure water. *J.Electrochem.Soc., 140(3), 804-10, (1993).*

[167] Takaaki, K, Y Akimasa, M Yasuji. Reaction mechanism of chemical vapor deposition using tetraethylorthosilicate and ozone at atmospheric pressure. *Jpn.J.Appl.Phys., 31, 2925-30, part 1, No.9A (1992).*

[168] Hishamune, Y. Covering semiconductor devices with silica films. *Jpn.Kokai Tokkyo Koho JP 01 82,634 [89 82,634] (Cl. H01L21/316), 28 Mar 1989, Appl. 87/242,113, 25 Sep 1989, 6pp.*

[169] Kurihara, H, M Saito, T Kozuki. Fabrication of nondoped silicate glass film with flat structure by O3-TEOS deposition. *Jpn.Kokai Tokkyo Koho JP 09 08,026 [97 08,026] (Cl. H01L21/316), 10 Jan 1997, JP Appl. 95/118,005, 19 Apr 1995, 8pp.*

[170] Homma, T. Future trends for interlayer dielectric films and their formation technologies in ULSI multilevel interconnections. *Mater.Chem.Phys., 41(4), 234-9, (1995).*

[171] Toris, K, S Saiton, Y Ohji. Preparation of lead zirconate thin films by reactive evaporation. *Jap.J.Appl.Physics, Part 1- nRegular Papers Short Notes & Review Papers; 33(9B), 5287-90, (1994).*

[172] Haluska IA. Formation of ceramic coatings on substrates in the presence of ozone. *Fr. Demande FR 2,697,852 (Cl. C23C16/22), 13 May 1994, Appl. 91/565, 18 Jan 1991, 23pp.*

[173] Nishida, SH. Method of forming zinc oxide light-shielding film. *Ger. offen DE 3,916,983 (Cl. C23C16/40), 30 Nov 1989, JP Appl. 88/125,792, 25 May 1988, 14pp.*

[174] Hollinger, G, D Gallet, M Gendry, MP Besland, J Joseph. Evidence for a new passivating indium rich phosphate prepared by UV/O_3 oxidation of indium phosphide. *Appl.Phys.Lett., 59(13) 1617-19, (1991).*

[175] Tawada, H, M Saitoh, CH Isobe. Method of heat-treating an oxide optical crystal and a heat treatment for carrying out the same. *Eur Pat. Appl. EP 407,945 (Cl. C30B29/28), 16 Jan 1991, JP Appl. 89/178,586, 11 Jul 1989, 23pp.*

[176] Tamada, H, A Yamada, M Saitoh. Lithiumniobate thin-film optical waveguide grown by liquid phase epitaxy and its application to second harmonic generation. *J. Appl.Phys., 70(5), 2536-41, (1991).*

[177] Sadamoto, M, Y Ashida, N Fukuda. Manufacture of solar cell modules. *Jpn.Kokai Tokkyo Koho JP 08 148,707 [96 148,707] (Cl. H01L31/04), 07 Jun 1996, Appl. 94/280,407, 15 Nov 1994, 14pp.*

[178] Bianchetti, MF, HR Canepa, S Larrondo, NE Walsel de Reca. Surface activation of photoconductive PbS films. *An Asoc.Quim.Argent.,* 84,(1), 73-7, (1996).

[179] Enomoto, R, M Asai. Adhesion-producing materials for electroless plating and printed circuits and methods for fabricating them using the materials. *Ger. offen DE 3,913,966 (Cl. C09J3/00), 09 Nov 1989, JP Appl. 88/104,044, 28 Apr 1988, 15pp.*

[180] Binz. *Real Encyclopedi der gesamten Helkunder* Bd. 18, S. 189, Berlin und Wien (1898).

[181] Lambert, BK. *Nature,* 181, 1202, (1958).

[182] Shepetinovskii, Z, I Mikashinovich. Second All-Union Conference "Ozone obtaining and application". Theses of reports. p. 157, Moscow (1991).

[183] *Eur. Pat.* Appl. EP 261 032 (1988).

[184] Carpendiae, J Freeberg. *J.Clin. Gastroenterol.* 17, 2, 142 (1993).

[185] Belianin, E, I Shmelev. *Ter. Arkh.*, 66, 2, 29 (1994).

[186] Sato. *Exp. Anim.*, 39, 2, 223 (1990).

[187] Grits, A Formichev. *Microbiologiya*, 59, 5, 831 (1990).

[188] Heindl. *Zentralbl.Hyg.Unwelmed.*, 194, 5/6, 464 (1993).

[189] Ancowski. *Complement Inflammation*, 7, 2, 57 (1990).

[190] Kontorshtikova, KN, NN Andreeva. Second All-Union Conference "Ozone obtaining and application. Theses reports, p. 157, Moscow (1991).

[191] Kitanova, S., G. Koynova, S. Rakovsky, V. Kagan. Xenobiotic Metabollism and Toxicity Workshop of Balkan Countries, P35, March 10-21st 1991, Novi Sad, Yugoslavia

[192] Rakovsky, S. Pross. I-st Bulg.-Russ. Sympos. "Free Radicals and Biostabilzers" I-st Bulg. - Russ. Symp. p.113, 23-26 November 1987, Sofia.

[193] Rakovsky, S. International Conference on "Regulation of Free Radical Reactions" - (Biomedical Aspects), Abstracts, No.140, 13-16 September, 1989.

[194] Rakovsky, S., R. Fotty, V. Ershov. International Conference on "Regulation of Free Radical Reactions" - (Biomedical Aspects), Abstracts, No.141, 13-16 September, 1989.

[195] Rakovsky, S., O Todorova, R Fotty, A. Terebenina, G. Borisov. International Conference on "Regulation of Free Radical Reactions" - (Biomedical Aspects), Abstracts, No.142, 13-16 September, 1989.

[196] Haruf, M, E Serbinova, A Erin, L Ulhin, V Savov, S Rakovsky, V Kagan. International Conference on "Regulation of Free Radical Reactions" - (Biomedical Aspects), Abstracts, No.198, 13-16 September, 1989.

[197] Aliev, MA, L. Ioffe. Second All-Union Conference "Ozone obtaining and application". Theses reports, p. 161 Moscow (1991).

[198] Krivatkina, SD. Second All-Union Conference "Ozone obtaining and application". Theses reports, p. 156, Moscow (1991).

[199] Gloor, M, B Lipphardt. *Z.Hautkr.*, 51, 3, 97 (1976).

[200] Sweet, F. *Science(Washington-D.C.)*, 209 (4459), 931 (1980).

[201] Richter, A. *J. Toxicol. Environ. Health*, 25, 3, 383 (1988).

[202] Shinriki, N. *Eur. Pat.* Appl. EP 601,891, Cl A61K9/08 (1994).

[203] Burleson, GR, TM Murray. *Appl. Microbiol.*, 29, 340 (1975).

[204] Karzenelson, E, G Koerner, N Biederman. *Appl. Environ. Microbiol*, 37, 715 (1979).

[205] J.Vaughn, J, YS Chen. *Appl. Environ. Microbiol.*, 53, 9, 2218 (1987).

[206] Lee, RE. *Semin. Oncol.*, 1, 254 (1974).

[207] Rokitansky. *Hospitalis*, 52, 643 (1982).

[208] Stroczynski, Z Antoszewski. *Pol. Tyd. Lek.*, 47 (42-43), 964 (1992).

[209] Turzynski, J Stroczynski. *Pol. Tyd. Lek.*,46 (37-39), 700-710 (1991).

[210] Rilling, S. *"Ozone" Science Engineering Intern.*, 7, 4, 257 (1985).

[211] Schulzs. *Lab. Anim.*, 20, 1, 41-48 (1986).

[212] Pribuda, S. *Semana Med.*, 123, 26, 10-269 (1963).

[213] Wolff. *Das medizinische Ozon. vfm Publications*, Heilderberg (1979).

[214] Payr. *Munchner Medizinische Wochenschrift*, 82, 220 (1935).

[215] Dorstewitz, H. *Kongres-bericht der artztlichen geselltschafen fur ozontherapie*, Baden-Baden (1981).

[216] S.H.Stadlaender, SH. *Erfahrungsheilkunde*, 30, 4, 277 (1981).

[217] Rokitansky. *Hospitalis*, 52, 711 (1982).

[218] Rotter, M, H Mittermayer. *Wien Klin. Wochenschr.*, 86, 24, 776 (1974).

[219] Washuttl, J. *Erfahrungsheilkunde*, 28, 811 (1979); 32, 815 (1983).

[220] Pryor, W. *Chem. Biol. Interact.*, 79, 41 (1991).

[221] Washuttl, J, R Viebahn. *"Ozone" Science Enginering Intern.*, 11, 4, 411 (1989).

[222] Sweet, F. *Science*, 202, 931 (1980).

[223] R.Viebahn, R, I Steiner. *"Ozone" Science Engineering Intern.*, 12, 65 (1989).
[224] Brinkman, R, H Lamberts. *Nature*, 181, 1202 (1958).
[225] Chow, C, M Mustafa. *Environ. Physiol. Biochem.*, 5, 142 (1975).
[226] Brickley, R, J Hackney, K Clark. *Arch. Environ. Health*, 30, 40 (1975).
[227] Goldstein, BD. *Arch. Environ. Health*, 26, 279 (1975).
[228] Vender, R. *Toxicol. Ind. Health*, 10, 1/2, 53 (1994).
[229] Lamberts, H, TS Veninga. *Lancet*, 1, 133 (1964).
[230] Schulz, E, G Moore, E Calabrese. *Bull. Environ. Contam. Toxicol.*, 26, 273 (1981).
[231] Chow, C, AL Tappel. *Arh. Environ. Health*, 26,205 (1973).
[232] Sato, S, M Kawakami. *Am. Rev. Resp. Dis.*, 113, 809 (1976).
[233] Chow, C, J Kanenko. *Environ. Res.*, 19, 49 (1979).
[234] Chan, PC, RJ Kindya. *J.Biol.Chem.*, 252, 23, 8537 (1977).
[235] A.Savino, A, ML Peterson. *Environ. Res.*, 15, 1, 65 (1978).
[236] A.Canada, A, G Airriess. *Nutrit Res.*, 7, 797 (1987).
[237] Murphy, S, H Davis. *Toxicol. Appl. Pharmacol.*, 6, 528 (1964).
[238] Giri, S, M Hollinger. *Environ. Res.*, 21, 467 (1980).
[239] Veninga, T, J Wagenaar. *Environ. Health Perspect.*, 39, 153 (1981).
[240] Shimasaki, H, T Takatori. *Biochem. Biophys. Res. Comm.*, 68, 1256 (1976).
[241] Jankovski, S, W Doroszkiewicz. *Complement Inflammation*, 7, 2, 57 (1990).
[242] Jegier, Z. *J.Am. Ing. Hyg. Assoc.*, 137, 6, 329 (1974).
[243] M.Kleinfeld, M, P Gielrnynb. *Am. J. Med. Sci.*, 231, 638 (1956).
[244] Laperwerff, M. *Aerospace Med.*, 34, 479 (1963).
[245] Trams, E, C Lauter. *Arch. Environ. Health.*, 24, 153 (1972).
[246] Berney, WB, RS Dyer. *U.S. Environ. Prot. Agency OFF Res. Dev., EPA*, 1, PB-264,233,555 (1976).
[247] Hackney, JD. From *Gov. Rep. Announce Index(U.S.)*, 77, 14, 98 (1977).
[248] Takanashi, Y, T Miura. *J. Toxicol. Environ. Health*, 15, 855 (1985).
[249] Kitanova, S., G. Koynova, S. Rakovsky, V. Kagan. *Xenobiotic Metabollism and Toxicity Workshop of Balkan Countries*, P34, March 10-21st 1991, Novi Sad, Yugoslavia
[250] Kitanova, S, G Koynova, S Rakovsky, V. Kagan. Pross. II-nd USSR Conference "*Ozone - Obtainment and Application*". Moscow, p. 168, 30.01-1.02.1991.
[251] Zelac, R, H Crompoy. *Environ. Res.*, 4, 325 (1971); 4, 262 (1971).
[252] Vaughan, T, W Moorman. *Toxicol. Appl. Pharmacol.*, 20, 404 (1971).
[253] Iwao, U. *Environ. Res.*, 48, 1, 78 (1989).
[254] Fairchild, E, S Grahm. *Toxicol. Appl. Pharmacol.*, 6, 607 (1964).
[255] Gordon, T, B Taylor, M Amdur. *Arch. Environ. Health*, 36,6, 284 (1981).
[256] Eskerv, M, M Scheuchenzuber. *Environ. Res.*, 40, 274 (1976).
[257] Criegee, R. *Rec. Chem. Prog.*, 18, 111 (1957).
[258] Menzel, DB. *Adv. Mod. Environ. Toxicol.*, 5 , 183 (1982).
[259] C.K.Chow, CK, MZ Hussain. *Exprl. Path.*, 25, 182 (1976).
[260] J.A.Last, JA, DB Greenberg. *Toxicol. Appl. Pharm.*, 51, 247 (1979).
[261] Hussain, ZM, CE Goss. *Life Sci.*, 18, 897, (1976).
[262] Veninga, TS. *Environ. Health Perspect.*, 39, 153 (1981).
[263] Chow, CK, CE Goss. *J.Toxicol. Environ. Health*, 3, 5/6, 877 (1977).
[264] Saymor. *Arch. Environ. Health*, 29, 3, 164 (1974).

[265] Mohammad, D Anthany. *J.Chest*, 66, 1, 165, suppl. pt 21 (1974).
[266] Summer, W. *Process Biochem*., 11, 7, 26 (1976).
[267] Plopper, CG. *Toxicol. Appl. Pharmacol*., 121, 1, 124 (1994).
[268] Derbal, E Kimberly. *Environ. Health Perspect. Suppl*., 102, 61 (1994).
[269] Kelly, FJ. *Hum.Exp. Toxicol*., 13, 6, 407 (1994).
[270] Drechsler, P, M Deborah. *Exp. Geront*., 30, 1, 65 (1995).
[271] Miller, FJ, JH Oventon, RH Kaskot. *Tox. and Appl. Pharm*., 79, 11 (1985).
[272] Stock, TH, DJ Kotchmar. *J.Air Pollut. Contr. Assoc*., 35, 1266 (1985).
[273] Dillar, C, N Urribarri. *Arch. Environ. Health*, 25, 6, 426 (1972).
[274] Mohammad, D Anthony. *J.Lab. Clin. Med*., 82, 3, 357 (1973).
[275] Peister, A. *Presse Term. Clim.*, 1, 58 (1977).
[276] Godstein, B, O Belchum. *Toxicol. Appl. Pharmacol*., 27, 2, 330 (1974).
[277] Elaine, S, D Dziedzic. *Toxicol. Lett*., 51, 125 (1990).
[278] Schesinger, RB, KE Driscoll. *J. Toxicol. Environ. Health*, 20,125 (1987).
[279] Abraham, WM, W Sielezak. *Eur. J. Respir. Dis*., 68, 114 (1986).
[280] W.M.Foster, WM, DL Costa, EG Langenbak, *J. Appl. Physiol*., 63, 996 (1987).
[281] Boorman, GA, LW Schwartz, DL Dungworth. *Lab. Invest*., 43, 108 (1980).
[282] K.Nikua, K, DW Wilson, SN Giri. *Am.J.Pathol*., 131, 373 (1988).
[283] Harkema, JR, CG Plopper, DM Hyde. *Am.J.Pathol.* ,128, 29, 44 (1987).
[284] Kirkpatrik, JR, RF Henderson. *Exp. Lung. Res*. 15, 1 (1989).
[285] Schlesinger, RB. *J.Toxicol. Environ. Health*, 21, 27 (1987).
[286] Lum, H, DG Plopper. *Exp. Lung Res*., 5, 61 (1983).
[287] Michael, WF. *J. Appl. Physiol*., 75, 5, 1938 (1993).
[288] Driscoll KE, RB Schleinzenger. *Toxicol. Appl. Pharmacol*., 93, 312 (1988).
[289] Kimura, I, E Godstein. *J. Infect. Dic*., 143, 247 (1981).
[290] Sherwood, RL, E Goldstein. *Environ. Res*., 41, 378 (1986).
[291] Bleavins, MR, HJ Whith. *Toxicologist*, 7, 129 (1987).
[292] Blevins, MR, D Dziedzic. *Toxicologist*, 8, 198 (1988).
[293] Sell, S. *Immunology, Immunopathology and Immunity*. Elsevier. New York (1987).
[294] Aranyi, C, SC Vana, PT Thomas. *J.Toxicol. Environ. Health,* 12, 55 (1983).
[295] Eskew, ML, WJ Scheuchenzuber, RW Scholz. *Environ. Res.*, 40, 274 (1986).
[296] Amoruso, MA, G With. *Life Sci*., 28, 20, 2215 (1981).
[297] Nikola, KJ, DW Wilson, DL Dungworth, CG Plopper. *Toxicol. Appl. Pharmacol*., 93, 394 (1989).
[298] Clark, T. *Toxicol. Letter*, 72,1-3, 279 (1994).
[299] Stephens, RJ, MF Sloan, G Reeman. *Am. J. Pathol*., 74, 31 (1973).
[300] Evans, MJ, N.P.Dekker. *Exp. Mol. Phatol*., 42, 366 (1985).
[301] Plopper, CG, CK Chow, DL Dunworth. *Exp. Mol. Pathol*., 29, 400 (1978).
[302] Nikula, KJ, DW Wilson, DL Dungworth, CG Plopper. *Am. J. Pathol*., 131,373 (1988).
[303] Hussain, MZ, CE Cross. *Life Sci*., 18, 897, 904 (1976).
[304] Last, JA, DB Greenberg. *Toxicol. Appl. Pharmacol.*, 51, 257 (1979).
[305] Fukase O, K Hashimoto. *Nippon Eiseigaku Zassi*., 37, 694 (1982).
[306] Kramer F. *Erfahrungsheilkunde,* 24, 120 (1975).
[307] Kramer F. *Ozonachrichten*, 2, 65 (1983).
[308] Turk, R. *Erfahrungsheilkunde*, 25, 177 (1976).
[309] Salzer, N, M Metka, M Shonbauer. *Erhfahrungsheilkunde*, 32, 9731 (1983).

[310] Werkmeister H. *Erfahrungsheilkunde*, 25, 180 (1976)
[311] Geabelin K. *Seifen, Oile, Fette, Wasche,* 112, 1, 1718 (1986).
[312] Eur. Pat. 222 309 !1987).
[313] Ger.Offen 3,806,203 Cl A 61 L9/01.
[314] US Pat. 5,266,275 Cl 422-116 A 61 L2/20.
[315] German Pat. 66 489, Cl A611 (1969).
[316] Gal G . *Intern. J. of Artifical Organs*, 15, 8,461 (1992).
[317] Hiromichi. JP 05329, 196(93,329,196) Cl A 61 L 2/18 (1993).

In: Progress in Monomers, Oligomers, Polymers, Composites … ISBN 978-1-60456-878-3
Editors: R. A. Pethrick, G.E. Zaikov et al.

Chapter 23

ELECTROSPUN BIODEGRADABLE AND BIOCOMPATIBLE NATURAL NANOFIBERS: *A DETAILED REVIEW*

A.K. Haghi**, *V. Mottaghitalab and M. Akbari
University of Guilan, Rasht, Iran

ABSTRACT

In this chapter, a detailed review on processing of biodegradable and biocompatible natural nanofibers is presented.

ABBREVIATIONS

$R =$	radius of the jet
$R' =$	slope of the jet surface
$Q =$	constant volume flow rate
$I =$	constant total current in the jet
$\sigma =$	surface charge density
$\upsilon =$	velocity
$E =$	z component of the electric field
$K =$	conductivity of the liquid
$\gamma =$	surface tension
$E_t =$	tangential components of the electric field at the surface
$E_n =$	normal components of the electric field at the surface
$g =$	gravity

* Correspondence to: University of Guilan, P.O.BOX 3756, Rasht, Iran, E-mail address: Haghi@guilan.ac.ir

$\tau_{zz} =$ axial viscous normal stress
$P =$ pressure
$t_n^e =$ normal traction on the surface of the jet due to electricity
$t_t^e =$ tangential traction on the surface of the jet due to electricity.
$\varepsilon =$ dielectric constants of the jet
$\bar{\varepsilon} =$ dielectric constants of the ambient air
$\eta =$ viscosity

1. Introduction

An emerging technology of manufacturing of thin natural fibers is based on the principle of electrospinning process. In conventional fiber spinning, the mechanical force is applied to the end of a jet. Whereas in the electrospinnig process the electric body force act on element of charged fluid. Electrospinning has emerged as a specialized processing technique for the formation of sub-micron fibers (typically between 100 nm and 1 μm in diameter), with high specific surface areas. Due to their high specific surface area, high porosity, and small pore size, the unique fibers have been suggested for wide range of applications.

Electrospinning of natural fibers offers unique capabilities for producing novel natural nanofibers and fabrics with controllable pore structure. Current research effort has focused in understanding the electrospinning of natural fibers in which the influence of different governing parameters are discussed.

Electrospinning [1–10] is an economical and simple method used in the preparation of polymer fibers. The fibers prepared via this method typically have diameters much smaller than is possible to attain using standard mechanical fiber-spinning technologies [11]. Electrospinning of polymer solutions has gain much attention in the last few years as a cheap and straightforward method to produce nanofibers [12-16]. Electrospinning differs from the traditional wet/dry fiber spinning in a number of ways, of which the most striking differences are the origin of the pulling force and the final fiber diameters. The mechanical pulling forces in the traditional industrial fiber spinning processes lead to fibers in the micrometer range and are contrasted in electrospinning by electrical pulling forces that enable the production of nanofibers [14]. Depending on the solution properties, the throughput of single-jet electrospinning systems ranges around 10 ml/min. This low fluid throughput may limit the industrial use of electrospinning. A stable cone-jet mode followed by the onset of the characteristic bending instability, which eventually leads to great reduction in the jet diameter, necessitate the low flow rate [6,17] When the diameters of polymer fiber materials are shrunk from micrometers (e.g. 10–100 mm) to submicrons or nanometers, there appear several amazing characteristics such as very large surface area to volume ratio (this ratio for a nanofiber can be as large as 103 times of that of a microfiber), flexibility in surface functionalities, and superior mechanical performance (e.g. stiffness and tensile strength) compared with any other known form of the material. These outstanding properties make the polymer nanofibers to be optimal candidates for many important applications[18]. These include filter media [19-21], composite materials [18, 22], biomedical applications (tissue engineering scaffolds [23-27], bandages [28], drug release systems [29-30]), protective clothing for the military [31-34], optoelectronic devices and semi-conductive materials [35-38], biosensor/chemosensor [39-42].

1.1. Electrospinning Setup

A schematic diagram to interpret electrospinning of polymer nanofibers is shown in Fig1. There are basically three components to fulfill the process: a high voltage supplier, a capillary tube with a pipette or needle of small diameter, and a metal collecting screen. In the electrospinning process a high voltage is used to create an electrically charged jet of polymer solution or melt out of the pipette. Before reaching the collecting screen, the solution jet evaporates or solidifies, and is collected as an interconnected web of small fibers [43-45]. One electrode is placed into the spinning solution/melt or needle and the other attached to the collector. In most cases, the collector is simply grounded. The electric field is subjected to the end of the capillary tube that contains the solution fluid held by its surface tension. This induces a charge on the surface of the liquid. Mutual charge repulsion and the contraction of the surface charges to the counter electrode cause a force directly opposite to the surface tension [46]. As the intensity of the electric field is increased, the hemispherical surface of the fluid at the tip of the capillary tube elongates to form a conical shape known as the Taylor cone [47]. Further increasing the electric field, a critical value is attained with which the repulsive electrostatic force overcomes the surface tension and the charged jet of the fluid is ejected from the tip of the Taylor cone [18, 48]. The jet exhibits bending instabilities due to repulsive forces between the charges carried with the jet. The jet extends through spiraling loops, as the loops increase in diameter the jet grows longer and thinner until it solidifies or collects on the target [49].

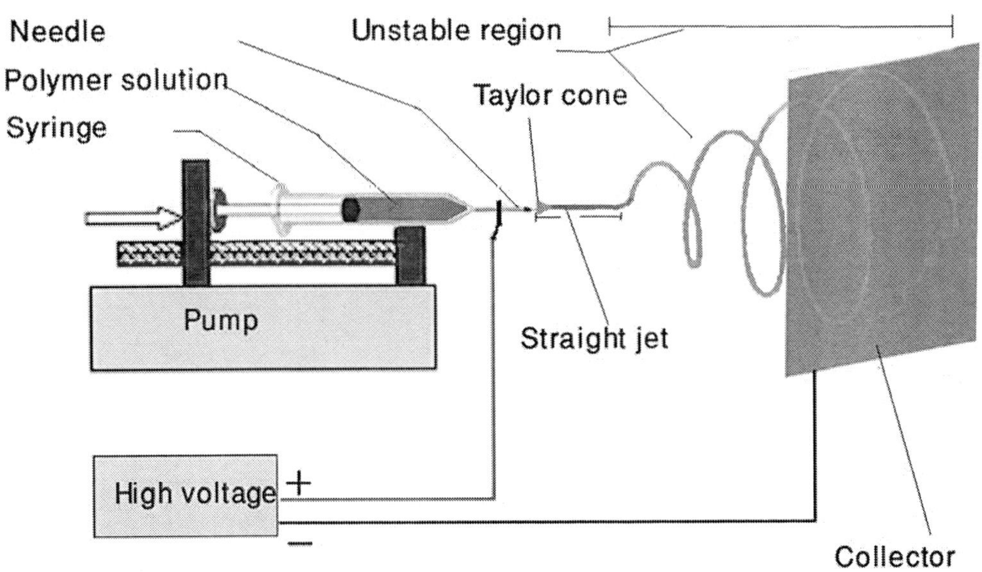

Figure 1.Shematic of electrospinning set up.

2. Effect of Systematic Parameters on Electrospun Nanofibers

It has been found that morphology such as fiber diameter and its uniformity of the electrospun polymer fibers are dependent on many processing parameters [50]. These parameters can be divided into three groups: a) solution properties, b) processing conditions, c) ambient conditions. Each of the parameters has been found to affect the morphology of the electrospun fibers.

2.1. Solution Properties

Solution parameters such as viscosity of solution, polymer concentration, molecular weight of polymer, electrical conductivity, elasticity and surface tension, those are which attribute to polymer and its solution characteristics has important effect on morphology [18].

2.1.1. Viscosity

The viscosity range of a different polymer solution which is spinnable is different. One of the most significant parameters influencing the fiber diameter is the solution viscosity. A higher viscosity results in a large fiber diameter. Figure 2 shows the representative images of beads and beaded fibers for solutions with a range of viscosity. Beads and beaded fibers are less likely to be formed for the more viscous solutions. The diameter of the beads become bigger and the average distance between beads on the fibers longer as the viscosity increases. Meanwhile, the shape of the beads gradually changes from spherical to spindle-like [51].

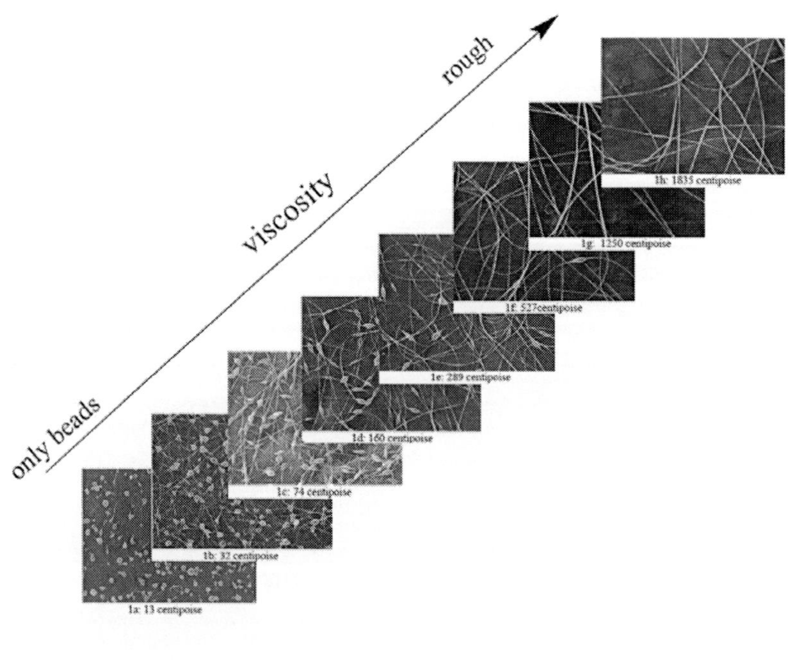

Figure 2. The morphology of beaded fibers versus solution viscosity [51].

2.1.2. Solution Concentration

As the solution concentration increase, a mixture of beads and fibers is obtained [52]. Higher solution concentration may result in fewer beads. The shape of the beads changed from spherical to spindle-like when the solution concentration varied from low to high levels [18,53]. The fiber distribution became gradually broader with increasing concentration, which is in consistency with the results obtained by Ryu et al. and Kidoaki et al. [54]. The fiber diameter increased with increasing solution concentration because the higher viscosity resisted the extension of the jet [50,55]. However, it is impossible to electrospin if the solution concentration or the corresponding viscosity is too high due to the difficulty in liquid jet formation [56-58].

2.1.3. Molecular Weight

Molecular weight also has a significant effect on the rheological and electrical properties such as viscosity, surface tension, conductivity and dielectric strength. It has been reported that too low molecular weight polymer tend to form beads rather than fibers and high molecular weight polymers give fibers with larger average diameter [59].

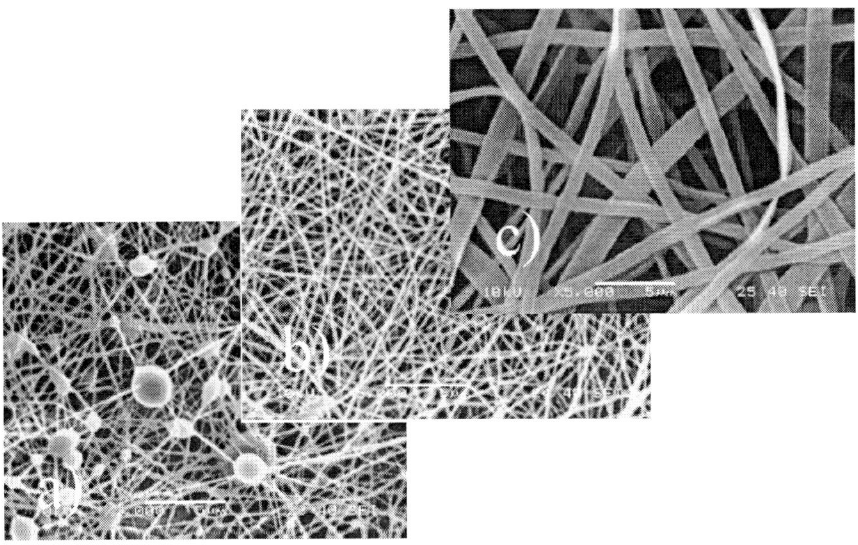

Figure 5. Photographs showing the typical structure in the electrospun polymer for various molecular weights. The samples were obtained at the center of the deposition area. (a) 9000–10,000 g/mol; (b) 13,000–23,000 g/mol; and (c) 31,000–50,000 g/mol (solution concentration: 25 wt.%) [58].

2.1.4. Surface Tension

The surface tension seems more likely to be a function of solvent compositions, but is negligibly dependent on the solution concentration. Different solvents may contribute different surface tensions. However, not necessarily a lower surface tension of a solvent will always be more suitable for electrospinning. Generally, surface tension determines the upper and lower boundaries of electrospinning window if all other variables are held constant. The formation of droplets, bead and fibers can be driven by the surface tension of solution and lower surface tension of the spinning solution helps electrospinning to occur at lower electric field [57].

2.1.5. Number of Entanglements

The above variables aren't independent of each other. Solution viscosity is a function of both concentration and polymer molecular weight. Also these two parameters affect on number of entanglements [52].

2.1.6. Solution Conductivity

Jun et al.[56] have shown that increasing solution conductivity by addition of a salt can significantly aid fiber formation (Figure 7). Addition of a salt, it has positive effect on the electrospinng number (Vq/ γR^2). Specifically, the electrical energy (Vq) increases. On the other hand, the change in the surface free energy due to change in surface tension is not expected significant. As the charges carried by the jet increased, higher elongation forces were imposed to the jet under the electrical field, resulting in smaller bead and thinner fiber diameters.

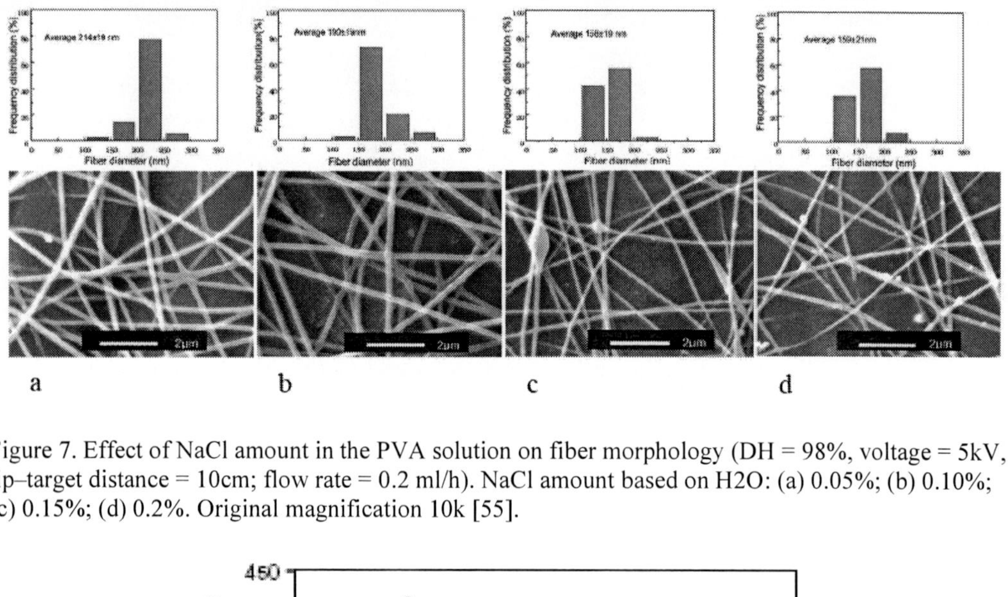

a b c d

Figure 7. Effect of NaCl amount in the PVA solution on fiber morphology (DH = 98%, voltage = 5kV, tip–target distance = 10cm; flow rate = 0.2 ml/h). NaCl amount based on H2O: (a) 0.05%; (b) 0.10%; (c) 0.15%; (d) 0.2%. Original magnification 10k [55].

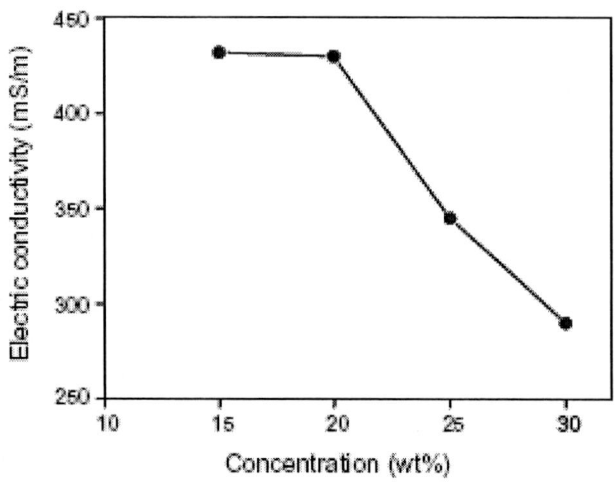

Figure 8. Electric conductivity as a function of polymer solution concentration [60].

2.1.7. Solution Conductivity

A comparison of the diameter of the electrospun fibers with the electrical conductivity of the solutions was shown in Figure 9. There was a significant drop in the diameter of the electrospun polymer fibers when the electrical conductivity of the solution increased. Beads were also observed due to low conductivity of the solution, which results in insufficient elongation of a jet by electrical force to produce uniform fiber.

Electrospun polymer nanofibers with the smallest fiber diameter were obtained with the highest electrical conductivity. This interprets that the drop in the size of the fibers was due to the increased electrical conductivity [50].

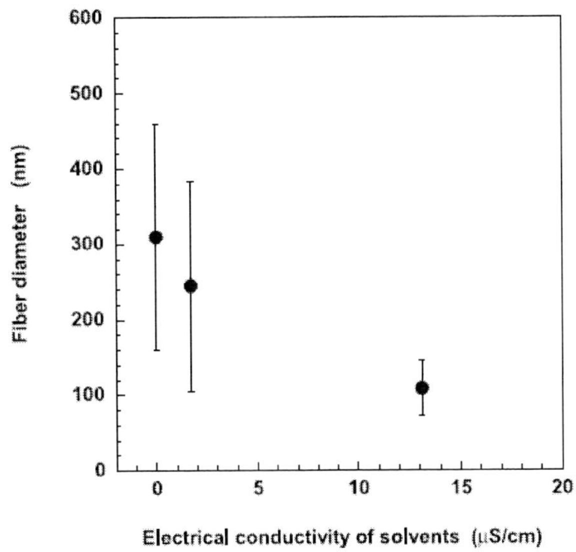

Figure 3. Solution conductivity effects on the diameter of the electrospun P(LLA-CL) (70/30 wt%) fibers[50].

2.1.8. Effect of Salt Addition

Figure 10 shows the SEM images of PAA nanofibers fabricated by electrospinning into solutions with different ionic strengths. Magnified PAA nanofibers are shown also. At the concentration of 1 M NaCl, no PAA nanofibers were synthesized. Bead structures were observed from all prepared PAA nanofibers. With increasing the concentration of NaCl, the relative viscosity slightly decreased. Variations in viscosity may be caused by the chain conformation change of PAA in solution upon adding NaCl. With increasing the ionic strength of the solution, the chain conformation of PAA changes from the extended linear conformation to the coil conformation because PAA is a polyelectrolyte. Although relative viscosity decreases slightly with increasing the amount of NaCl [61].

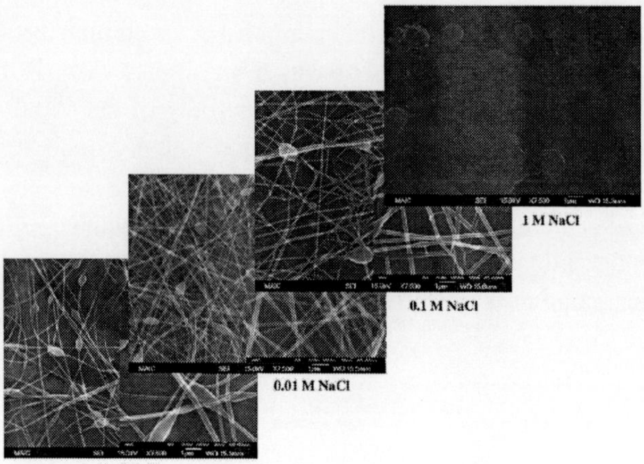

Figure 10. The SEM images of PAA nanofibers with different concentrations of NaCl [61].

The effect of NaCl addition on the morphology of electrospun PVA fibers was shown in Figure 11. Even a little sodium chloride added in the solution made its conductivity increase sharply. Net charge density carried by the jet in the electrospinning process can be affected by the conductivity of the solution. PVA fiber diameters were gradually decreased from 214 ± 19nm to 159 ± 21nm with increasing content of NaCl from 0.05% to 0.2%, for the higher net charge density increased the electrical force exerted on the jet and led to decreased fiber diameter. When 0.2% NaCl was added in PVA solution, many small particles, perhaps sodium chloride crystals were observed on the SEM micrograph [Figure 11(d)].

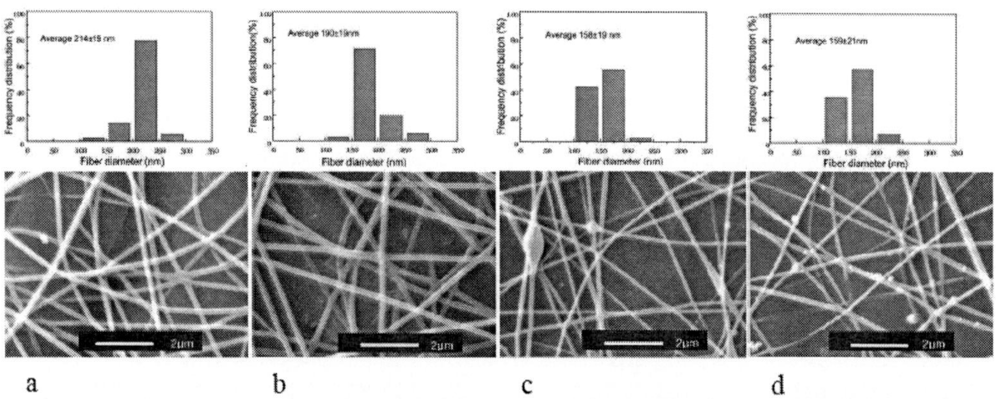

Figure 11. Effect of NaCl amount in the PVA solution on fiber morphology (DH = 98%, voltage = 5kV, tip–target distance = 10cm; flow rate = 0.2 ml/h). NaCl amount based on H2O: (a) 0.05%; (b) 0.10%; (c) 0.15%; (d) 0.2%. Original magnification 10k.[56].

2.1.9. Solvent

The effects of solvents and their properties on electro-spinnability of the as-prepared polystyrene (PS) solutions and the morphological appearance of the as-spun PS fibers were investigated qualitatively by means of a scanning electron microscope (SEM) in [62]. The eighteen solvents used were benzene,t-butylacetate, carbontetrachloride, chlorobenzene, chloroform, cyclohexane, decahydronaphthalene (decalin), 1, 2-dichloroethane, dimethylformamide (DMF), 1,4-dioxane, ethylacetate, ethylbenzene, hexane, methylethylketone (MEK), nitrobenzene, tetrahydrofuran (THF), 1,2, 3,4-tetrahydronaphthalene (tetralin), and toluene. DMF was the best solvent to dissolve PS pellets within six hours, while others were found to dissolve PS pellets within one to three days, with the exception of hexane which was not able to dissolve PS pellets after seven days. Only the PS solutions in 1,2-dichloroethane, DMF, ethylacetate, MEK, and THF could produce fibers with high enough productivity, while the PS solutions in benzene, cyclohexane, decalin, ethylbenzene, nitrobenzene, and tetralin were not spinnable [62].

2.2. Processing Condition

2.1.1. Applied Voltage

In the case of electrospinning, the electric current due to the ionic conduction of charge in the polymer solution is usually assumed small enough to be negligible. The only mechanism of charge transport is the flow of polymer from the tip to the target. Thus, an increase in the electrospinning current generally reflects an increase in the mass flow rate from the capillary tip to the grounded target when all other variables (conductivity, dielectric constant, and flow rate of solution to the capillary tip) are held constant [16].

With the increase of the electrical potential the resulting nanofibers became rougher. It was already reported that a diameter of electrospun fibers was not significantly affected by an applied voltage This voltage effects was particularly diminished when the polymer concentration was low. According to past works, higher voltage was reported to induce not only larger diameter but also smaller diameter. Applied voltage may affect some factors such as mass of polymer fed out from a tip of needle, elongation level of a jet by an electrical force, morphology of a jet (a single or multiple jets), etc. A balance among these factors may determine a final diameter of electrospun fibers. It is also noted that beaded fibers have been found to be electrospun with too high level of applied voltage. Although voltage effects show different tendencies, the voltage did not show a significant role in controlling the fiber morphology.[59]

A series of experiments were carried out when the applied voltage was varied from 5 to 13kV and the tip to target distance was held at 15cm. Results were shown in Figure 12. There was a slightly increase in average fiber diameter with increasing applied electric field. A considerable amount of thin fibers with diameters below 150nm were found when the applied voltage is above 10kV. A narrow distribution of fiber diameters was observed at a lower voltage of 5kV, while broad distribution in the fiber diameter was obtained at higher applied voltages of 10–13kV. Increasing the applied voltage, i.e., increasing the electric field strength will increase the electrostatic repulsive force on the fluid jet which favors the thinner fiber formation. On the other hand, the solution will be removed from the capillary tip more quickly as jet is ejected from Taylor cone. This results in the increase of the fiber diameter. Corona discharge was observed at voltages above 13kV, making electrospinning impossible [56]. Also an increase in voltage from 9.2 kV to 25 kV did not induce a significant change in the mean diameter of the fibers [54]. The diameter of electrospun PAN fibers did not change significantly over the range of applied voltage for the various solution concentrations in the experimental region [56].

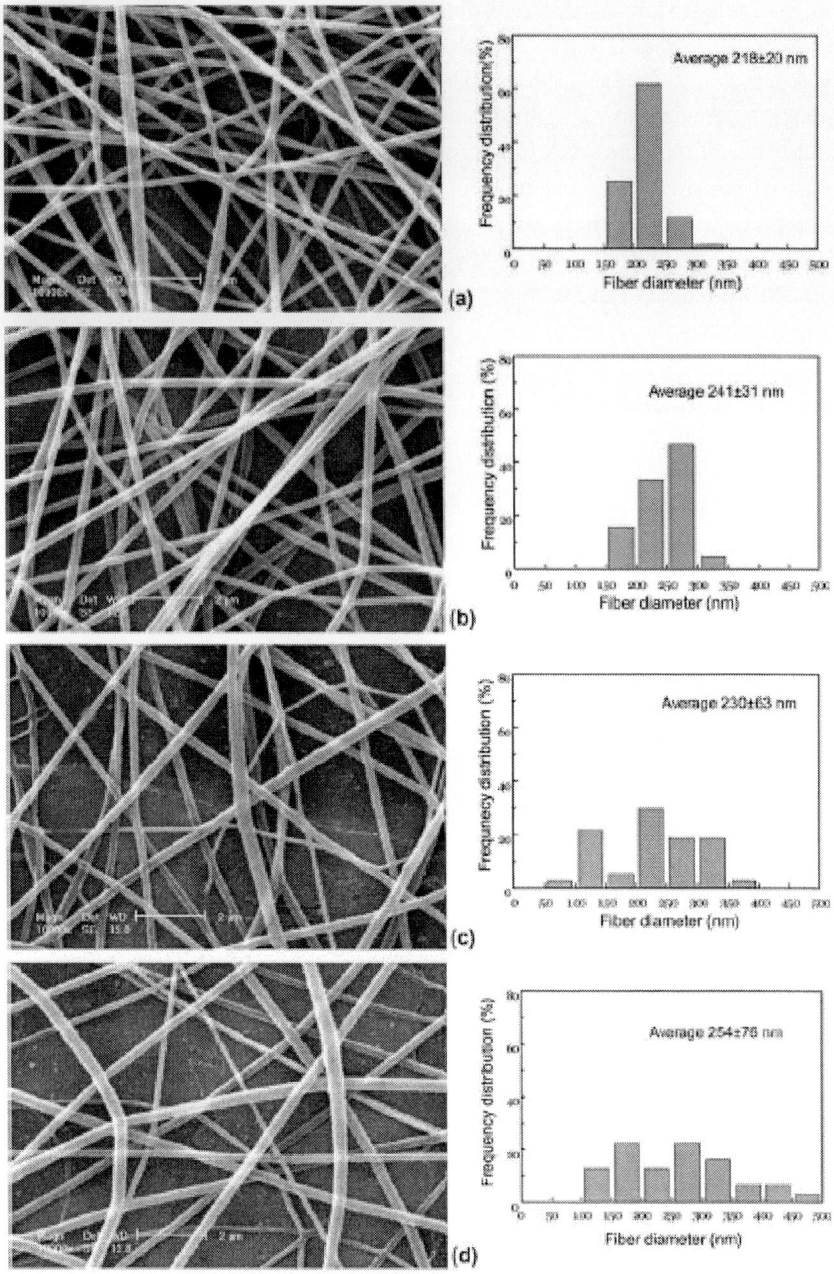

Figure 12. Effect of voltage on morphology and fiber diameter distribution from a 7.4% PVA/water solution (DH = 98%, tip–target distance = 15cm, flow rate = 0.2 ml/h). Voltage: (a) 5kV; (b) 8kV; (c)10kV; (d) 13kV. Original magnification 10kx [56]·.

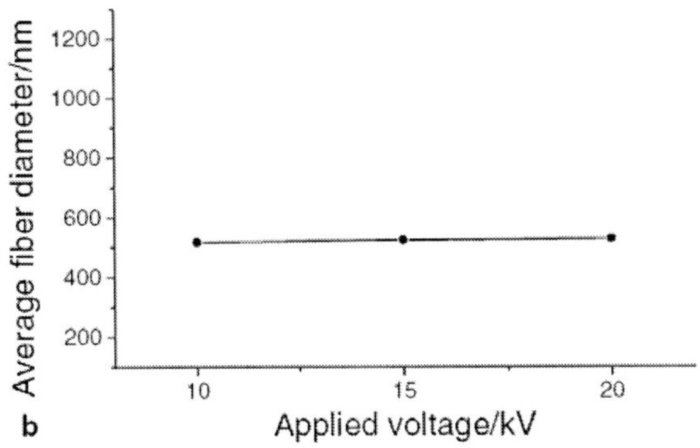

Figure 13. Main effect plots of applied voltage (b) on average fiber diameter [55].

SEM micrographs of nanofibers electrospun at different electrospinning voltage from a constant polymer concentration of 5wt% are shown in Figure 14a. A fiber diameter tended to decrease with increasing electrospinning voltage, although the influence was not as great as that of polymer concentration. Figure 14b shows fiber diameter as a function of electrospinning voltage. The increase of the electrospinning voltage causes increase of the electrostatic stress on the jet, which may be analogous to increase of the draw rate in conventional fiber spinning [64].

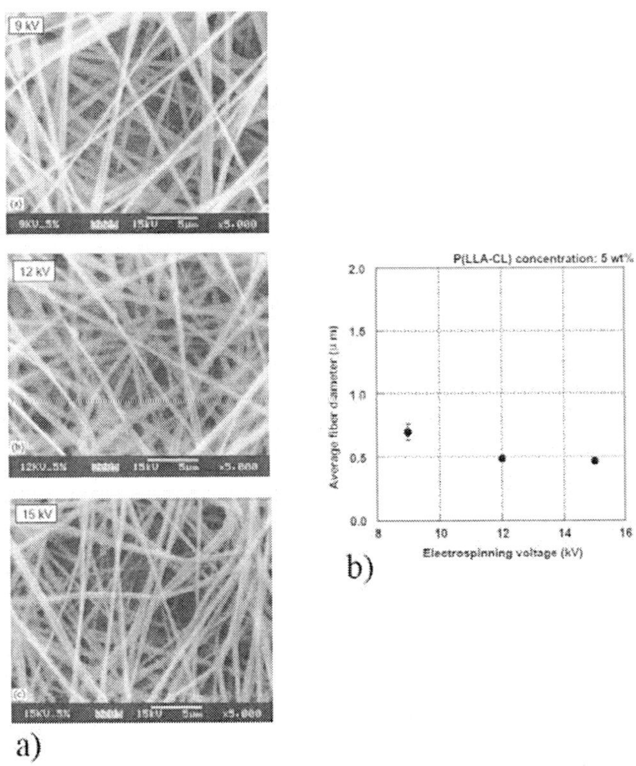

Figure 14.a) SEM micrographs of P(LLA-CL) fibers electrospun from a polymer concentration of 5wt% at different applied voltage: (a) 9 kV; (b) 12 kV and (c) 15 kV.b) Relation between fiber diameter and applied voltage in the electrospinning with 5wt% P(LLA-CL) solution [64].

2.2.2. Feed Rate

The morphological structure can be slightly changed by changing the solution flow rate as shown in Figure 15. At the flow rate of 0.3 ml/h, a few of big beads were observed on the fibers. The flow rate could affect electrospinning process. When the flow rate exceeded a critical value, the delivery rate of the solution jet to the capillary tip exceeded the rate at which the solution was removed from the tip by the electric forces. This shift in the mass-balance resulted in sustained but unstable jet and fibers with big beads were formed [56].

Figure 16 shows that the diameter of the electrospun HM-PLLA fibers was not significantly changed with varied volume feed rate. The influence due to the volume feed rate also diminished when the polymer concentration is low.

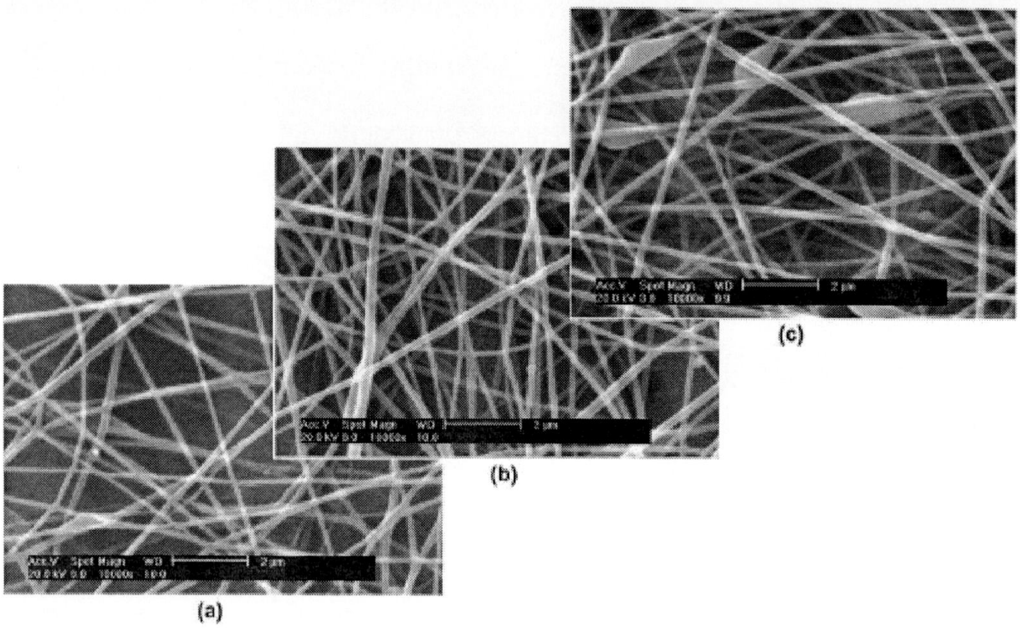

Figure 15. Effect of flow rate of 7% PVA water solution on fiber morphology (DH = 98%, voltage = 8kV, tip–target distance = 15cm).Flow rate: (a) 0.1 ml/h; (b) 0.2 ml/h; (c) 0.3 ml/h. Original magnification 10k.[56].

The solution's electrical conductivity, were found as dominant parameters to control the morphology of electrospun polymer fibers [9]. In the case of low-molecular- weight liquid, when a high electrical force is applied, formation of droplets can occur. A theory proposed by Rayleigh explained this phenomenon. As evaporation of a droplet takes places, the droplet decreases in size. Therefore the charge density of its surface is increased. This increase in charge density due to Coulomb repulsion overcomes the surface tension of the droplet and causes the droplet to split into smaller droplets. However, in the case of a polymer solution (high molecular weight liquid) the emerging jet does not break up into droplets, but is stabilized and forms a string of beads connected by a fiber. As the concentration is increased, a string of connected beads is seen, and with further increase there is reduced bead formation until only smooth fibers are formed. And sometimes spindle-like beads can form due to the extension causing by the electrostatic stress. The changing of fiber morphology can probably

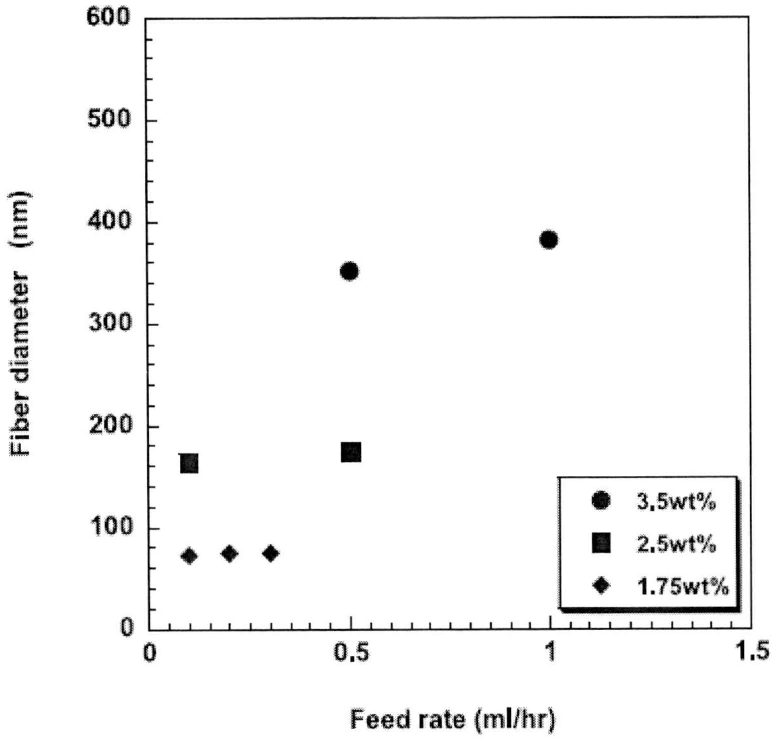

Figure 16. Volume feed rate effects on the diameter of the PLLA (Mw: 300 K) fibers electrospun from solutions with different polymer concentration [50].

be attributed to a competition between surface tension and viscosity. As concentration was increased, the viscosity of the polymer solution increased. The surface tension attempted to reduce surface area per unit mass, thereby caused the formation of beads/spheres. Viscoelastic forces resisted the formation of beads and allowed for the formation of smooth fibers. Therefore formation of beads at lower polymer solution concentration (low viscosity) occurred where surface tension had a greater influence than the viscoelastic force. However, bead formation was reduced and finally eliminated at higher polymer solution concentration, where viscoelastic forces had a greater influence in comparison with surface tension. But when the concentration was too high, high viscosity and rapid evaporation of solvent made the extension of jet more difficult, thicker and ununiform fibers were formed [55].

2.2.3. Distance of Needle Tip to Collector

Tip–target distance had no significant effect on the electrospun fiber morphology of fully hydrolyzed PVA, as shown in Figure 17. The micrographs were undistinguishable for electrospinning at 8–15cm of tip–target distance. It was assumed that solution jets were elongated and solidified quickly after they flowed out of the needle tip because of the high conductivity of fully hydrolyzed PVA used [56].

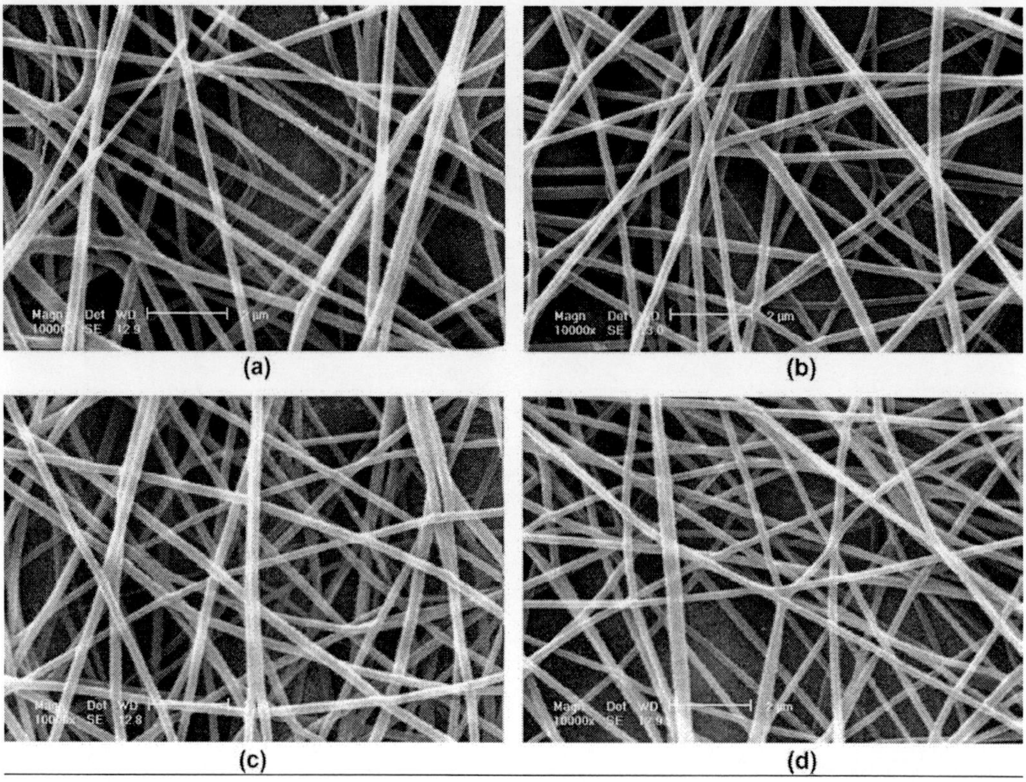

Figure 17. Effect of tip–target distance on fiber morphology from a 7.4% PVA/water solution (DH = 98%, voltage = 5kV, flow rate = 0.2 ml/h). Tip–target distance: (a) 8cm; (b) 10cm; (c) 12cm; (d) 15cm. Original magnification 10k· [56].

Based on the processing parameter studies, all the parameters effects on the morphology of the electrospun nanofibers were summarized in a processing map (Figure 18). Suitable level of processing parameters must be optimized to electrospin polymers into nanofibers with desired morphology and the parameters levels are dependent on properties of polymers and solvents used in each of electrospinning process. Understanding of the concept how each of processing parameter affect the morphology of the electrospun nanofibers is essential. All the parameters have been divided into two groups; i.e. one with parameters which affect the mass of polymer fed out from a tip of needle, and the other with parameters which affect an electrical force during electrospinning. Polymer concentration, applied voltage and volume feed rate were considered to affect the mass of polymer. Polymer concentration and feed rate directly reflect to the mass of polymer. Increased polymer concentration and feed rate tend to bring more mass of polymer into the polymer jet. It is noteworthy that the minimum polymer concentration to electrospin uniform fibers was determined by the molecular weight of polymer. High molecular weight of polymer provides enough level of solution viscosity to produce a uniform jet during electrospinning even when polymer concentration is relatively low. Applied voltage reflects to force to pull a solution out from the needle hence higher applied voltage causes more solution coming out. On the other hand, it was considered that solution electrical conductivity and applied voltage affect a charge density thus an electrical force, which acts to elongate a jet during electrospinning. Hence, higher solution electrical conductivity and applied voltage increase the jet elongation. Therefore, it is summarized that electrospun fibers with smaller diameter can be produced with lower polymer concentration, feed rate and applied voltage when the effects of mass of polymer dominates to determine the final diameter of electrospun fibers, while smaller diameter of fibers can be electrospun with higher solution electrical conductivity and applied voltage when the effects of the jet elongation is dominant. For both cases, non-uniformed/ beaded fibers were found if the parameters were either too high or too low. In fact, applied voltage affects both the polymer mass and the jet elongation, however, the effects is not as dominant as the other parameters for controlling the morphology of electrospun fibers. It must be noted that polymer concentration, molecular weight and solution electrical conductivity play a primal role in determining the morphology of electrospun fibers. Polymer fibers with smaller diameter can be electrospun with higher electrical conductivity of solution and lower polymer concentration which can be further decreased by higher molecular weight of polymer [50].

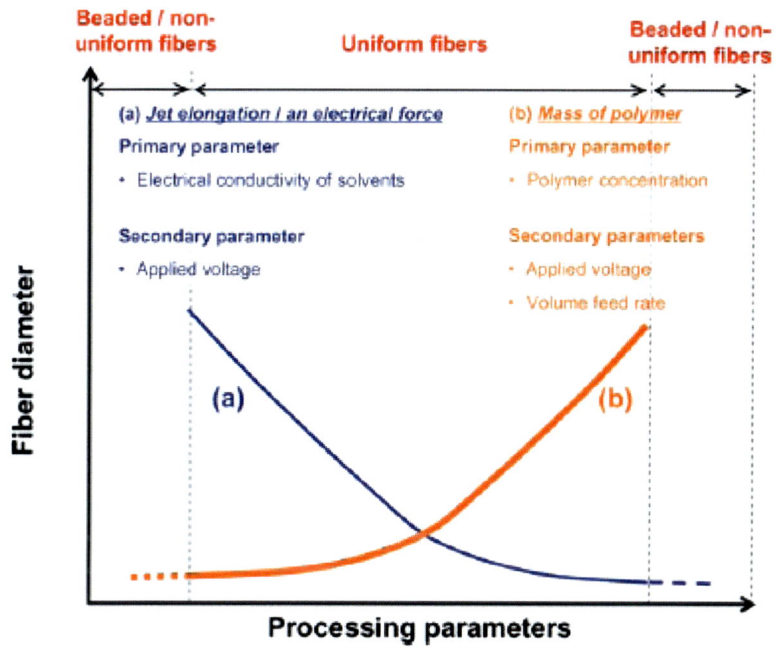

Figure 18. Processing map obtained based on the systematic parameter study: (a) jet elongation/an electrical force (affected by electrical conductivity of solvents, applied voltage), (b) mass of polymer (affected by polymer concentration, applied voltage, volume feed rate) [50].

3. Theory and Modeling

Though easily realizable in the laboratory, electrospinning is a complex phenomenon to analyze because of the coupling between the electric field and the deformation of the fluid, the latter in turn determined by the rheology of the material. Typically, electrospinning has two stages. In the first, the polymer jet issues from a nozzle and thins steadily and smoothly downstream. In the second stage, the thin thread becomes unstable to a non-axisymmetric instability and spirals violently in large loops. The enormously increased contour length produces a very large stretch ratio and a nanoscale diameter.[9] For the steady stretching in stage one, Spivak and Dzenis[65] published a simple model that assumes the electric field to be uniform and constant, unaffected by the charges carried by the jet. .(Reneker et al. modeled the viscoelasticity of the jet by a linear Maxwell equation [9]. Hohman *et al.*[67,68] developed a slender-body theory for electrospinning that couples jet stretching, charge transport, and the electric field. The model encounters difficulties, however, with the boundary condition at the nozzle.

For stage two, the bending instability has been carefully documented by two groups (Reneker *et al.*[69,70] Shin *et al.*[71]); each has proposed a theory for the instability. Reneker *et al.* modeled the polymer jet by a linear Maxwell equation. Like-charge repulsion generates a bending force that destabilizes the jet. Hohman *et al.*[67] built an electrohydrodynamic instability theory, and predicted that under favorable conditions, a nonaxisymmetric instability prevails over the familiar Rayleigh instability and a varicose instability due to electric charges. In theoretical work to date, the rheology of the polymer jet has been represented by a Newtonian viscosity [67,68] a power-law viscosity [66] and the linear Maxwell equation.[9,69,71].

The jet is governed by four steady-state equations representing the conservation of mass and electric charges, the linear momentum balance, and Coulomb's law for the **E** field [9]. Mass conservation requires that

$$\pi R^2 \upsilon = Q \tag{1}$$

where Q is a constant volume flow rate. Charge conservation may be expressed by

$$\pi R^2 KE + 2\pi R \upsilon\sigma = I \tag{2}$$

where E is the z component of the electric field, K is the conductivity of the liquid, and I is the constant total current in the jet. The momentum equation is formulated by considering the forces on a short segment of the jet (Figure 19):

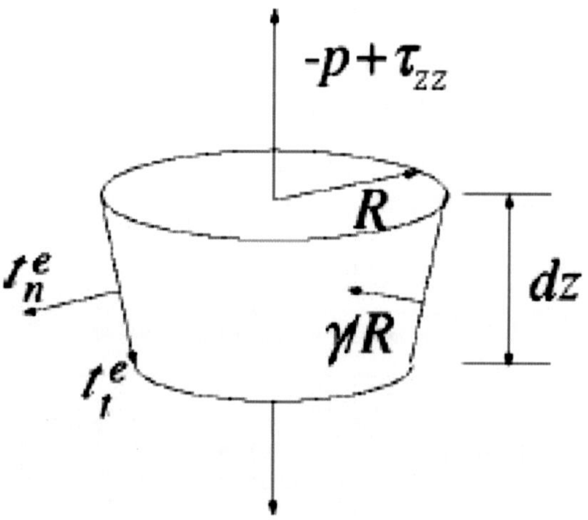

Figure 19. Momentum balance on a short section of the jet.[9]

$$\frac{d}{dz}(\pi R^2 \rho \upsilon^2) = \pi R^2 \rho g + \frac{d}{dz}[\pi R(-P+\tau_{ZZ})] + \frac{\gamma}{R}.2\pi RR' + 2\pi R(t_t^e - t_n^e R'), \tag{3}$$

Where τ_{zz} is the axial viscous normal stress, p is the pressure, γ is the surface tension, and t_t^e and t_n^e are the tangential and normal tractions on the surface of the jet due to electricity. The prime indicates derivative with respect to z, and R' is the slope of the jet surface. The ambient pressure has been set to zero. The electrostatic tractions are determined by the surface charge density and the electric field:

$$t_n^e = \left\| \frac{\varepsilon}{2}(E_n^2 - E_t^2) \right\| \approx \frac{\sigma^2}{2\bar{\varepsilon}} - \frac{\varepsilon' - \varepsilon}{2} E^2, \tag{4}$$

$$t_t^e = \sigma E_t \approx \sigma E, \tag{5}$$

where ε and $\bar{\varepsilon}$ are the dielectric constants of the jet and the ambient air, respectively, E_n and E_t are the normal and tangential components of the electric field at the surface, and $\| * \|$ indicates the jump of a quantity across the surface of the jet. We have used the jump conditions for E_n and E_t : $\| \varepsilon E_n \| = \bar{\varepsilon}\bar{E} - \varepsilon E_n = \sigma$, $\| E_t \| = \bar{E}_t - E_t = 0$, and assumed that εE_n<<(see Ganan-Calvo [13])and $E_t \approx E$. The overbar indicates quantities in the surrounding air. The pressure $p(z)$ is determined by the radial momentum balance, and applying the normal force balance at the jet surface leads to

$$-p + \tau_{rr} = t_n^e - \frac{\gamma}{R}, \tag{6}$$

Inserting Eqs. (4)-(6) into Eq. (3) yields:

$$\rho \upsilon \upsilon' = \rho g + \frac{3}{R^2}\frac{d}{dz}(\eta R^2 \upsilon') + \frac{\gamma R'}{R^2} + \frac{\sigma\sigma'}{\bar{\varepsilon}} + (\varepsilon - \bar{\varepsilon})EE' + \frac{2\sigma E}{R}, \tag{7}$$

4. Natural Fibers

The fibers in modern textile manufacture can be classified into two groups (a) natural (b) man-made fibers. A vast classification of fibers illustrated in figure 20. Natural fibers are those are provided by nature in a ready made fibrous form. Natural fibers can be subdivided into three main classes, according to the nature of their source.

a) vegetable fibers
b) animal fibers
c) mineral fibers

Vegetable fibers include the most important of all textile fibers cotton – together with flax, hemp, jute and other fibers which have produced by plants. They are base on cellulose, the material used by nature as a structural material in the plant world.

Animal fibers include wool and other hair-like fibers, and fibers such as silk, produced as filaments by cocoon-spinning creatures. These animal fibers are based on proteins, the complex substances from which much of the animal body is made [72].

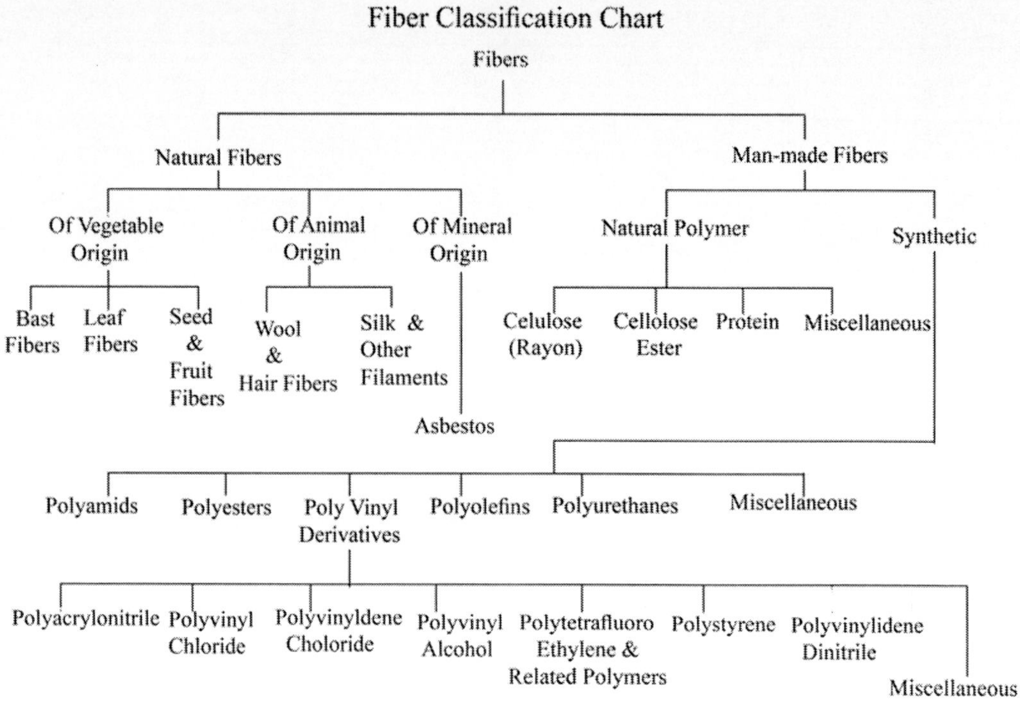

Figure 20. A vast classification chart for fibers[72].

5. El Ecreospinng of Silk Fibers

5.1. Introduction

The silk fiber protein is synthesized by the silk gland cells and stored in the lumen of silk gland .Subsequently; it is converted into silk fibers. Each strand of silk fiber is a double structure with 2 fibroin filaments covered by sericin (Figure 21).

In the fibroin, alanine and glycine together account for 70 % of the total composition, whereas in the sericin they make up about 15%. The chief component of sericin is another amino acid, serine (30% of the total).

- Silk is the only natural fibre which exists as a continuous filament. Each *Bombyx mori* cocoon can yield up to 1600 meters of filament. These can be easily joined together using the adhesive qualities of sericin to form a theoretically endless filament.

- The silk fiber's triangular cross-section gives it excellent light reflection capability.
- The silk fiber is smooth, unlike those of wool, cotton and others. This is one of the reasons why silk fabrics are so lustrous and soft.
- Silk can absorb up to 30 % of its weight in moisture without creating a damp feeling. When moisture is absorbed it generates 'wetting-heat' which helps to explain why silk is comfortable to wear next to the skin.
- Silk has a tenacity of approximately 4.8 grams per denier, slightly less than that of nylon.
- Silk has poor resistance to ultraviolet light and for this reason is only recommended for those curtains that are lined or not exposed to direct sunlight [73].

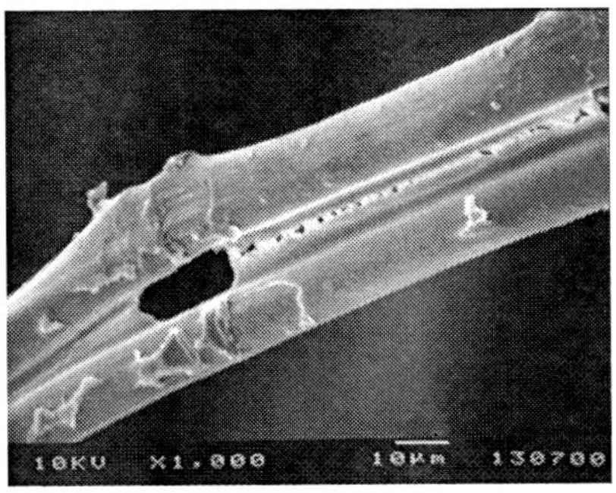

Figure 21.Silk micrograph (ITF) [73].

5.2. Crystal Structure of Silk (Fibroin) In Various Stages of Electrospinning

The physiological properties of SF matrices strongly depend on its molecular conformation and surface texture. SF exhibits at least three crystalline forms: silk I, silk II, and alpha-helix. All three conformations can be formed from the appropriate preparation conditions and each is interchangeable under certain conditions. Crystallization of SF involved the conformational transition that can be easily induced by simple physical (thermal), mechanical or chemical treatments. The most common method to convert the distorted conformation (random coil or silk I) of SF into the more stable β-sheet (silk II) conformation is to treat the SF film with an organic solvent. It is well known that organic solvents, particularly methanol, are highly effective in the crystallizing SF from a distorted conformation to α-sheet.

Infrared spectroscopy (IR) has been often applied to study the molecular conformation of silk fibroin fibers or films. The sensitive absorption bands on the IR spectrum are located in the spectral regions of ~1625 cm^{-1} (amide I), ~1528 cm^{-1} (amide II), ~1230 cm^{-1} (amide III), and ~700 cm−1 (amide V). To characterize the structure of SF matrices, the ATR-IR spectra in the amide I and II regions is examined. The IR spectra of SF matrices are shown in Figure 22(a)–(c). The as-spun SF-N matrix was characterized by absorption bands at 1651 cm^{-1} (amide I) and 1528 cm^{-1} (amide II), attributed to the random coil conformation, as shown in Figure 22a. The methanol-treated SF-N matrix showed strong β -sheet absorptions at 1622 and 1514 cm^{-1} within a methanol-treating time of 10 min, indicating that the random coil conformation of the SF nanofibers rapidly converted into β-sheet structure (Figure 22b) [75].

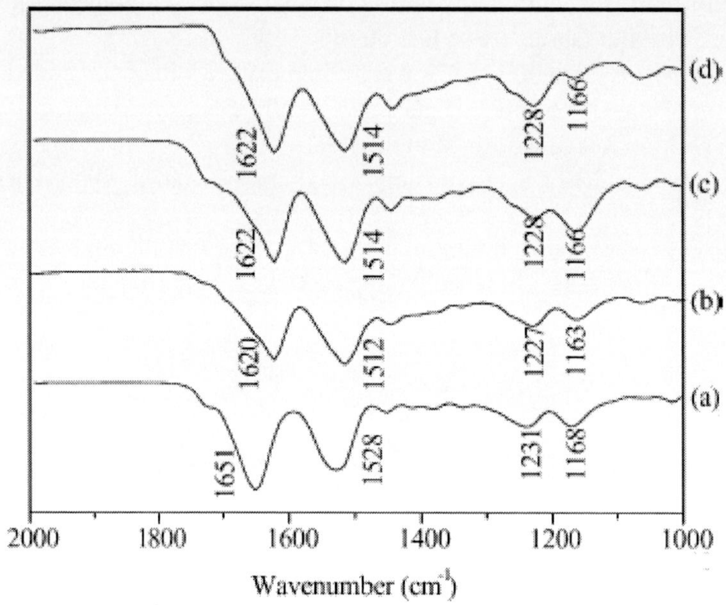

Figure 22. ATR-IR spectra of the as-prepared and methanol-treated SF matrices: (a) as-spun SF-N; (b) methanol-treated SF-N; (c) as-cast SF-F; (d) SF-M [75].

On the contrary, both the as-prepared SF-F and SF-M matrices showed strong β-sheet absorptions at 1622 and 1514 cm^{-1} without methanol treatment, as shown in Figure 22c and d. The raw SF had highly ordered intermolecular hydrogen bonds (β -sheet), and the SF film took a mainly β-sheet conformation, when cast from the formic acid solution.

Solid-state 13C NMR has been shown to be a more effective analytical tool for demonstrating the formation of β-sheets in polypeptides and proteins, because the isotropic 13C NMR chemical shifts of carbon atoms in proteins are sensitive to the β-sheet's secondary structure. It is well established that SF conformations are dependent upon the species of silkworms and conditions of the sample preparation. In particular,it has been reported that fibroin from *Bombyx mori* adopts two dimorphic structures, silk I and silk II . The silk II form is identified by the ^{13}C chemical shifts of glycine (Gly), serine (Ser), and alanine (Ala) that are indicative of β-sheets, while the silk I form produces chemical shifts that are associated with a loose helix or distorted β-turn. However, when compared with silk II, the less stable silk I shows a relatively unresolved structure, and the conformation of the soluble form of SF rapidly undergoes a transition to the insoluble silk II conformation. In ^{13}C CP/MAS NMR structural analyses of *B. mori* silk fibroins, the two crystalline forms of silk fibroins, silk I and silk II (β-sheet), have been distinguished by the conformation-dependent ^{13}C chemical shifts of the respective amino acid residues Figure 22 shows ^{13}C CP/MAS NMR spectra (a) and the expanded ^{13}C NMR spectra of the methyl alanine region (b) of SF matrices. The dotted arrow indicates the chemical shift of alanine in the β-sheet conformation. solid-state 13CNMR spectra of the as-prepared SF-N and SFF matrices, together with that of the SF-M. Assignments were made according to the literature. The chemical shifts of the Ala residue in

Figure 22a were 17.6 ppm for AlaC_β, 49.9 ppm for AlaC_α, and 173.5 ppm for Ala C=O carbon. The observed ^{13}C NMR chemical shifts of the peaks suggest that the Ala residue of the SF-N matrix takes a mainly non β-sheet conformation (random coil and silk I). The vertical dashed line and arrow in Figure 22 show the chemical shift of Ala in the β-sheet conformation. A shoulder was detected at ~ 20.4 ppm and assigned to Ala C_β in a β-sheet conformation, indicating that as-spun SF-N take some β-sheet conformations. The β-sheet structure of as-spun SF nanofibers could be formed partially by the elongational forces during the electrospinning process. However, the characteristic resonances of SF-F matrix, especially the Ala C_β at 20.4 ppm and Ala C_α at 48.9 ppm in are similar with that of the SF-M matrix [75].

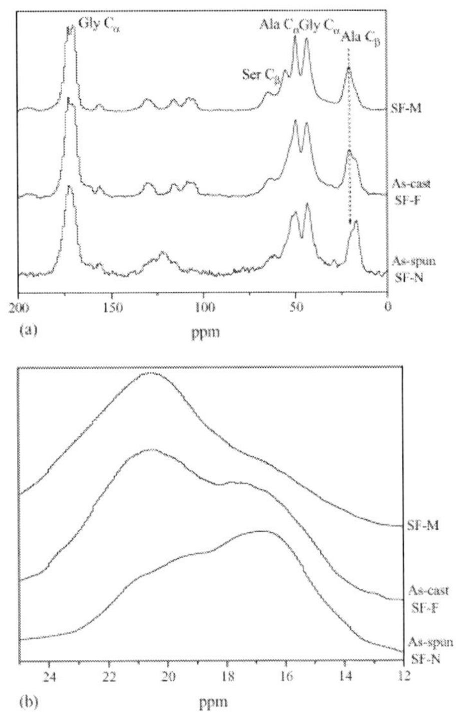

Figure 22. ^{13}C CP/MAS NMR spectra (a) and the expanded 13C NMR spectra of the methyl alanine region (b) of SF matrices. The dotted arrow indicates the chemical shift of alanine in the β-sheet conformation.[75]

Shimizu has reported that there are two types of crystal formation during solidification of liquid fibroin (by losing water) inside the silk gland. One of them corresponds to the crystal form of fibrous fibroin. The fibroin molecular chain which constitutes the spatial crystal form is considered to be the intermediate form during the course of shifting the molecular chain of liquid silk to the fibrous fibroin molecule chain, that is, β –chain. It is therefore named as α –type of fibroin .The condition for the appearance of a type thereafter include solidification when there is no stress from liquid silk and the temperature being less than 40°C. It becomes β type when the temperature is above 50°C.Both α and β types of none stretched are insoluble in water and α type is softer than β type [76].

β Structure of Silk

β structure is the structure where the polypeptide chain is elongated. The structure can be the type where all the molecular chain run in the same direction and form a parallel pleated sheet or the type where the molecular chain run in the alternate direction and form anti parallel chain pleated sheet. In the case of β–structure, there are three important features, namely the period that is repeating period of the polypeptide chain, the spacing of the molecular chain in the sheet and the distance between the sheets [76].

5.3. Spinning Dope Preparation for Electrospinning

5.3.1. Degumming

The natural gum, sericin is normally left on the silk during reeling, throwing and weaving. It acts as a size which protects the fiber from mechanical injury. The gum is removed from finished yarns or fabrics, usually by boiling whit soap and water [73].The fibers were heated in an aqueous Na_2CO_3 (0.02 M) or 0.5 %(w/w) $NaHCO_3$ and rinsed with water to extract sericin.[77 , 75].

5.3.2. Dissolving of Fibroin

The extracted fibers (degummed silk) were then dissolved in 50% $CaCl_2$ [77] (100°C, then cooled) or in 9.3 M LiBr solution at 60°C yielding a solution. The solution pours into regenerated cellulose dialysis tubing to carry out dialysis against 1000 ml deionized water (for 48 h at 23 °C). Another way reported by Min et al. [76,75] degummed silk (SF) is dissolved in ternary $CaCl_2/CH_3CH_2OH/H_2O$ (1:2:8 in molar ratio) at 70°C for 6 h and then dialyzed with cellulose tubular membrane. After these stages the regenerated silk fibroin sponge is obtained by lyophilization. The silk sponge solution is electrospun in formic acid (98-100%).

5.4.Elecrospinning of Silk Fibroin

Electrospinning generally produces non-woven matrices with randomly arranged, ultrathin fibers that have nanometerscale diameters. Figure 23 shows a SEM micrograph (magnification 500×) of the woven SF microfiber and as-spun SF nanofiber matrix. From the image analysis, the SF nanofibers have an average diameter of 80 nm and their diameters ranged from 30 to 120 nm, while the diameter of the SF microfiber was 11 nm. The average diameter of SF nanofibers corresponds to about two orders of magnitude smaller than a SF microfiber [75].

Figure 24. SEM image of woven SF microfibers and nonwoven SF nanofibers [75].

5.4.1.Effect of Silk Polymer Concentration on Fiber Diameter

Silk concentration plays a major role in fiber diameter. No fibers were formed at less than 5% silk concentration for any electric field and spinning distances. Figures. 24 and 25 show the morphology of fibers obtained at the electric fields of 3 and 4 kV/cm, respectively, at silk/formic acid concentrations of 5, 8, 10, 12, 15, and 19.5% with a constant tip-to-collection plate distance of 7 cm. At 8% concentration less than 30 nm diameter fibers were formed with beads (drops of polymer over the woven mesh) and they were not uniform and were branched off (Figure 25). At 10% concentration with 5 cm spinning distance and 2, 3 and 4 kV/cm electric fields, drops were formed instead of fibers. Continuous fibers were obtained above 12% regardless of electric field and distance (Figures 24 and 25). At 19.5%, the average fiber diameter was much larger than that of fibers spun at lower concentrations. In the short distance as well as low concentration (10%), the solution reaches the collection plate before the solvent fully evaporates. This explains the formation of droplets and beads at the low concentration and distance. Fewer beads were observed in electrospun fibers at higher concentration. Increase in the regenerated silk concentration in the formic acid increases the solution viscosity. At low concentrations beads are form instead of fibers and at high concentrations the formation of continuous fibers are prohibited because of inability to maintain the flow of the solution at the tip of the needle resulting in the formation of larger fibers. Continuous nanofibers were obtained above 12% regardless of electric field and distance and at higher concentration of 19.5% the average fiber diameter was larger than at lower concentrations [77].

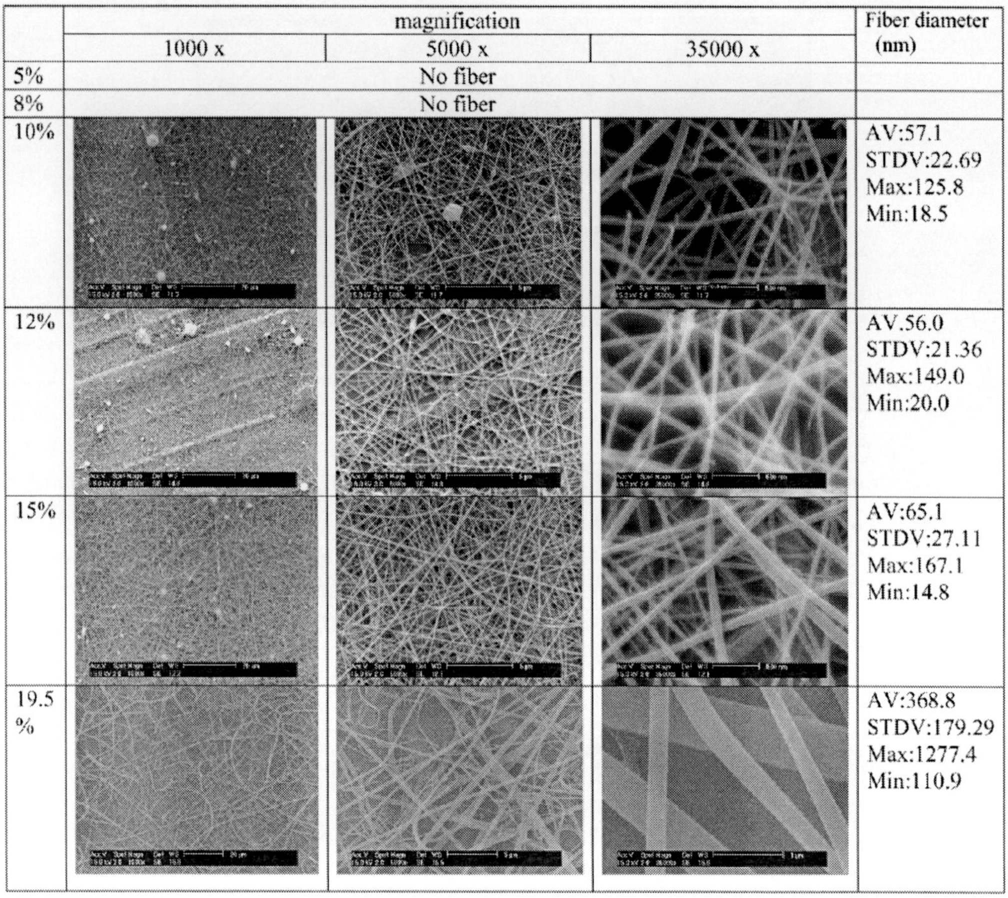

Figure 24. The morphology of fibers at electric field of 3 kV/cm at concentrations from 5 to 19.5% with a constant spinning distance of 7 cm. The figure also shows the average, standard deviation, maximum and minimum values of the fiber diameter [77].

The SEM micrographs of nanofibers electrospun from the SF solutions with different concentrations or viscosities ranged from 3% to 15% by weight were shown in Figure 26. At the concentration below 9% by weight, beads or beaded fibers were generated by electrospinning. The continuous nanofibers can be obtained at the concentration above 12% by weight, and this concentration appears to correspond to the onset of significant chain entanglements in the viscosity data shown in Figure 26. Therefore, it can be concluded that extensive chain entanglements are necessary to produce the continuous fibers by electrospinning [75].

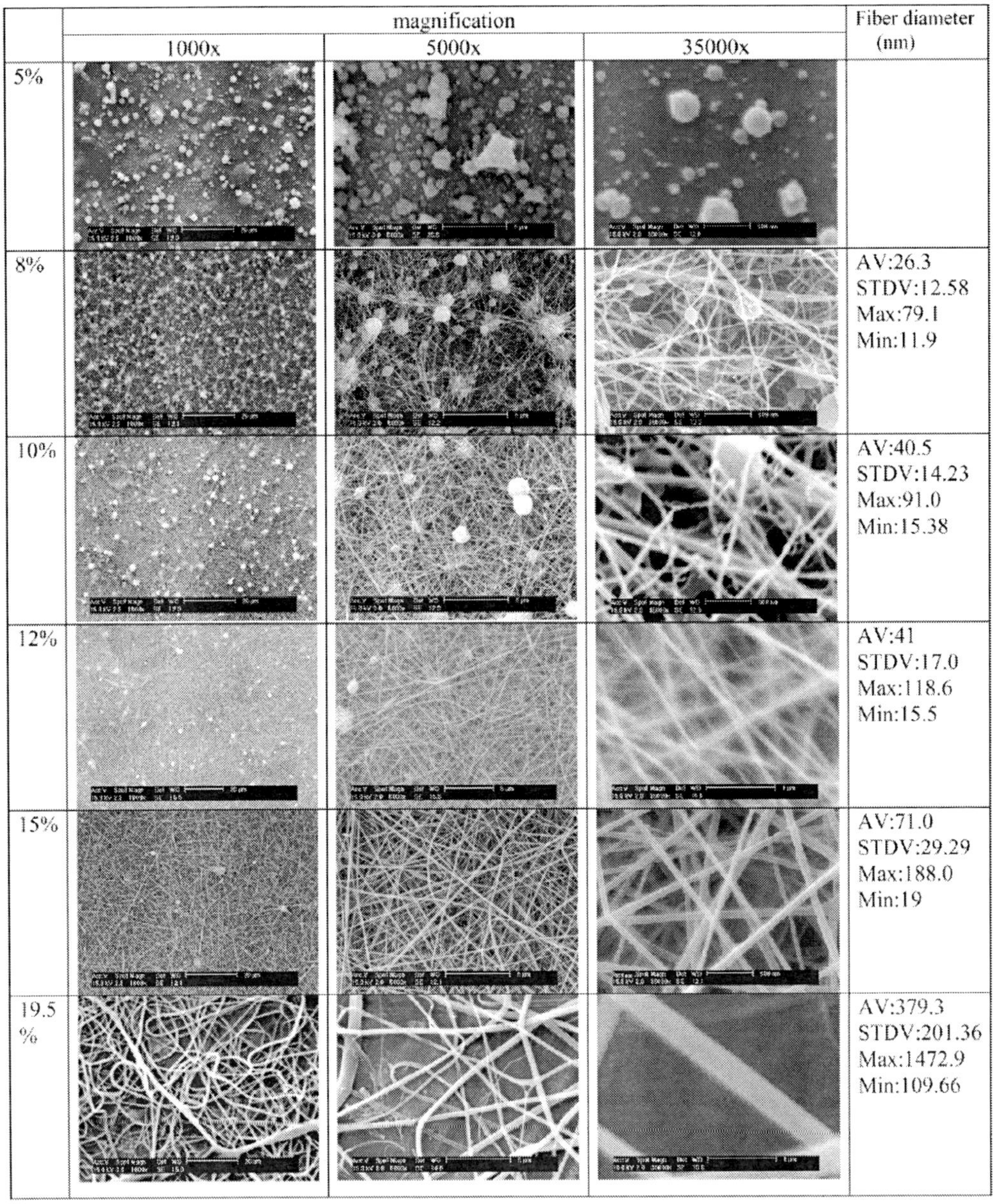

Figure 25. The morphology of fibers at electric field of 4 kV/cm at concentrations from 5 to 19.5% with a constant spinning distance of 7 cm.[77].

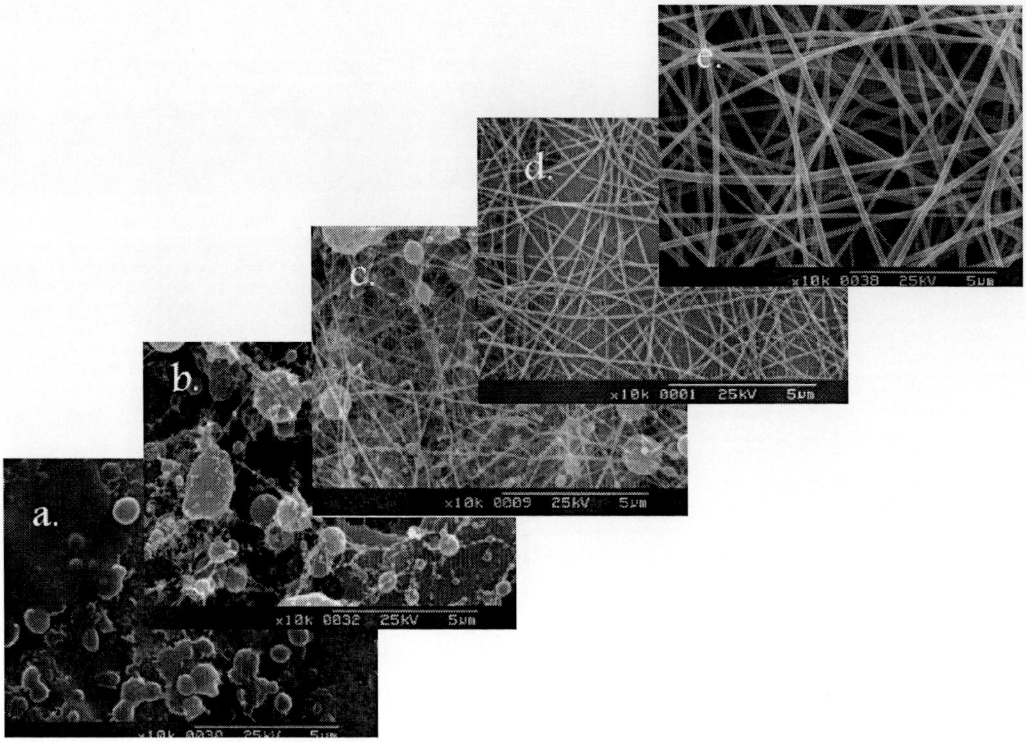

Figure 26. SEM micrographs of electrospun SF nanofibers with concentration of (a) 3%, (b) 6%, (c) 9%, (d) 12%, or (e) 15% by weight.[75]

5.4.2. Effect of Voltage and Spinning Distance on Morphology and Diameter

Figure 27 shows the relationship between mean fiber diameter and electric field with concentration of 15% at spinning distances of 5, 7 and 10 cm. The mean fiber diameter obtained at 2 kV/cm is larger than other electric fields. The effect of two factors, concentration and electric field on fiber diameter was investigated by two-way analysis of variance. The interaction effect between two factors is also obtained from this analysis [77].

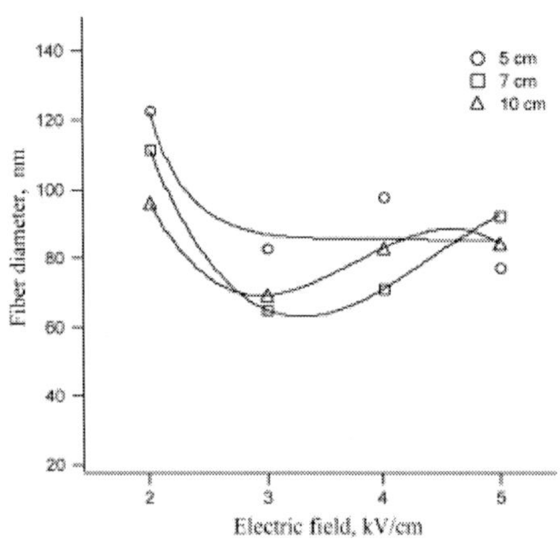

Figure 27. The relationship between mean fiber diameter and electric field with concentration of 15% at spinning distances of 5, 7 and 10 cm. (Please refer to text for explanation) [77].

Figure 28 shows that the concentration apparently has more effect on the fiber diameter than electricfield. The multiple regression analysis was carried out to evaluate the contribution of concentration and electric field on the fiber diameter [77].

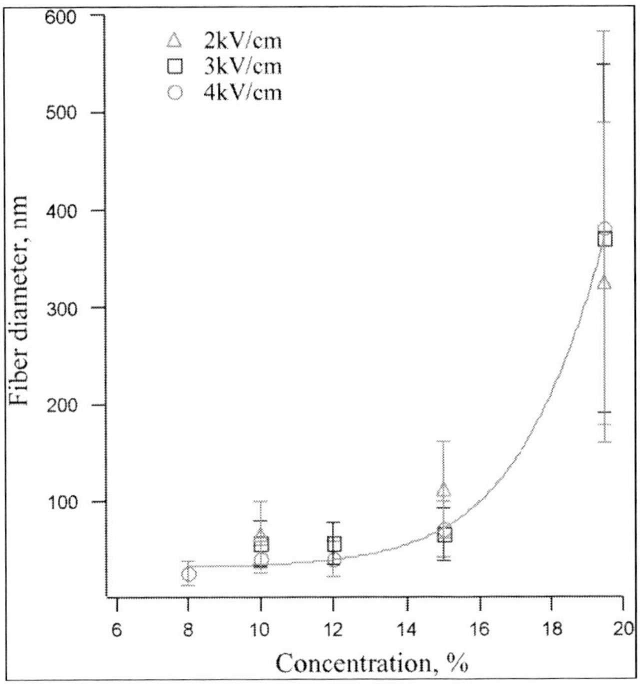

Figure 28. The relationship between the fiber diameter and the concentration at three electric fields (2, 3, 4 kV/cm)[77].

Sukigara et al.[78] used RSM analysis (Response Surface Methodology) to the experimental results to develop a processing window which will produce nanoscale regenerated silk fibers by electrospinning process. RSM is used in situations where several variables influence a feature (called the response) of the system. The steps in the procedure are described briefly as follows.

1. Identification of variables ζ_1; ζ_2; ζ_3... for response η:
2. Calculation of corresponding coded variables x_1; x_2; x_3... by using the following equation.

$$x_i = \frac{\zeta_i - [\zeta_{Ai} + \zeta_{Bi}]/2}{[\zeta_{Ai} + \zeta_{Bi}]/2} \quad , \tag{1}$$

where ζ_{Ai} and ζ_{Bi} refer to the high and low levels of the variables ζ_i; respectively.

3. Determination of the empirical model by multiple regression analysis to generate theoretical responses ($\hat{y}$): The second-order model is widely used in RSM. The general equation for response h of the second-order model is given by:

$$\eta = \beta_o + \sum_{i=1}^{k} \beta_i x_i + \sum_{i=1}^{k} \beta_{ii} x_i^2 + \sum_{i<j=2}\sum^{k} \beta_{ij} x_i x_j , \qquad (2)$$

where k is the number of factors, x_i are the coded variables and β are coefficients.

4. Calculation of the coefficients b to fit the experimental data as close as possible.

The relationship between the response and the variables is visualized by a response surface or contour plot to see the relative influence of the parameters, to find an optimum parameter combination, and to predict experimental results for other parameter. for two variables , When k = 2; the empirical model from the general Eq. (2) becomes

$$y = \beta_o x_1 + \beta_1 x_1 + \beta_2 x_2 + \beta_{11} x_1^2 + \beta_{22} x_2^2 + \beta_{12} x_1 x_2 + \varepsilon , \qquad (3)$$

RSM procedure to optimize the process parameters for the electrospinning silk is shown in Figure 29.

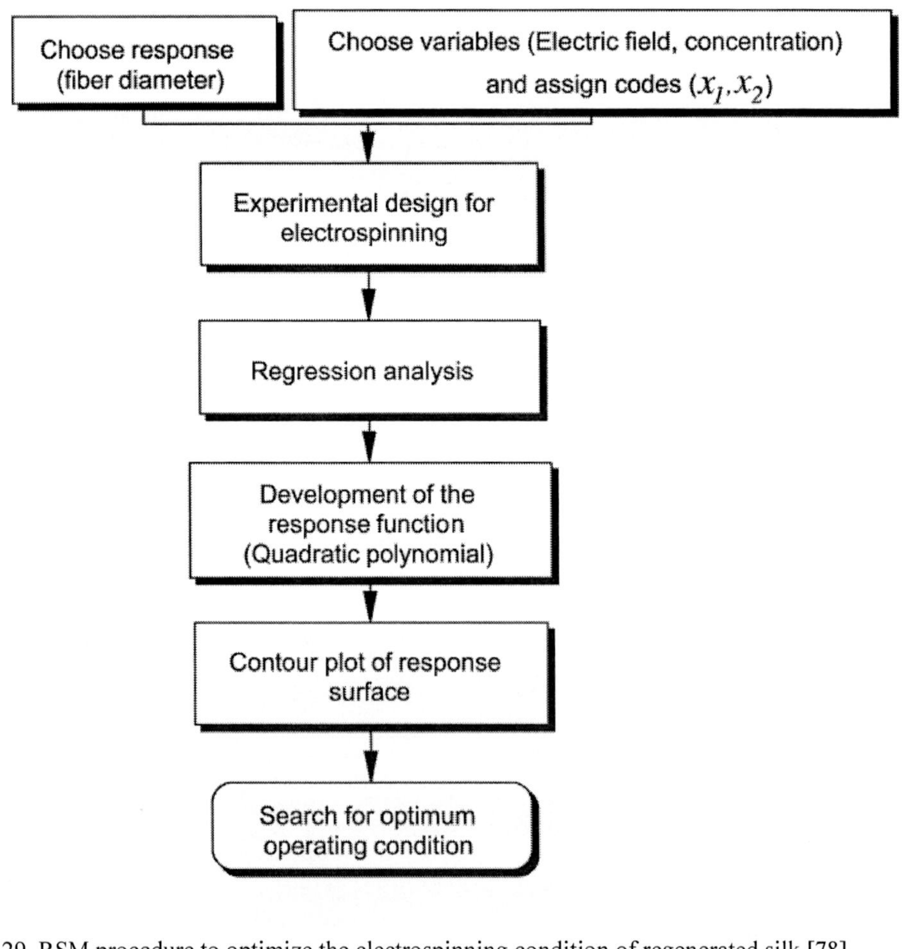

Figure 29. RSM procedure to optimize the electrospinning condition of regenerated silk.[78]

Sukigara et al. [78] designed a factorial experiment by using two factors (electric field and concentration). For a quadratic model, experimentsmust be performed for at least three levels of each factor. These levels are best chosen equally spaced. The two factors (silk concentration and electric field) and three levels resulted in nine possible combinations of factor settings. A schematic of the experimental design is shown in Figure 30(A) and (B).

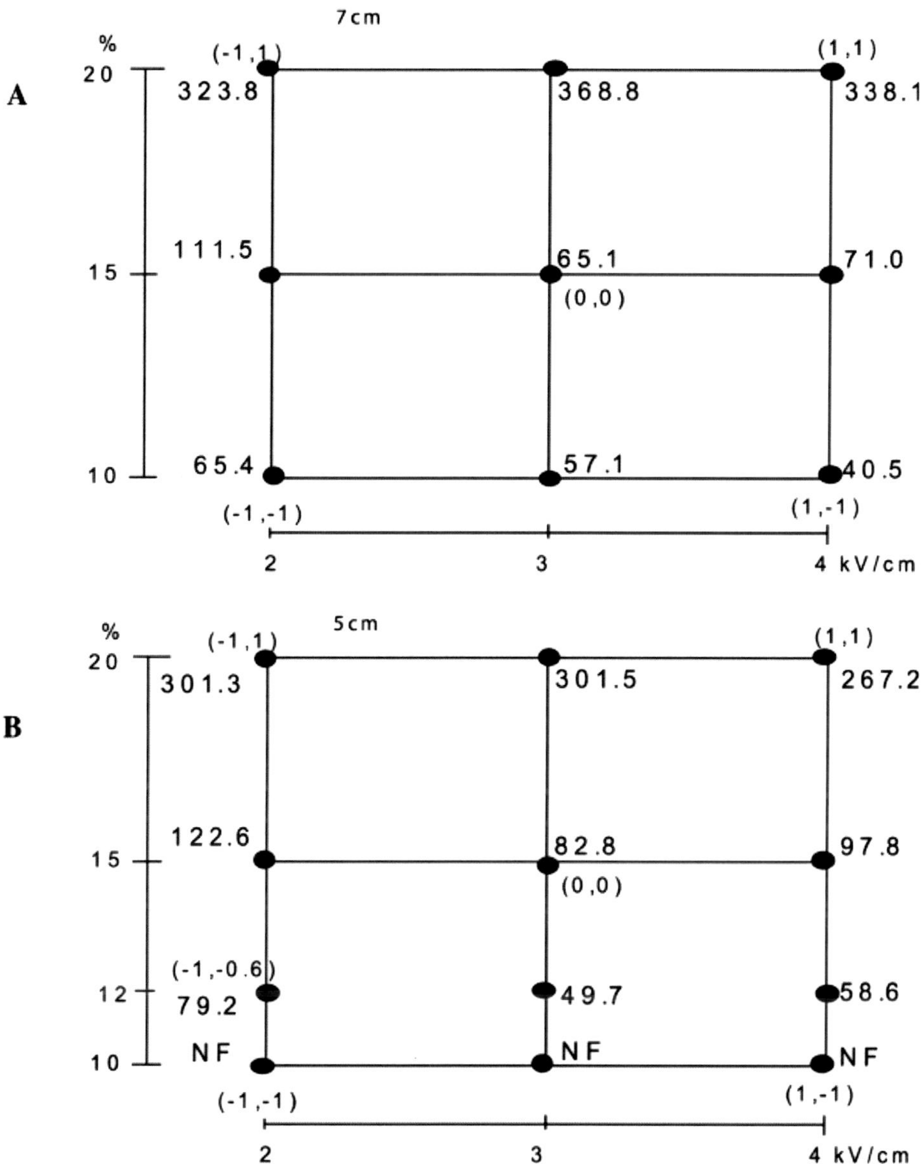

Figure 30. Experimental design (A) spinning distance 7 cm and (B) spinning distance 5 cm. The values at the coordinate points show the mean fiber diameter (nm) of 100 measurements and coded values are shown in the brackets (electric field and concentration). NF: no fiber formation [78].

5.5. Characterization

The conformational changes of the secondary structure of silk fibroin, which occur during the electrospinning process, were analyzed by Raman spectroscopy. The spectra for pristine, degummed and electrospun fibers are shown in Figure 31(A). Pristine and degummed silk fibroins display characteristic conformational bands in the range 1650–1667 and 1241–1279 cm^{-1} which correspond to amide I and complex amide III, respectively. In this study, the amide I (random coil) pristine band was observed at 1665 cm^{-1} and the amide III (β-sheet) pristine band at 1231 cm^{-1}. These well-defined bands were chosen because they give a clear indication of changes in the secondary structures from random to β-sheet. The degummed silk also shows absorption bands at thesewavelengths. No significant spectral changes were observed indicating that the fibroin conformation is unchanged during the degumming process. The Raman spectra of the electrospun fiber are essentially the same as that of the pristine and degummed fibers although minor bands and some differences in peak intensities appear. This shows that the electrospinning process preserves the natural conformation of the fibroin. Figure 31(B) shows the amide I (1665 cm^{-1}, random) to amide III (1228 cm^{-1}, β-sheet) ratio of the electrospun fiber is less than that of the pristine fiber. This means that the electrospun fiber has a higher β-sheet content than the pristine fiber.

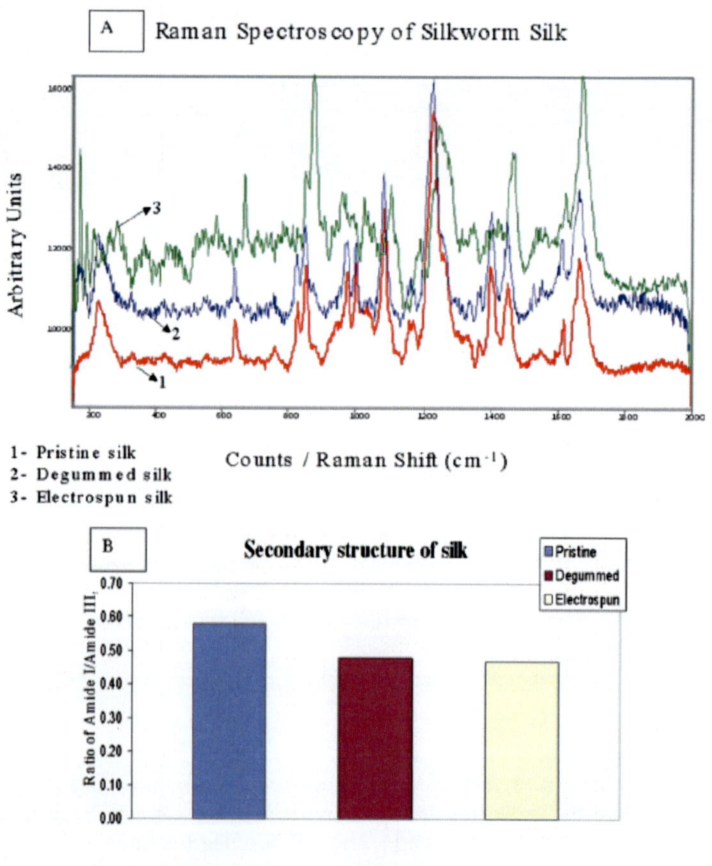

Figure 31. (A) Raman spectroscopy of pristine, degummed and electrospun nonwoven silk mat. (B) Secondary structural compositions of silk fibroin showing the fraction of Amide I to Amide III conformations [79].

6. ELCTROSPINNING OF CELLULOSE AND CELLULOSE ACETATE

Cellulose is the most abundant naturally occurring polysaccharide formed out of glucose-based repeat units, connected by 1,4-beta-glucosidic linkages. Cellulose and its derivatives are widely used as tough versatile materials. Cellulose nitrate, cellulose acetate (CA) and cellulose xanthate (rayon) can be easily molded or drawn into fibers for textile applications, for designing composite materials (safety glass), as thermoplastics etc [80].

6.1. Electrospinning of CTA Solution

CTA was dissolved in MC or solvent mixtures of MC/EtOH (90 / 10, 85 / 15, and 80 / 20, v/v). The needle (ID=0.495 mm) was connected to the high voltage supply (Chungpa EMT, CPS-40K03), which can generate positive DC voltages up to 40 kV. The distance between the needle tip and the collecting target was 10 cm. Positive voltage applied to polymer solutions were 15 kV. CA solutions were delivered via a syringe (20 mL) with a mass flow rate of 1mL/h. The optimum concentration of CTA solution for fiber formation was 5 wt.%. All electrospinnings were carried out at room temperature [81].

Figure 32 shows SEM photographs of ultrafine CTA fibers electrospun from a 5 wt.% CTA solution in MC. CTA fibers showed a mixed pattern of flat ribbons and round shapes [Figure 32(a)]. These flat ribbons were wrinkled or twisted in an irregular way. At a higher magnification (×20,000), it was found that ultrafine CTA fibers had isolated circular shape pores with a narrow size distribution in the range of 50–100 nm [Figure 32(b)].The porous structure was induced by phase separation resulting from the rapid evaporation of MC during the electrospinning process. The polymer-rich phase formed the fiber matrix and solvent-rich phase gave rise to the pores. Figure32(c) shows a SEM image for the fractured cross-section of the porous CTA fiber. Isolated pores were also observed inside the fiber, which were larger than those in the fiber surface. In liquid–liquid demixing of polymer solution, the resulting polymer morphology is dependent on the phase separation mechanism. The structure formed by the spinodal decomposition (SD) mechanism shows interconnected pores, whereas nucleation and growth (NG) mechanism mainly results in isolated spherical pores [81].

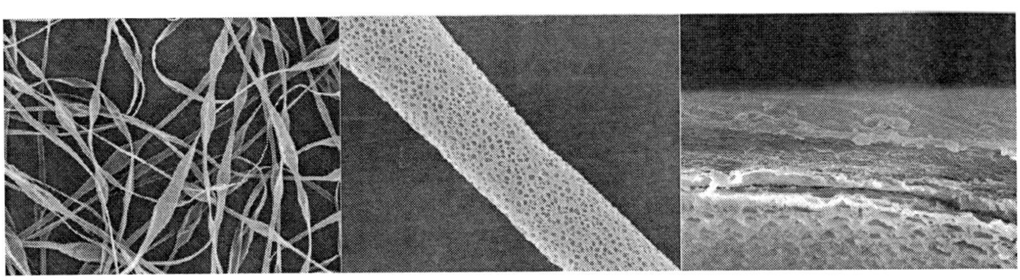

Figure 32. SEM images of electrospun CTA fibers: (a) magnification ×1000,(b) magnification ×20,000, and (c) fractured cross-section of porous CTA fibers[81].

Figure 33 shows SEM photographs of ultrafine CTA fibers electrospun from different ratios of MC/ EtOH (90 / 10, 85/ 15, and 80/20 v/v). All ultrafine CTA fibers were electrospun with the same electrospinning condition for a CTA solution in MC. It was found that the addition of EtOH changed not only the pore structure but also the fiber diameter. In the case of MC/EtOH (90 / 10), the resulting CTA fibers had interconnected pores in the range of 200–500 nm, indicating phase separation proceeded according to SD mechanism. However, non-porous corrugated fibers were obtained from the other mixed solvents [Fgure 33(b) and (c)] [81].

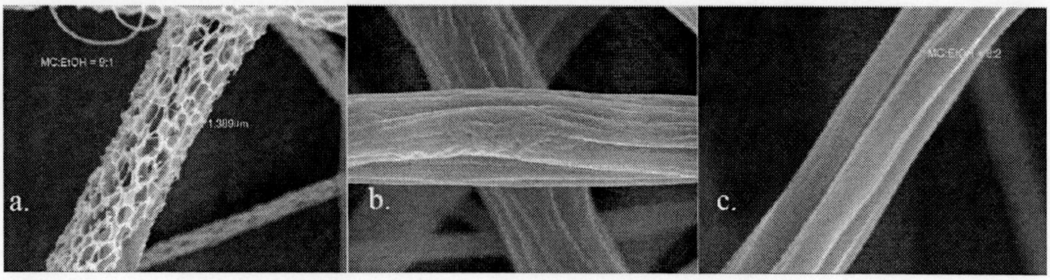

Figure 33. SEM images of ultrafine CTA fibers electrospun using different volume ratios of MC/EtOH: (a) 90/ 10, (b) 85/ 15, and (c) 80/ 20 [81].

7. Concluding Remarks

Nanotechnology has become in recent years a topic of great interest to scientists and engineers, and is now established as prioritized research area in many countries. The reduction of the size to the nano-meter range brings an array of new possibilities in terms of material properties, in particular with respect to achievable surface to volume ratios. Electrospinning of natural fibers is a novel process for producing superfine fibers by forcing a solution through a spinnerette with an electric field. A comprehensive review on this technique has been made in this paper. Based on this review, many challenges exist in the electrospinning process of natural fibers, and a number of fundamental questions remain open. The electrospinning technique provides an inexpensive and easy way to produce natural nanofibers on low basis weight, small fiber diameter and pore size. It is hoped that this chapter will pave the way toward a better understanding of the application of electrospinning of natural fibers.

Nevertheless, Among the biodegradable and biocompatible polymers, SF was extensively studied as one of the candidate materials for biomedical applications, because it has several distinctive biological properties including good biocompatibility, biodegradability, and minimal inflammatory reaction. One of the promising applications of SF in biomedical engineering is as scaffolding materials for tissue engineering. It was reported that SF matrices could be useful for the culture of fibroblasts and osteoblasts as well as stem cells, and could enhance the adhesion, growth, and differentiation of the cells in a manner similar to that of collagen matrices. nonwoven matrices of electrospun nanofibers could be prepared from a regenerated SF solution, and that the matrices were effective in the cell attachment and spreading of normal human keratinocytes and fibroblasts [82].

REFERENCES

[1] YQ. Wan, Q. Guo, N. Pan. 2004. Thermo-electro-hydrodynamic model for electrospinning process. *Int J Nonlinear Sci Num Simul*,5.pp5–8.

[2] JH. He, YQ. Wan, JY. Yu. 2004. Allometric scaling and instability in electrospinning. *Int J Nonlinear Sci Num Simul*,5(3).pp.243–52.

[3] JH.He, YQ. Wan, JY. Yu . 2004.Application of vibration technology to polymer electrospinning. *Int J Nonlinear Sci Num Simul*,5(3).pp.253–61.

[4] JH He, YQ Wan. 2004. Allometric scaling for voltage and current in electrospinning. *Polymer,45*(19).pp-6731–4.

[5] X-H .Qin, YQ ,Wan, JH .He . 2004. Effect of LiCl on electrospinning of PAN polymer solution: theoretical analysis and experimental verification. *Polymer,45*(18).pp.6409–13.

[6] SA. Therona, E. Zussmana, AL. Yarin. 2004. Experimental investigation of the governing parameters in the electrospinning of polymer solutions. *Polymer*,45.pp.2017–30.

[7] MM .Demir, I .Yilgor, E. Yilgor, B Erman. 2002. Electrospinning of polyurethane fibers. *Polymer,*43.pp.3303–9.

[8] AM. Ganan-Calvo. 1999. The surface charge in electrospraying: its nature and its universal scaling laws. *J Aerosol Sci*;30(7):863–72.

[9] JJ. Feng. 2003. Stretching of a straight electrically charged viscoelastic jet. *J Non-Newtonian Fluid Mech*, 116.pp.55–70.

[10] YM .Shin, MM. Hohman, MP. Brenner, GC. Rutledge. 2001. Experimental characterization of electrospinning: the electrically forced jet and instabilities. *Polymer,* 42.pp.9955–67.

[11] J. He, Y. Wan, J-Y.Yu. 2005. Scaling law in electrospinning: relationship between electric current and solution flow rate. *Polymer*, 46.PP. 2799–2801.

[12] PK. Baumgarten. 1971. Electrostatic spinning of acrylic microfibers. *J Colloid Interface Sci*,36.pp.71–79.

[13] J .Doshi, DH. Reneker. 1995. Electrospinning process and applications of electrospun fibers. *J Electrostat*,35.pp.151–60.

[14] DH. Reneker, AL. Yarin, H. Fong, S. Koombhongse. 2000. Bending instability of electrical charged liquid jets of polymer solutions in electrospinning. *J Appl Phys*,87.pp.4531– 4547.

[15] M. Bognitzki, W. Czando, T. Frese, A. Schaper, M Hellwig, M. Steinhart, A. Greiner, JH. Wendorff. 2001 Nanostructured fibers via electrospinning. *Adv Mater*,13.pp.70–72.

[16] E. Zussman, A. Theron, AL. Yarin. 2003. Formation of nanofiber crossbars in electrospinning. *Appl Phys Lett*,82.pp.973–975.

[17] S.A. Theron, A.L. Yarin, E. Zussman, E. Kroll. 2005. Multiple jets in elctrospinning: experiment and modeling. *Polymer, 46*.pp . 2889–2899.

[18] Z-M Huang, Y.-Z. Zhang, M. Kotak, S. Ramakrishna. 2003. A review on polymer nanofibers by electrospinning and their applications in nanocomposites. *Composites Science and Technology*, 63 .pp.2223–2253

[19] D. Emig, A. Klimmek, E. Raabe. *US Patent* 6395046. 2002.

[20] D. Groitzsch, E. Fahrbach. *US patent* 4,618,524, 1986.

[21] HL. Schreuder-Gibson, P. Gibson, K Senecal, M Sennett, J. Walker, W. Yeomans, et al. 2002. Protective textile materials based on electrospun nanofibers. *Journal of Advanced Materials,34*(3).pp. 44–55.

[22] M.M. Bergshoef, G.J. Vancso. 1999. Transparent nanocomposites with ultrathin, electrospun nylon-4,6 fiber reinforcement. *Adv. Mater*, 11.pp.1362–1365.

[23] Z. Ma, M. Kotaki, R. Inai, S. Ramakrishna. 2005. Potential of nanofiber matrix as tissue engineering scaffolds. *Tissue Eng*, 11 .pp.101.

[24] C.Y. Xu, R. Inai, M. Kotaki, S. Ramakrishna. 2004 .Aligned biodegradable nanofibrous structure: a potential scaffold for blood vessel engineering. *Biomaterials*, 25 .pp. 877.

[25] H. Yoshimoto, Y.M. Shin, H. Terai, J.P. Vacanti. 2003. A biodegradable nanofiber scaffold by electrospinning and its potential for bone tissue engineering. *Biomaterials,* 24 .pp. 2077.

[26] M. Shin, O. Ishii, T. Sued, J.P. Vacanti. 2004. Contractile cardiac grafts using a novel nanofibrous mesh. *Biomaterials* ,25 .pp. 3717.

[27] Z Ma, M. Kotaki, Thomas Yong, Wei He, S. Ramakrishna.2005 .Surface engineering of electrospun polyethylene terephthalate (PET) nanofibers towards development of a new material for blood vessel engineering. *Biomaterials*, 26 .pp.2527–2536.

[28] HJ. Jin, S. Fridrikh, GC. Rutledge, D Kaplan. 2002. Electrospinning Bombyx mori silkwith poly(ethylene oxide). *Abstracts of Papers American Chemical Society*,224(1–2).pp408.

[29] Y.K. Luu, K. Kim, B.S. Hsiao, B. Chu, M. Hadjiargyrou.2003. Development of a nanostructured DNA delivery scaffold via electrospinning of PLGA and PLA–PEG block copolymers, *J. Control. Release*, 89 .pp. 341.

[30] G. Verreck, I. Chun, J. Rosenblatt, J. Peeters, A.V. Dijck, J. Mensch, M. Noppe, M.E. Brewster. 2003. Incorporation of drugs in an amorphous state into electrospun nanofibers composed of a waterinsoluble, nonbiodegradable polymer. *J. Control. Release,* 92 .pp.349.

[31] PW. Gibson, HL. Schreuder-Gibson, D. Riven . 1999.Electrospun fiber mats: transport properties. AIChE J,45(1).pp.190–195.

[32] PW. Gibson, HL. Schreuder-Gibson, C. Pentheny. 1998. Electrospinning technology: direct application of tailorable ultrathin membranes.*J of Coated Fabrics*,28.pp.63.

[33] PW. Gibson, HL. Schreuder-Gibson, D. Rivin. 2001. Transport properties of porous membranes based on electrospun nanofibers. *Colloids and Surfaces A: Physicochemical and Engineering Aspects*,187.pp.469–481.

[34] HL. Schreuder-Gibson, PW. Gibson, K Senecal, M. Sennett, J .Walker, W Yeomans, et al. 2002.Protective textile materials based on electrospun nanofibers. *Journal of Advanced Materials*,.34(3).pp.44–55.

[35] ID Norris, MM Shaker, FK Ko, AG Macdiarmid. 2000.Electrostatic fabrication of ultrafine conducting fibers: polyaniline/polyethylene oxide blends. *Synthetic Metals,*114(2).pp.109–114.

[36] C.M. Waters, T.J. Noakes, C. I. Pavery . *US patent*,5088807, 1992.

[37] K.J. Senecal, L. Samuelson, M. Sennett, G.H. Schreuder. US patent application publication 0045547, 2001.

[38] K.J. Senecal, DP. Ziegler, J. He, R. Mosurkal, H. Schreuder-Gibson, LA. 2002. Samuelson Photoelectric response from nanofibrous membranes. *Materials Research Society Symposium Proceedings*,708.pp.285–9.

[39] X. Wang, C. Drew, S.H. Lee, K.J. Senecal, J. Kumar, L.A. Samuelson. 2002. Electrospun nanofibrous membranes for highly sensitive optical sensors. *Nano Lett*, 11. pp.1273–1275.

[40] B. Ding, J.H. Kim, Y. Miyazaki, S.M. Shiratori. 2004. Electrospun nanofibrous membranes coated quartz crystal microbalance as gas sensor for NH3 detection, Sens. Actuators *B: Chem*, 101 .pp. 373–380.

[41] P.I. Gouma. 2003. Nanostructured polymorphic oxides for advanced chemosensors. *Rev. Adv. Mater. Sci*, 5 .pp. 147–154.

[42] K. Sawicka, P. Goum, S Simon. 2005. Electrospun biocomposite nanofibers for urea biosensing. *Sensors and Actuators* B, 108. pp. 585 – 588.

[43] JM Deitzel, J Kleinmeyer, JK Hirvonen, et al. 2001. Controlled deposition of electrospun poly(ethylene oxide) fibers. *Polymer*,42.pp.8163–8170.

[44] H Fong, DH. Reneker. 2001. Electrospinning and formation of nanofibers. In: Salem DR, editor. Structure formation in polymeric fibers. *Munich: Hanser*. pp. 225–246.

[45] K. Fujihara,_, M. Kotak, S. Ramakrishn. 2005. Guided bone regeneration membrane made of polycaprolactone/calcium carbonate composite nano-fibers. *Biomaterials*, 26 .pp.4139– 4147.

[46] X. Fang, DH. Reneker.1997. DNA fibers by electrospinning. *J Macromolecular Sci-Phys*,B36 .pp. 169–173.

[47] GI. Taylor. 1969. Electrically driven jets. *Proc R Soc London, Ser* A ,313.pp.453–475.

[48] El-R. Kenawy , G. L. Bowlin , K. Mansfield , J. Layman , D. G. Simpsonc, E. H. Sanders., G. E. Wnek. 2002 . Release of tetracycline hydrochloride from electrospun poly(ethylene-co-vinylacetate), poly(lactic acid), and a blend . *Journal of Controlled Release*, 81 .pp. 57–64

[49] S. F. Fennessey, J. R.. Farris. 2004 . Fabrication of aligned and molecularly oriented electrospun polyacrylonitrile nanofibers and the mechanical behavior of their twisted yarns . *Polymer*, 45.pp. 4217– 4225.

[50] S-H. Tan, R. Inai, M. Kotaki, S. Ramakrishna. 2005. Systematic parameter study for ultra-fine fiber fabrication viaelectrospinning process. *Polymer,* 46 .pp. 6128–6134.

[51] H. Fong, I. Chun, D.H. Reneker. 1999. Beaded nanofibers formed during electrospinning *.Polymer, 40* .pp.4585– 4592.

[52] S. L. Shenoy, W. D. Bates, H. L. Frisch, G. E. Wnek. 2005.Role of chain entanglements on fiber formation during electrospinning of polymer solutions: good solvent, non-specific polymer–polymer interaction limit . *Polymer ,46* .pp. 3372–3384.

[53] H Fong, DH Reneker. 1999. Elastomeric nanofibers of styrene-butadiene-styrene triblockcopolym er. *J Polym Sci: Part B Polym Phys*,37(24).pp.3488–3493.

[54] S. Kidoaki, K. Kwon, T. Matsuda.2005. Mesoscopic spatial designs of nano- and microfiber meshes fortissue-engineering matrix and scaffold based on newly devised multilayering and mixing electrospinning techniques . *Biomaterials,* 26 .pp. 37–46.

[55] S.Y. Gu, J. Ren, G.J. Vancso .2005. Process optimization and empirical modeling for electrospun polyacrylonitrile (PAN) nanofiber precursor of carbon nanofibers. European Polymer *Journal,* xxx .pp.xxx–xxx.

[56] CH. Zhang , X. Yuan , L. Wu , Y. Han, Jing Sheng. 2005. Study on morphology of electrospun poly(vinyl alcohol) mats. *European Polymer Journal* ,41 .pp. 423–432.

[57] X. Geng, O-H. Kwon, J. Jang. 2005. Electrospinning of chitosan dissolved in concentrated acetic acid solution. *Biomaterials,* 26 .pp. 5427–5432

[58] A. Koski, K. Yim, S. Shivkumar. 2004. Effect of molecular weight on fibrous PVA produced by electrospinning. *Mater Lett*,58.pp.493–497.

[59] JM Deitzel, J Kleinmeyer, D Harris, T. N. Beck. 2001. The effect of processing variables on the morphology of electrospun nanofibers and textiles. *Polymer*,42.pp.261–272.

[60] Y. J. Ryu, Hak Yong Kim , K. H. Lee, H. C. Park , D. R. Lee.2003 . Transport properties of electrospun nylon 6 nonwoven mats. *European Polymer Journal*, 39 .pp. 1883–1889.

[61] B. Kim, H. Park, S-H. Lee, W. M. Sigmund.2005 .Poly(acrylic acid) nanofibers by electrospinning. *Materials Letters* ,59.pp. 829–832.

[62] T. Jarusuwannapoom , W. Hongrojjanawiwat, S Jitjaicham, L. Wannatong, M. Nithitanakul , C. Pattamaprom , P. Koombhongse , R. Rangkupan , P. Supaphol . 2005. Effect of solvents on electro-spinnability of polystyrene solutions and morphological appearance of resulting electrospun polystyrene fibers. *European Polymer Journal* ,41 .pp. 409–421.

[63] C.J. Buchko, L.C. Chen, Y. Shen, D.C. Martin. 1999. Processing and microstructural characterization of porous biocompatible protein polymer thin films. *Polymer*,40.pp.7397–407.

[64] X.M. Moa, C.Y. Xub, M. Kotakib, S. Ramakrishna .2004.Electrospun P(LLA-CL) nanofiber: a biomimetic extracellular matrixfor smooth muscle cell and endothelial cell proliferation. *Biomaterials* ,25 .pp. 1883–1890

[65] A. F. Spivak and Y. A. Dzenis.1998.Asymptotic decay of radius of a weakly conductive viscous jet in an external electric field. *Appl. Phys. Lett*, 73.pp.3067

[66] D.H. Reneker, A.L. Yarin, H. Fong, S. Koombhongse. 2000. Bending instability of electrically charged liquid jets of polymer solutions in electrospinning, *J. Appl. Phys*, 87.pp. 4531–4547.

[67] M. M. Hohman, M. Shin, G. Rutledge, and M. P. Brenner. 2001 .Electrospinning and electrically forced jets: I. Stability theory. *Phys. Fluids* 13.pp. 2201.

[68] M. M. Hohman, M. Shin, G. Rutledge, and M. P. Brenner.2001.Electrospinning and electrically forced jets: II. *Applications.Phys. Fluids* 13.pp. 2221.

[69] D. H. Reneker, A. L. Yarin, H. Fong, and S. Koombhongse. . 2000.Bending instability of electrically charged liquid jets of polymer solutions in electrospinning.*J. Appl. Phys*,87.pp. 4531 .

[70] A. L. Yarin, S. Koombhongse, and D. H. Reneker. 2001. Bending instability in electrospinning of nanofibers. *J. Appl. Phys*, 89.pp.3018 .

[71] Y. M. Shin, M. M. Hohman, M. P. Brenner, and G. C. Rutledge. 2001. Electrospinning: A whipping fluid jet generates submicron polymer *fibers''Appl. Phys. Lett*,78.pp. 1149.

[72] J. G. Cook.1964. Handbook of Textile Fibers : I. Natural Fibers. Third Edition .Merrow publisher.

[73] R. R. Franck.2001.Silk, mohair,cashmere and other luxury fibres. Woodhead Publishing.

[74] D Kaplan, WW Adams, B Farmer, C Viney, editors. . 1994. Silk polymers: materials science and biotechnology. Washington DC: *ACS Symposium Series*, p. 544.

[75] B-M. Min., L. Jeong., Y.S. Nam., J-M. Kim,J. Y. Kim, W.H. Park .2004. Formation of silk fibroin matrices with different texture and its cellular response to normal human keratinocytes. International *Journal of Biological Macromolecules*, 34.pp.281–288

[76] N. Hojo,2000.Structure of Silk Yarn Part A. *Biological and Physical Aspects.Science Publisher* . USA.

[77] S. Sukigara, M. Gandhi, J. Ayutsede, M. Micklus, F. Ko. 2003 .Regeneration of Bombyx mori silk by electrospinning: part *1:Processing Parameters and Geometric Properties.Polymer*, 44.pp. 5721–5727.

[78] S. Sukigara, M Gandhi, J. Ayutsede, M. Micklus, F. Ko. 2004 . Regeneration of Bombyx mori silk by electrospinning.Part 2. Process optimization and empirical modeling using response surface methodology . *Polymer, 45* .pp. 3701–3708.

[79] J. Ayutsede, M. Gandhi, S. Sukigara, M. Micklus, H-E. Chen,F. Ko. 2005. Regeneration of Bombyx mori silk by electrospinning. Part 3:characterization of electrospun nonwoven mat. *Polymer*, 46 .pp. 1625–1634

[80] E. Entchevaa, H. Biena, L. Yina, C-Y. Chunga, M. Farrella,Y. Kostovc.2004. Functional cardiac cell constructs on cellulose-based scaffolding. *Biomaterials*, 25 .pp. 5753 –5762.

[81] S. O. Han, W. K. Son, J. H. Youk, T. S. Lee, W. H. Park,2005, Ultrafine porous fibers electrospun from cellulose *triacetateMaterials Letters*, xx .pp. xxx – xxx.

[82] K. E. Park, S. Y. Jung, S. J. Lee, B-M Min, W. H. Park .2006 . Biomimetic nanofibrous scaffolds: Preparation and characterization of chitin/silk fibroin blend nanofibers . International *Journal of Biological Macromolecules* ,38 .pp. 165–173.

In: Progress in Monomers, Oligomers, Polymers, Composites ... ISBN 978-1-60456-878-3
Editors: R. A. Pethrick, G.E. Zaikov et al.

Chapter 24

POLYMERIC NANOCOMPOSITES BASED ON THE ORGANOMODIFIED LAYERED SILICATES: STRUCTURE, MANUFACTURE AND PROPERTIES

***A.K. Mikitaev* [1,*], *A.Yu. Bedanokov* [2] *and M.A. Mikitaev* [3]**

[1]Kabardino – Balkarian State University, Nalchik, Russian Federation
[2]D.I.Mendeleev Russian University for Chemical Technology, Moscow, Russian Federation
[3]L.Ya.Karpov Research Institute, Moscow, Russian Federation

ABSTRACT

The analytic review of recent advances in the field of synthesis and study of the structures and properties of the polymeric nanocomposites on the basis of the organomodified layered silicates is given

Keywords: montmorillonite, organoclays, nanocomposites

1. INTRODUCTION

The polymeric nanocomposites are the polymers filled with nanoparticles which interact with the polymeric matrix on the molecular level in contrary to the macrointeraction in composite materials. Mentioned nanointeraction results in high adhesion hardness of the polymeric matrix to the nanoparticles [1,52].

Usual nanoparticle is less than 100 nanometers in any dimension, 1 nanometer being the billionth part of a meter [1,2].

The analysis of the reported studies tells that the investigations in the field of the polymeric nanocomposite materials are very promising.

* Correspondence to: 360004, Nalchik, Chernyshevskogo street, 173

The first notion of the polymeric nanocomposites was given in patent in 1950 [3]. Blumstain pointed in 1961 [4] that polymeric clay – based nanocomposites had increased thermal stability. It was demonstrated using the data of the thermogravimetric analysis that the polymethylmetacrylate intercalated into the Na^+ - methylmetacrylate possessed the temperature of destruction 40-50°C higher than the initial sample.

This branch of the polymeric chemistry did not attract much attention until 1990 when the group of scientists from the Toyota Concern working on the polyamide – based nanocomposites [5-9] found two – times increase in the elasticity modulus using only 4.7weight% of the inorganic compound and 100°C increase in the temperature of destruction, both discoveries widely extending the area of application of the polyamide. The polymeric nanocomposites based on the layered silicates began being intensively studied in state, academic and industrial laboratories all over the world only after that.

2. Structure of the Layered Silicates

The study of the polymeric nanocomposites on the basis of the modified layered silicates (broadly distributed and well – known as various clays) is of much interest. The natural layered inorganic structures used in producing the polymeric nanocomposites are the montmorillonite [10,11,12], hectorite [13], vermiculite [14], kaolin, saponine [15] and others. The sizes of inorganic layers are about 220 and 1 nanometers in length and width respectively [16,17].

The perspective ones are the bentonite breeds of clays which include at least 70% of the minerals from the montmorillonite group.

Montmorillonite $(Na,K,Ca)(Al,Fe,Mg)[(Si,Al)_4O_{10}](OH)_2 \times nH_2O$, named after the province Montmorillion in France, is the high – dispersed layered aluminous silicate of white or gray color in which appears the excess negative charge due to the non-stoichiometric replacements of the cations of the crystal lattice, charge being balanced by the exchange cations from the interlayer space. The main feature of the montmorillonite is its ability to adsorb ions, generally cations, and to exchange them. It produces plastic masses with water and may enlarge itself 10 times. Montmorillonite enters the bentonite clays (the term "bentonite" is given after the place Benton in USA).

The inorganic layers of clays arrange the complexes with the gaps called layers or galleries. The isomorphic replacement within the layers (such as Mg^{2+} replacing Al^{3+} in octahedral structure or Al^{3+} replacing Si^{4+} in tetrahedral one) generates the negative charges which electrostatically are compensated by the cations of the alkali or alkali-earth metals located in the galleries (Fig.1) [18]. The bentonite is very hydrophilic because of this. The water penetrates the interlayer space of the montmorillonite, hydrates its surface and exchanges cations what results in the swelling of the bentonite. The further dilution of the bentonite in water results in the viscous suspension with bold tixotropic properties.

The more pronounced cation – exchange and adsorption properties are observed in the bentonites montmorillonite of which contains predominantly exchange cations of sodium.

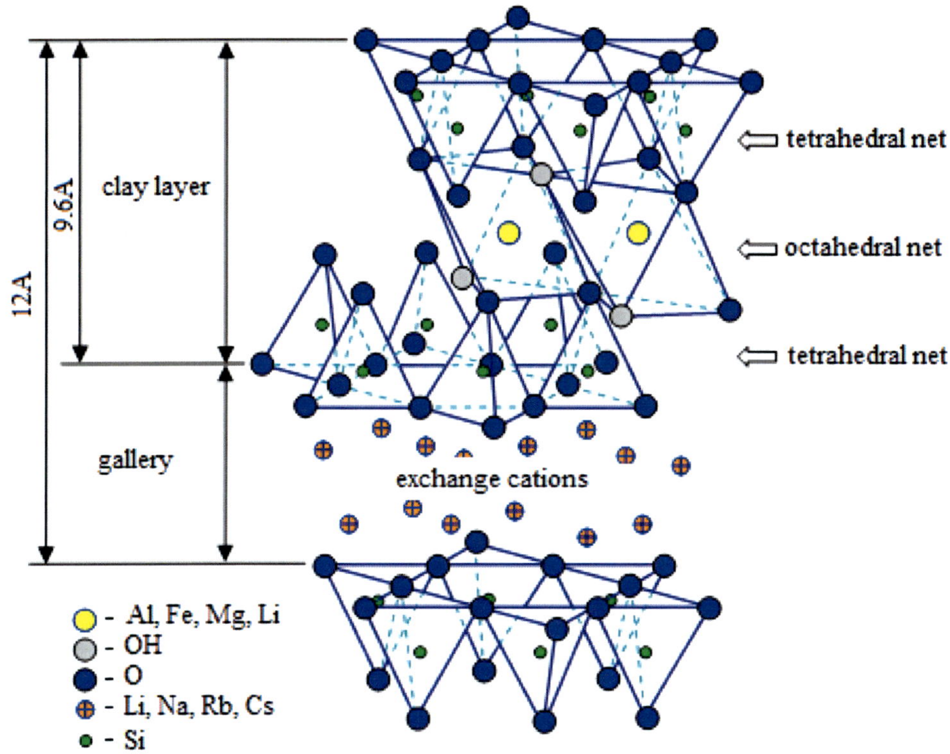

Figure 1.The structure of the layered silicate

3. Modification of the Layered Silicates

Layered silicates possess quite interesting properties – sharp drop of hardness at wetting, swelling at watering, dilution at dynamical influences and shrinking at drying.

The hydrophility of aluminous silicates is the reason of their incompatibility with the organic polymeric matrix and is the first hurdle need to be overridden at producing the polymeric nanocomposites.

One way to solve this problem is to modify the clay by the organic substance. The modified clay (organoclay) has at least two advantages: (1) it can be well dispersed in polymeric matrix [19] and (2) it interacts with the polymeric chain [13].

The modification of the aluminous silicates can be done with the replacement of the inorganic cations inside the galleries by the organic ones. The replacing by the cationic surface – active agents like bulk ammonium and phosphonium ions increases the room between the layers, decreases the surface energy of clay and makes the surface of the clay hydrophobic. The clays modified such a way are more compatible with the polymers and form the layered polymeric nanocomposites [52]. One can use the non-ionic modifiers besides the organic ones which link themselves to the clay surface through the hydrogen bond. Organoclays produced with help of non – ionic modifiers in some cases become more chemically stable than the organoclays produced with help of cationic modifiers (Fig.2.I) [20].

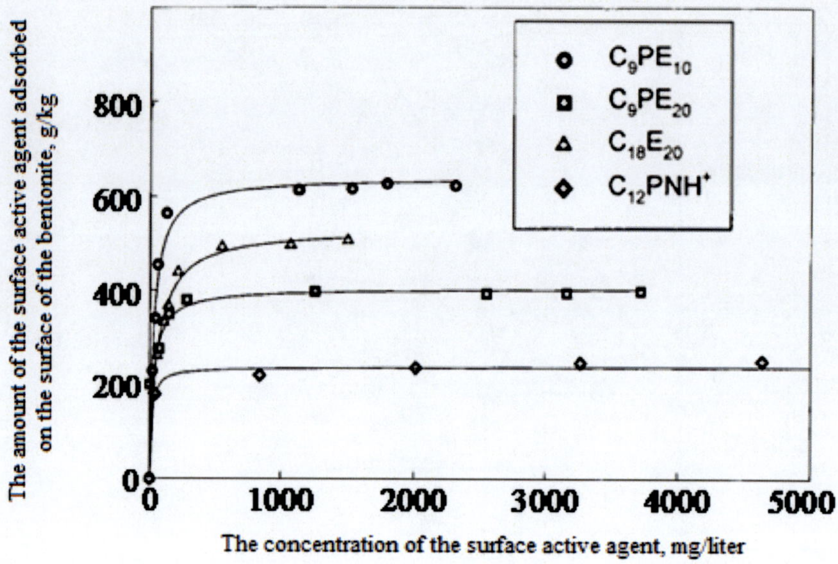

Figure 2.I. The adsorption of different modifiers on the clay surface

The least degree of desorption is observed for non – ionic interaction between the clay surface and organic modifier (Fig.2.II). The hydrogen bonds between the ethylenoxide grouping and the surface of the clay apparently make these organoclays more chemically stable than organoclays produced with non-ionic mechanism.

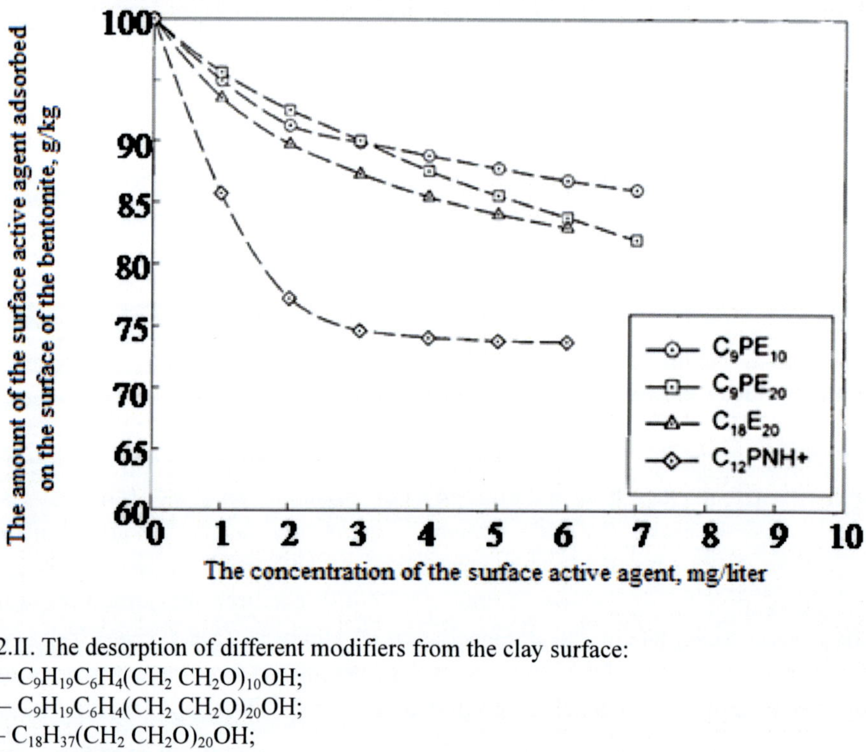

Figure 2.II. The desorption of different modifiers from the clay surface:
C_9PE_{10} – $C_9H_{19}C_6H_4(CH_2\,CH_2O)_{10}OH$;
C_9PE_{20} – $C_9H_{19}C_6H_4(CH_2\,CH_2O)_{20}OH$;
$C_{18}E_{20}$ – $C_{18}H_{37}(CH_2\,CH_2O)_{20}OH$;
C12PNH+ – C12H25C6H4NH+Cl-.

4. Structure of the Polymeric Nanocomposites on the Basis of the Montmorillonite

The study of the distribution of the organoclay in the polymeric matrix is of great importance because the properties of composites obtained are in the direct relation from the degree of the distribution.

According Giannelis [21], the process of the formation of the nanocomposite goes in several intermediate stages (Fig.3). The formation of the tactoid happens on the first stage – the polymer surrounds the agglomerations of the organoclay. The polymer penetrates the interlayer space of the organoclay on the second stage. Here the gap between the layers may reach 2-3 nanometers [22]. The further separation of the layers, third stage, results in partial dissolution and disorientation of the layers. Exfoliation is observed when polymer shifts the clay layers on more than 8-10 nanometers.

All mentioned structures may be present in real polymeric nanocomposites in dependence from the degree of distribution of the organoclay in the polymeric matrix. Exfoliated structure is the result of the extreme distribution of the organoclay. The excess of the organoclay or bad dispersing may born the agglomerates of the organoclays in the polymeric matrix what finds experimental confirmation in the X-ray analysis [11,12,21,23].

In the following subsections we describe a number of specific methods used at studying the structure of the polymeric nanocomposites.

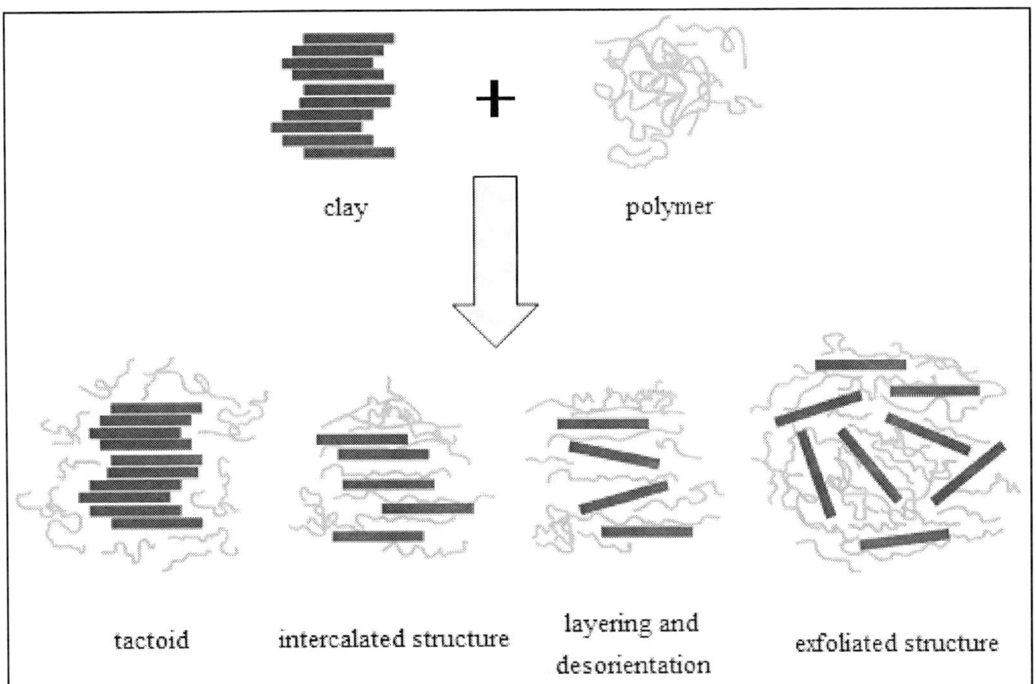

Figure 3. The schematic formation of the polymeric nanocomposite [24]

4.1. Determination of the Interlayer Space

The X-ray determination of the interlayer distance in the initial and modified layered silicates as well as in final polymeric nanocomposite is one of the main methods of studying the structure of the nanocomposite on the basis of the layered silicate. The peak in the small – angle diapason (2θ=6-8°C) is characteristic for pure clays and responds to the order of the structure of the silicate. This peak drifts to the smaller values of the angle 2θ in organomodified clays. If clay particles are uniformly distributed in the bulk of the polymeric matrix then this peak disappears, what witnesses on the disordering in the structure of the layered silicate. If the amount of the clay exceeds the certain limit of its distribution in the polymeric matrix, then the peak reappears again. This regularity was demonstrated on the instance of the polybutylenterephtalate (Fig.4) [11].

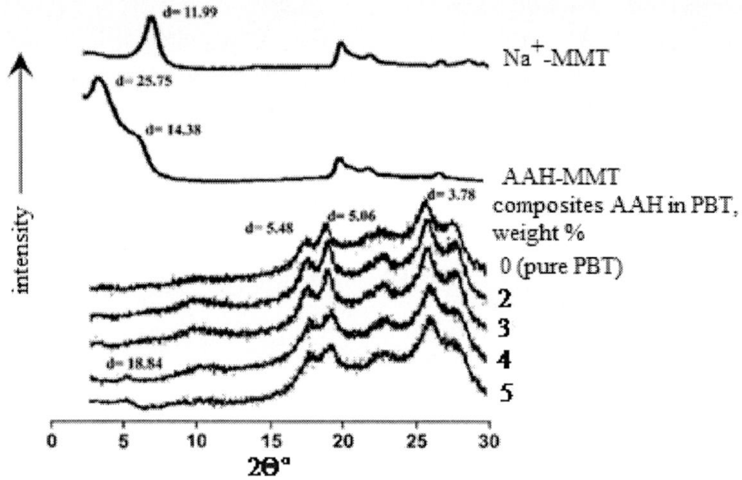

Figure 4. The data of the X-ray analysis for clay, organoclay and nanocomposite PBT/organoclay

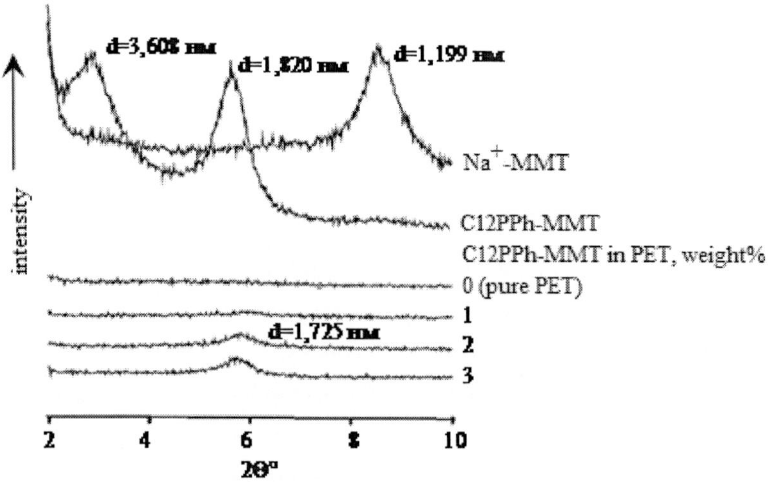

Figure 5. The data of the X-ray analysis for clay, organoclay and nanocomposite PET/organoclay

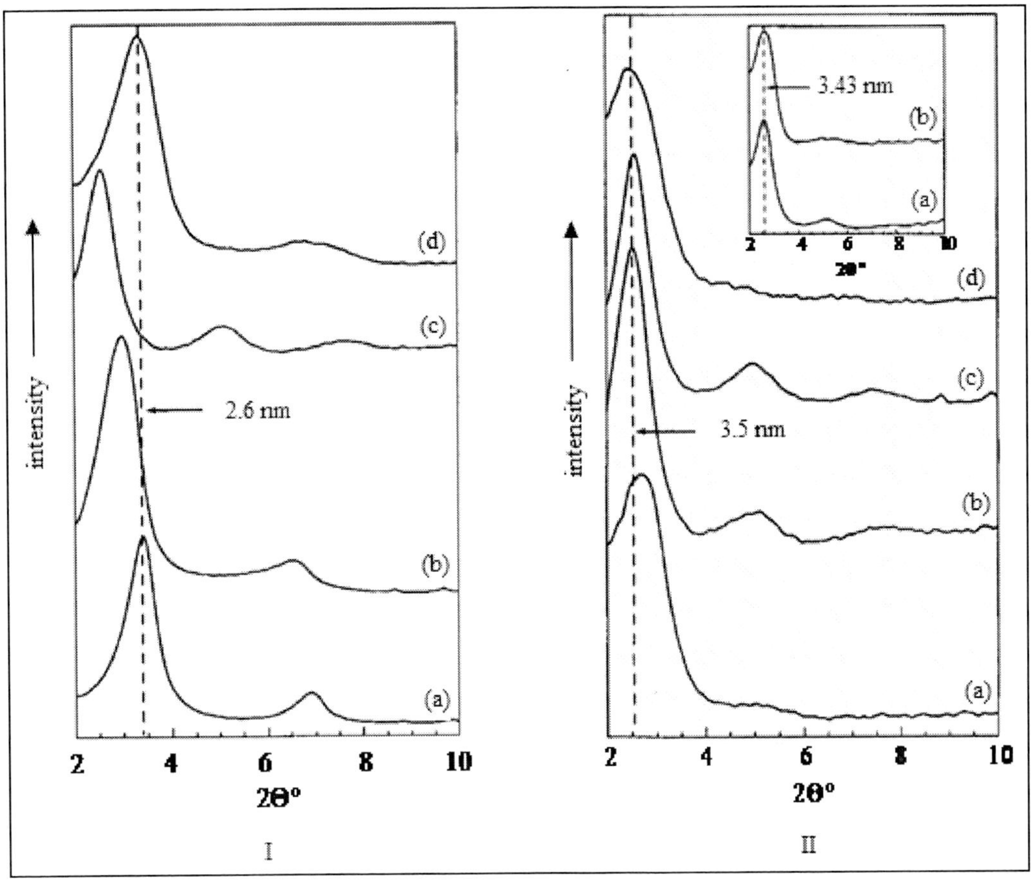

Figure 6. The data of the X-ray analysis for:
I. (a) – dimetyldioctadecylammonium (DMDODA) – hectorite;
(b) – 50% polystyrene (PS) / 50% DMDODA – hectorite;
(c) – 75% polyethylmetacrylate (PEM) / 25% DMDODA;
(d) – 50% PS/ 50 % DMDODA – hectorite after 24 hours of etching in cyclohexane.
II. mix of PS, PEM and ofganoclay:
(a) – 23.8% PS / 71.2% PEM / 5% DMDODA – hectorite;
(b) – 21.2% PS / 63.8% PEM / 15% DMDODA – hectorite;
(c) – 18.2% PS / 54.8% PEM / 27% DMDODA – hectorite;
(d) – 21.2% PS / 63.8% PEM / 15% DMDODA – hectorite after 24 hours of etching in cyclohexane.

The knowledge of the angle 2θ helps to define the size of the pack of the aluminous silicate consisting of the clay layer and interlayer space. The size of such pack increases in a row from initial silicate to polymeric nanocomposite according to the increase in the interlayer space. The average size of that pack for montmorillonite is 1.2-1.5 nanometers but for organomodified one varies in the range of 1.8-3.5 nanometers.

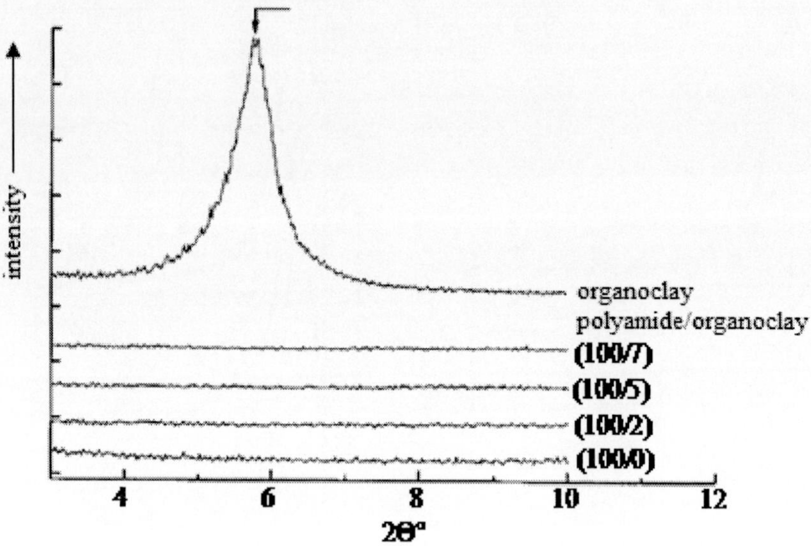

Figure 7. The data of the X-ray analysis for organoclay and nanocomposite polyamide acid / organoclay

Summing up we conclude that comparing the data of the X-ray analysis for the organoclay and nanocomposite allows for the determination of the optimal clay amount need be added to the composite. The data from the scanning tunneling (STM) and transmission electron (TEM) microscopes [27, 28] can be used as well.

4.2. The Degree of the Distribution of the Clay Particles in the Polymeric Matrix

The two structures, namely the intercalated and exfoliated ones, could be distinguished with the respect to the degree of the distribution of the clay particles, Fig.8. One should note that clay layers are quite flexible though they are shown straight in the figure. The formation of the intercalated or exfoliated structures depends on many factors, for example the method of the production of the nanocomposite or the nature of the clay etc [29].

The TEM images of the surface of the nanocomposites can help to find out the degree of the distribution of the nanosized clay particles, see plots (a) to (d) in the Fig.9.

The smooth surface tells about the uniform distribution of the organoclay particles. The surface of the nanocomposite becomes deformed with the increasing amount of the organoclay, see plots (a) to (d) in the Fig.10. Probably, this is due to the influence of the clay agglomerates [30,31].

Also one can use the STM images to judge on the degree of the distribution of the organoclay in the nanocomposite, Figs 11 and 12. If the content of the organoclay is 2-3weight% then the clay layers are separated by the polymeric layer of 4 to 10 nanometers width, Fig.11. If the content of the organoclay reaches 4-5weight% then the majority of the clay becomes well distributed however the agglomerates of 4 to 8 nanometers may appear.

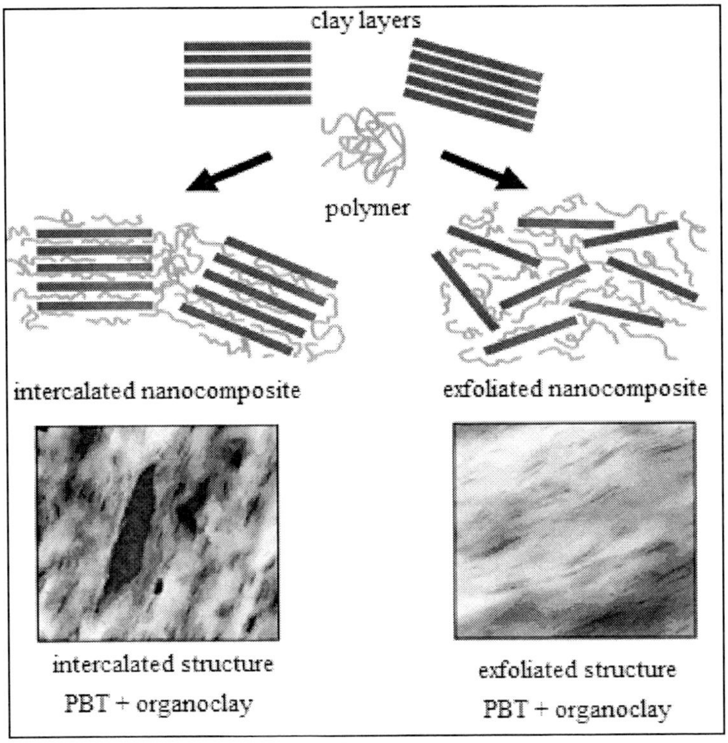

Figure 8. The formation of the intercalated and exfoliated structures of the nanocomposite

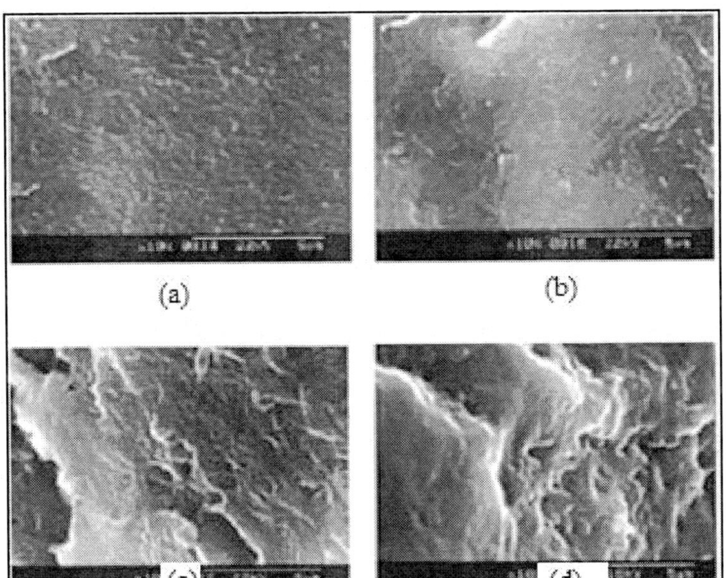

Figure 9. The images from scanning electron microscope for the nanocomposite surfaces:
(a) – pure PBT;
(b) – 3weight% of organoclay in PBT;
(c) – 4weight% of organoclay in PBT;
(d) – 5weight% of organoclay in PBT.

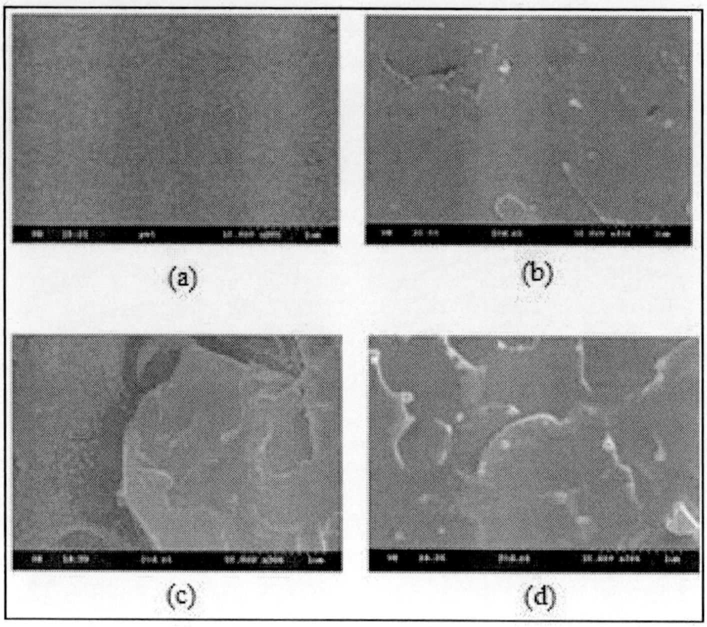

Figure 10. The images from scanning electron microscope for the nanocomposite surfaces:
(a) – pure PET;
(b) – 3weight% of organoclay in PET;
(c) – 4weight% of organoclay in PET;
(d) – 5weight% of organoclay in PET.

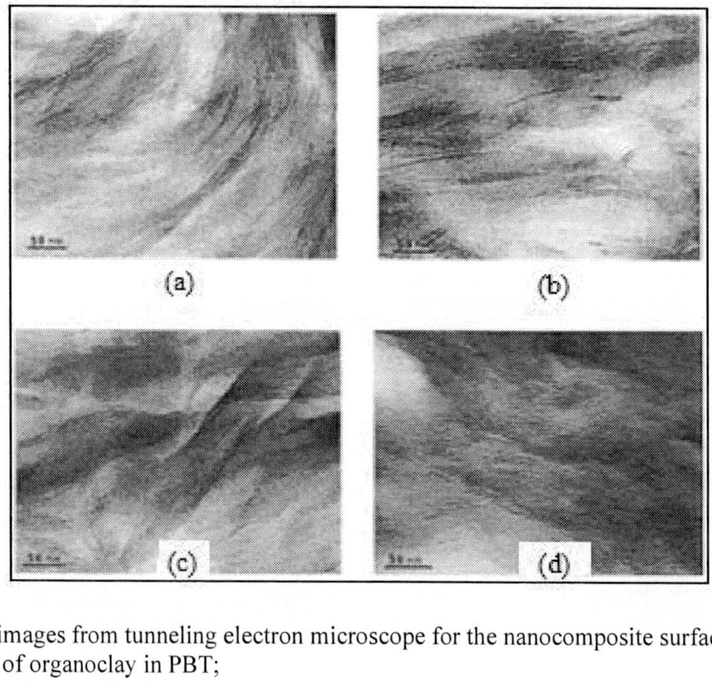

Figure 11. The images from tunneling electron microscope for the nanocomposite surfaces:
(a) – 2weight% of organoclay in PBT;
(b) – 3weight% of organoclay in PBT;
(c) – 4weight% of organoclay in PBT;
(d) – 5weight% of organoclay in PBT.

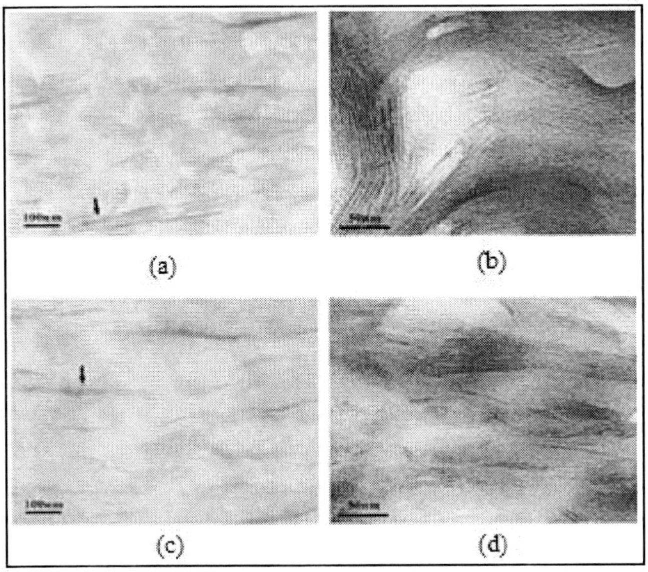

Figure 12. The images from tunneling electron microscope for the nanocomposite surfaces:
(a) – 1weight% of organoclay in PET;
(b) – 2weight% of organoclay in PET;
(c) – 3weight% of organoclay in PET;
(d) – 4weight% of organoclay in PET.

So the involvement of the X-ray analysis and the use of the microscopy data tell that the nanocomposite consists of the exfoliated clay at the low content (below 3weight%) of the organoclay.

5. Production of the Polymeric Nanocomposites on the Basis of the Aluminous Silicates

Different groups of authors [32-35] offer following methods for obtaining nanocomposites on the basis of the organoclays: (1) in the process of the synthesis of the polymer [33,36,37], (2) in the melt [38,39], (3) in the solution [40-46] and (4) in the sol-gel process [47-50].

The most popular ones are the methods of producing in melt and during the process of the synthesis of the polymer.

The producing of the polymeric nanocomposite in situ is the intercalation of the monomer into the clay layers. The monomer migrates through the organoclay galleries and the polymerization happens inside the layers [19,51], Fig.13.

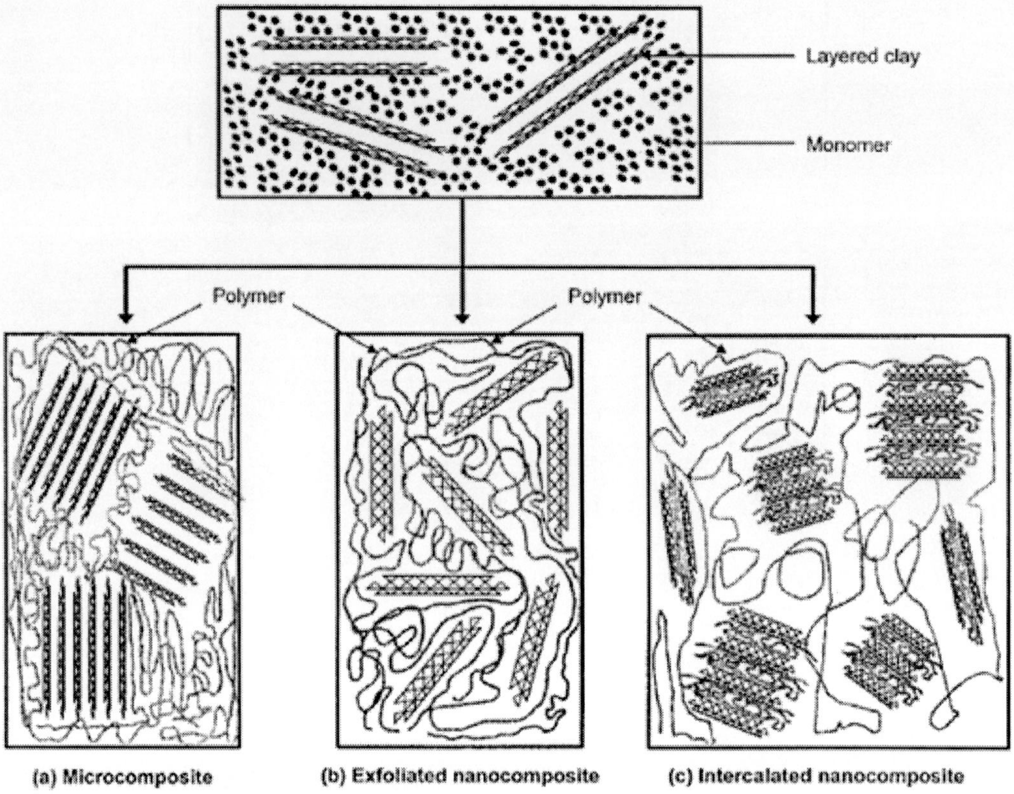

Figure 13. The production of the nanocomposite in situ:
(a) – microcomposite;
(b) – exfoliated nanocomposite;
(c) – intercalated nanocomposite [51].

The polymerization may be initiated by the heat, irradiation or other source. Obviously, the best results on the degree of the distribution of the clay particles in the polymeric matrix must emerge if using given method. This is associated with the fact that the separation of the clay layers happens in the very process of the inclusion of the monomer in the interlayer space. In other words, the force responsible for the separation of the clay layers is the growth of the polymeric chain whereas the main factor for reaching the necessary degree of the clay distribution in solution or melt is just satisfactory mixing. The most favorable condition for synthesizing the nanocomposites is the vacuuming or the flow of the inert gas. Besides, one has to use the fast speeds of mixing for satisfactory dispersing of the organoclay in the polymeric matrix.

The method of obtaining the polymeric nanocomposites in melt (or the method of extrusion) is the mixing of the polymer melted with the organoclay. The polymeric chains lose the considerable amount of the conformational entropy during the intercalation. The probable motive force for this process is the important contribution of the enthalpy of the interaction between the polymer and organoclay at mixing. One should add that polymeric nanocomposite on the basis of the organoclays could be successively produced by the extrusion [22]. The advantage of the extrusion method is the absence of any solvents what excludes the unnecessary leaks. Moreover the speed of the process is several times more and the technical side is simpler. The extrusion method is the best one in the industrial scales of production of the polymeric nanocomposites what acquires the lesser source expenses and easier technological scheme.

If one produces the polymer – silicate nanocomposite in solution then the organosilicate swells in the polar solvent such as toluene or N,N-dimethylformamide. Then the added is the solution of the polymer which enters the interlayer space of the silicate. The removing of the solvent by means of evaporation in vacuum happens after that. The main advantage of the given method is that "polymer – layered silicate" might be produced from the polymer of low polarity or even non-polar one. However this method is not widely used in industry because of much solvent consumption [52].

The sol-gel technologies find application at producing nanocomposites on the basis of the various ceramics and polymers. The initial compounds in these technologies are the alcoholates of specific elements and organic oligomers.

The alcoholates are firstly hydrolyzed and obtained hydroxides being polycondensated then. The ceramics from the inorganic 3D net is formed as a result. Also the method of synthesis exists in which the polymerization and the formation of the inorganic glass happen simultaneously. The application of the nanocomposites on the basis of ceramics and polymers as special hard defensive coverages and like optic fibers [53] is possible .

6. Properties of the Polymeric Nanocomposites

Many investigations in physics, chemistry and biology have shown that the jump from macroobjects to the particles of 1-10 nanometers results in the qualitative transformations in both separate phases and systems from them [54].

One can improve the thermal stability and mechanical properties of the polymers by inserting the organoclay particles into the polymeric matrix. It can be done by means of joining the complexes of properties of both the organic and inorganic substances, that is combining the light weight, flexibility and plasticity of former and durability, heat stability and chemical resistance of latter.

Nanocomposites demonstrate essential change in properties if compared to the non-filled polymers. So, if one introduces modified layered silicates in the range of 2 to 10weight% into the polymeric matrix then he observes the change in mechanical (tensile, compression, bending and overall strength), barrier (penetrability and stability to the solvent impact), optical and other properties. The increased heat and flame resistance even at low filler content is among the interesting properties too. The formation of the thermal isolation and negligible penetrability of the charred polymer to the flame provide for the advantages of using these materials.

The organoclay as a nanoaddition to the polymers may change the temperature of the destruction, refractoriness, rigidity and rupture strength. The nanocomposites also possess the increased rigidity modulus, decreased coefficient of the heat expansion, low gas-penetrability, increased stability to the solvent impact and offer broad range of the barrier properties [54]. In Table 1 we gather the characteristics of the nylon-6 and its derivative containing 4.7weight% of the organomodified montmorillonite.

Table 1. The properties of the nylon-6 and composite based on it [54]

	Rigidity modulus, GPa	Tensile strength, MPa	Temperature of the deformation, °C	Impact viscosity, kJ/m^2	Water consumption, weight %	Coefficient of the thermal expansion (x,y)
Nylon-6	1.11	68.6	65	6.21	0.87	13×10^{-5}
Nanocomposite	1.87	97.2	152	6.06	0.51	6.3×10^{-5}

In the paper "Polymeric nanocomposites based on the organomodified layered silicates: structure, manufacture and properties" by Mikitaev A.K., Bedanokov A.Yu. and Mikitaev M.A.

It is important that the temperature of the deformation of the nanocomposite increases on 87°C.

The thermal properties of the polymeric nanocomposites with the varying organoclay content are collected in Table 2.

The inclusion of the organoclay into the polybutyleneterephtalate leads to the increase in the glass transition temperature (T_g) from 27 to 33 degrees centigrade if the amount of the clay raises from zero to 2weight%. That temperature does not change with the further increase of the organoclay content. The increase in T_g may be the result of two reasons [56-59]. The first is the dispersion of the small amount of the organoclay in the free volume of the polymer, and the second is the limiting of the mobility of the segments of the polymeric chain due to its interlocking between the layers of the organoclay.

Table 2. The main properties of the polymeric nanocomposites

Property	Composition								
	Polybutyleneterephtalate + AAX-montmorillonite					Polyethyleneterephtalate + C_{12}PPh-montmorillonite			
	Organoclay content, %								
	0	2	3	4	5	0	1	2	3
Viscosity, dliter/g	0.84	1.16	0.77	0.88	0.86	1.02	1.26	0.98	1.23
T_g, °C	27	33	34	33	33	---	---	---	---
T_m, °C	222	230	230	229	231	245	247	245	246
T_d, °C	371	390	388	390	389	370	375	384	386
W_{tR}^{600c}, %	1	6	7	7	9	1	8	15	21
Strength limit, MPa	41	50	60	53	49	46	58	68	71
Rigidity modulus, GPa	1.37	1.66	1.76	1.80	1.86	2.21	2.88	3.31	4.10
Relative enlargement, %	5	7	6	7	7	3	3	3	3

In the paper "Polymeric nanocomposites based on the organomodified layered silicates: structure, manufacture and properties" by Mikitaev A.K., Bedanokov A.Yu. and Mikitaev M.A.

The same as the T_g, the melting temperature T_m increases from the 222 to 230°C if the organoclay content raises from 0 to 2weight% and stays constant up to 5weight%, see Table 2. This increase might be the consequence of both complex multilayer structure of the nanocomposite and interaction between the organoclay and polymeric chain [60,61]. Similar regularities have been observed in other polymeric nanocomposites also.

The thermal stability of the nanocomposites polybutyleneterephtalate, briefly PBT, (or polyethyleleterephtalate) / organoclay determined by the thermogravimetric analysis is presented in Table 3 and in Figs 14 and 15 [11,12].

The temperature of the destruction, T_D, increases with the organoclay content up to 350°C in case of the composite PBT / organoclay. The thermogravimetrical curves for pure and composite PBTs have similar shapes below 350°C. The values of temperature T_D depends on the amount of organoclay above 350°C. The organoclay added becomes a barrier for volatile products being formed during the destruction [61,62]. Such example of the improvement of the thermal stability was studied in papers [63,64]. The mass of the remnant at 600°C increases with organoclay content.

Following obtained data authors draw the conclusion that the optimal results for thermal properties are being obtained if 2weight% of the organoclay is added [11,12,19].

The great number of studies on the polymeric composite organoclay – based materials show [11,12,13,19] that the inclusion of the inorganic component into the organic polymer improves the thermal stability of the latter, see Tables 3 and 4.

Table 3. The thermal properties of the fibers from PET with varying organoclay content

Organoclay content, weight %	η_{inh}^{a}	T_m (°C)	ΔH_m^{b} (J/g)	$T^{i}_{D}{}^{c}$ (°C)	Wt_R^{600d} (%)
0 (pure PET)	1.02	245	32	370	1
1	1.26	247	32	375	8
2	0.98	245	33	384	15
3	1.23	246	32	386	21

In the paper "Polymeric nanocomposites based on the organomodified layered silicates: structure, manufacture and properties" by Mikitaev A.K., Bedanokov A.Yu. and Mikitaev M.A.

a – viscosities were measured at 30°C using 0.1 gram of polymer on 100 milliliters of solution in mix phenol / tetrachlorineethane (50/50);

b – change in enthalpy of melting;

c – initial temperature of decomposition;

d – weight percentage of the coke remnant at 600°C.

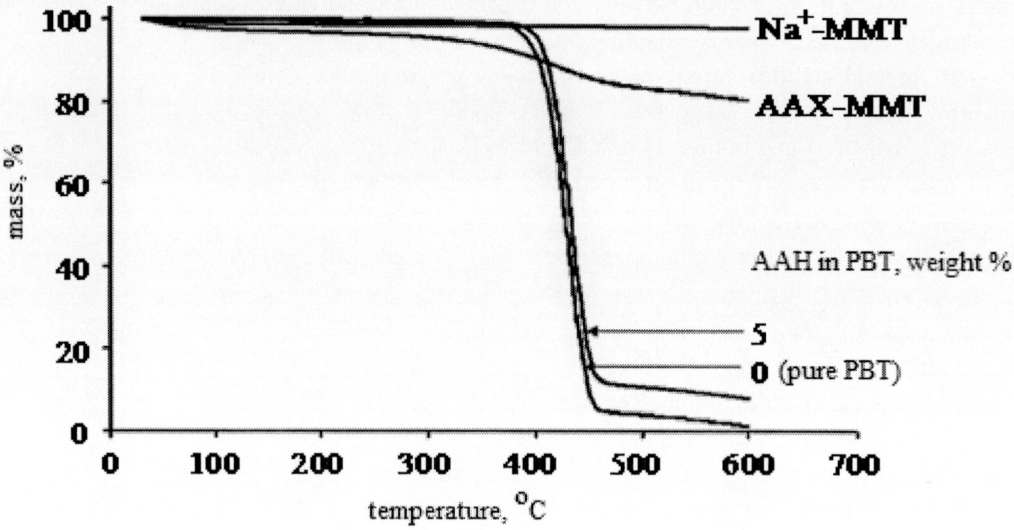

Figure 14. The thermogravimetrical curves for the montmorillonites, PBT and nanocomposites PBT / organoclay

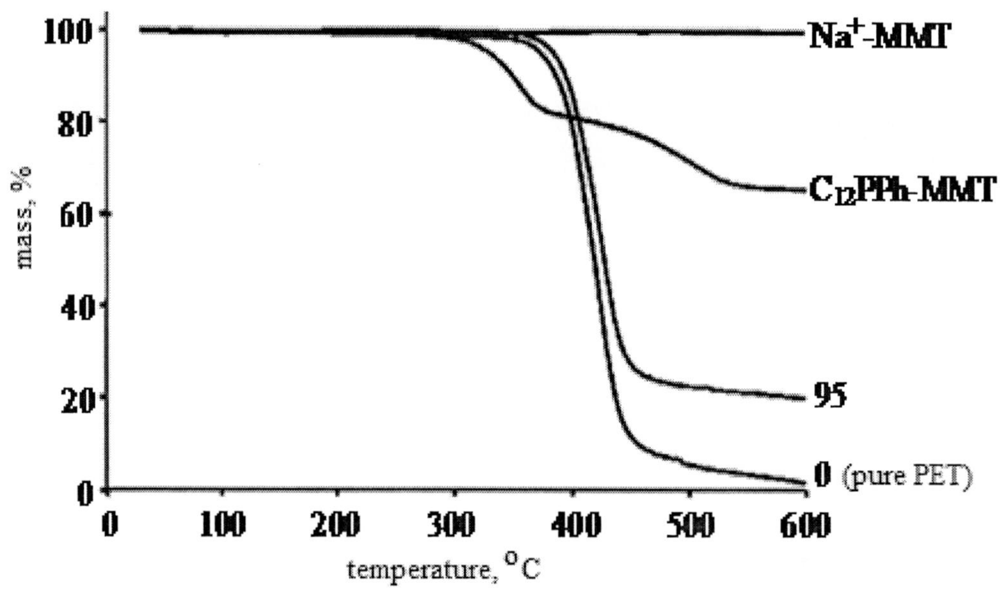

Figure 15. The thermogravimetrical curves for the montmorillonites, PET and nanocomposites PET / organoclay

The values of the melting temperature increase from 222 to 230°C if the amount of the organoclay added reaches 2weight% and then stay constant. This effect can be explained by both thermal isolation of the clay and interaction between the polymeric chain and organoclay [43,64]. The increase in the glass transition temperature also occurs what can be a consequence of several reasons [55,56,58]. One of the main among them is the limited motion of the segments of the polymeric chain in the galleries within the organoclay.

Table 4. The basic properties of the nanocomposite based on PBT with varying organoclay content

Organoclay content, weight %	I.V.[a]	T_g	T_m (°C)	$T_D^{i\ b}$ (°C)	Wt_R^{600c} (%)
0 (pure PBT)	0,84	27	222	371	1
2	1,16	33	230	390	6
3	0,77	34	230	388	7
4	0,88	33	229	390	7
5	0,86	33	231	389	9

In the paper "Polymeric nanocomposites based on the organomodified layered silicates: structure, manufacture and properties" by Mikitaev A.K., Bedanokov A.Yu. and Mikitaev M.A.

a – viscosities were measured at 30°C using 0.1 gram of polymer on 100 milliliters of solution in mix phenol / tetrachlorineethane (50/50);

b – initial temperature of the weight loss;

c – weight percentage of the coke remnant at 600°C.

If the organoclay content in the polymeric matrix of the PBT reaches 2weight% then both the temperature of the destruction increases and the amount of the coke remnant increase at 660°C and then both stay practically unchanged with the further increase of the organoclay content up to 5weight%. The loss of the weight due to the destruction of the polymer in pure PBT and its composites looks familiar in all cases below 350°C. The amount of the organoclay added becomes important above that temperature because the very clays possess good thermal stability and make thermal protection by their layers and form a barrier preventing the volatile products of the decomposition to fly off [43,60]. Such instance of the improvement in thermal properties was observed in many polymeric composites [64-68]. The weight of the coke remnant increases with the rising organoclay content up to 2weight% and stays constant after that. The increase of the remnant may be linked with the high thermal stability of the organoclay itself. Also it is worth noticing that the polymeric chain closed in interlayer space of the organoclay has fewer degrees of oscillatory motion at heating due to the limited interlayer space and the formation of the abundant intermolecular bonds between the polymeric chain and the clay surface. And the best result is obtained at 2-3weight% content of the organoclay added to the polymer.

If one considers the influence of the organoclay added to the polyetheleneterephtalate, briefly PET, [69] then the temperature of the destruction increases on 16°C at optimal amount of organoclay of 3weight%. The coke remnant at 600°C again increases with the rising organoclay content, see Table 3.

Regarding the change in the temperature of the destruction in the cases of PET and PBT versus the organoclay content one can note that both trends look similar. However the coke remnant considerably increases would the tripheyldodecylphosphonium cation be present within the clay. The melting temperature does not increase in case of the organoclay added into the PET in contrary to the case of PBT. Apparently this may be explained by the more crystallinity of the PBT and the growth of the degree of crystallinity with the organoclay content.

It becomes obvious after analyzing the above results that the introduction of the organoclay into the polymer increases the thermal stability of the latter according the (1) thermal isolating effect from the clay layers and (2) barrier effect in relation to the volatile products of destruction.

The studying of the mechanical properties of the nanocomposites, see Table 2, have shown that the limit of the tensile strength increases with the organoclay added up to 3weight% for the majority of the composites. Further addition of the organoclay, up to 5weight%, results in the decreasing limit of the tensile strength. We explain this by the fact that agglomerates appear in the nanocomposite when the organoclay content exceeds the 3weight% value [61,70,71]. The proof for the formation of the agglomerates have been obtained from the X-ray study and using the data from electron microscopes.

Nevertheless the rigidity modulus increases with the amount of the organoclay added into the polymeric matrix, the resistance of the clay itself being the explanation for that. The oriented polymeric chains in the clay layers also participate in the increase of the rigidity modulus [72]. The percentage of enlargement at breaking became 6-7weight% for all mixes.

Using data of the Table 2 we explain the improvement in the mechanical properties of the nanocomposites with added organoclay up to 3weight% by the good degree of distribution of the organoclay within the polymeric matrix. The degree of the improvement also depends on the interaction between the polymeric chain and clay layers.

The study of the influence of the degree of the extract of fibers on the mechanical properties has shown that the limit of strength and the rigidity modulus both increase in PBT whereas they decrease in nanocomposites, Table 5. This can be explained by the breaking of the bonds between the organoclay and PBT at greater degree of extract. Such phenomena have been observed in numerous polymeric composites [73-75].

The first notions on the lowered flammability of the polymeric nanocomposites on the organoclay basis appeared in 1976 in the patent on the composite based on the nylon-6 [5]. The serious papers in the field were absent till the 1995 [76].

The use of the calorimeter is very effective for studying the refractoriness of the polymers. It can help at measuring the heat release, the carbon monoxide depletion and others. The speed of the heat release is one of the most important parameters defining the refractoriness [77]. The data on the flame resistance in various polymer / organoclay systems such as layered nanocomposite nylon-6 / organoclay, intercalated nanocomposites

Table 5. The ability to stretch of the nanocomposites PBT / organoclay at varying degrees of extract

Organoclay content, weight %	Limit of strength, MPa			Rigidity modulus, GPa		
	DR=1	DR=3	DR=6	DR=1	DR=3	DR=6
0 (pure PBT)	41	50	52	1.37	1.49	1.52
3	60	35	29	1.76	1.46	1.39

In the paper "Polymeric nanocomposites based on the organomodified layered silicates: structure, manufacture and properties" by Mikitaev A.K., Bedanokov A.Yu. and Mikitaev M.A.

Table 6. Calorimetric data

Sample	remnant (%)±0.5	Peak of the HRR (Δ%) (kW/m^2)	Middle of the HRR (Δ%) (kW/m^2)	Average value H_c (MJ/kg)	Average value SEA (m^2/kg)	Average CO left (kg/kg)
Nylon-6	1	1010	603	27	197	0.01
Nylon-6 / organoclay, 2%, delaminated	3	686 (32%)	390 (35%)	27	271	0.01
Nylon-6 / organoclay, 5%, delaminated	6	378 (63%)	304 (50%)	27	296	0.02
Polystyrene	0	1120	703	29	1460	0.09
PS / organoclay, 3%, bad mixing	3	1080	715	29	1840	0.09
PS / organoclay, 3%, intercalated / delaminated	4	567 (48%)	444 (38%)	27	1730	0.08
PS w/DBDPO/Sb_2O_3, 30%	3	491 (56%)	318 (54%)	11	2580	0.14
Polypropylene	0	1525	536	39	704	0.02
PP / organoclay, 2%, intercalated	5	450 (70%)	322 (40%)	44	1028	0.02

In the paper "Polymeric nanocomposites based on the organomodified layered silicates: structure, manufacture and properties" by Mikitaev A.K., Bedanokov A.Yu. and Mikitaev M.A.

H_c – heat of combustion;

SEA – specific extinguishing area;

DBDPO – dekabrominediphenyloxide;

HRR – speed of the heat release

polystyrene (or polypropylene) / organoclay were given in paper [78] in where the lowered flammability was reported, see Table 6. And the lowered flammability have been observed in systems with low organoclay content, namely in range from 2 to 5weight%.

The curve of the heat release for the polypropylene and the nanocomposite on its basis (organoclay content varying from 2 to 4weight%) is given in the Fig.16 from which one can see that the speed of the heat release for the nanocomposite enriched with the 4weight% organoclay (the interlayer distance 3.5 nanometers) is 75% less than for pure polypropylene.

The comparison of the experimental data for the nanocomposites on the basis of the nylon-6, polypropylene and polystyrene gathered in Table 7 show that the heat of combustion, the smoke release and the amount of the carbon monoxide are almost constant at varying organoclay content. So we conclude that the source for the increased refractoriness of these materials is the stability of the solid phase and not the influence of the vapor phase. The data for the polystyrene with the 30% of the dekabrominediphenyloxide and Sb_2O_3 are given in Table 6 as the proof of the influence of the vapor phase of bromine. The incomplete combustion of the polymeric material in the latter case results in low value of the heat of the combustion and high quantity of the carbon monoxide released [79].

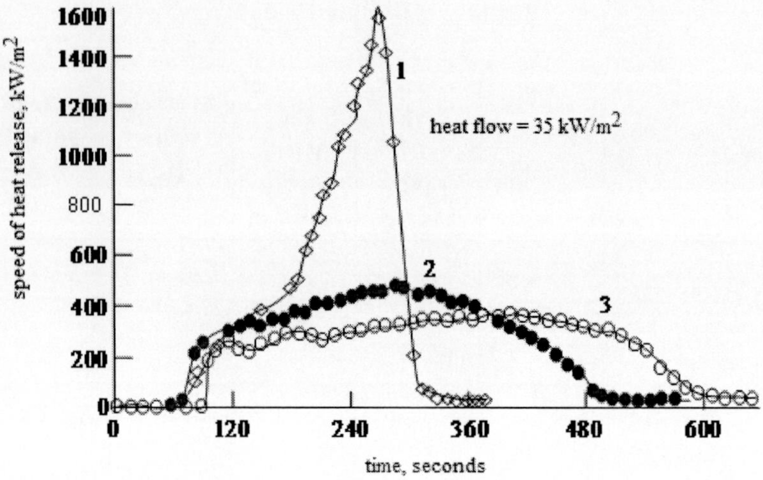

Figure 16. The speed of the heat release for:
1 – pure polypropylene;
2 – nanocomposite with 2weight% of organoclay;
3 – nanocomposite with 4weight% of organoclay.

One should note that the mechanism for the increased fire resistance of the polymeric nanocomposites on the basis of the organoclays is not, in fact, clear at all. The formation of the barrier from the clay layers during the combustion at their collapse is supposed to be the main mechanism. That barrier slows down the combustion [80]. In our paper we study the influence of the nanocomposite structure on the refractoriness. The layered structure of the nanocomposite expresses higher refractoriness comparing to that in intercalated nanocomposite, see Fig.17.

The data on the polymeric polystyrene – based nanocomposites presented in Fig.17 are for (1) initial ammoniumfluorine hectorite and (2) quaternary ammonium montmorillonite. The intercalated nanocomposite was produced in first case whereas the layered-intercalated nanocomposite was produced in the second one. But because the chemical nature and the morphology of the organoclay used was quite different it is very difficult to draw a unique conclusion about the flame resistance in polymeric nanocomposites produced. Nonetheless, one should point out that good results of the same quality were obtained for both layered and intercalated structures when studying the aliphatic groupings of the polyimide nanocomposites based on these clays. The better refractoriness is observed in case of polystyrene embedded in layered nanocomposite while intercalated polystyrene – based nanocomposite (with MMT) also exhibits increased refractoriness.

As one can see the from the Fig.18, the speed of the heat release for the nanocomposite produced in nitrogen atmosphere at 170°C is much lower than for other samples. Probably, the reason for the low refractoriness of the nanocomposite produced in extruder without the vacuuming at 180°C is the influence of the high temperature and of the oxygen from the air what can lead to the destruction of the polymer in such conditions of the synthesis.

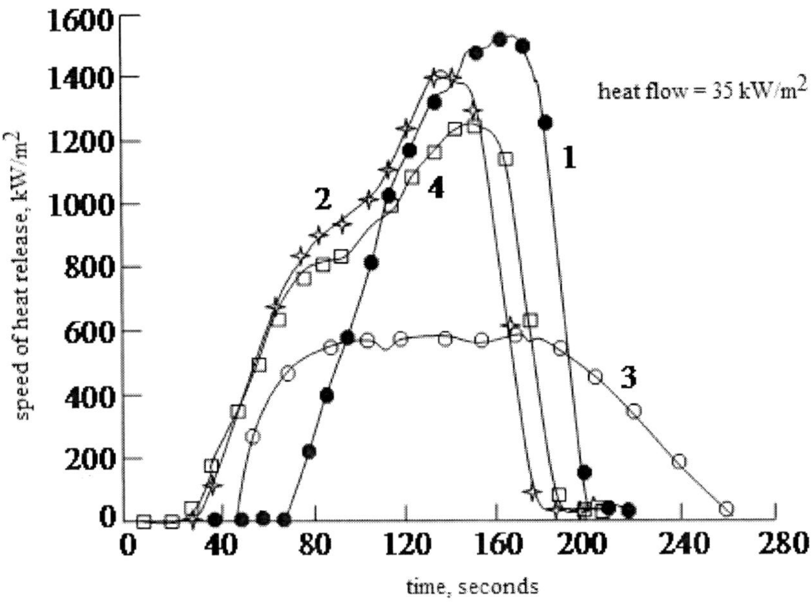

Figure 17. The speed of the heat release for:
1 – pure polystyrene (PS);
2 – PS mixed with 3weight% of Na^+ MMT;
3 – intercalated / delaminated PS (3weight% 2C18-MMT) extruded at 170°C;
4 – intercalated PS (3weight% C14-FH) extruded at 170°C.

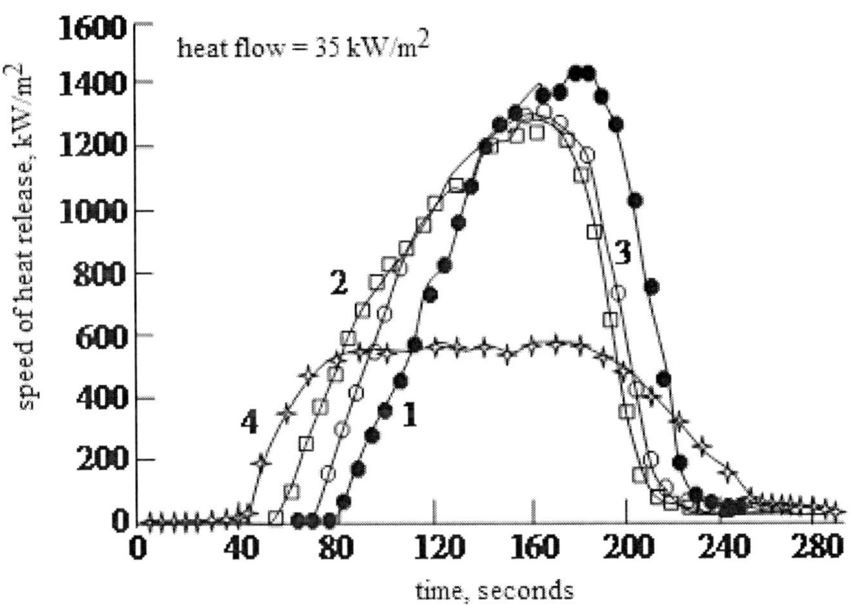

Figure 18. The speed of the heat release for:
1 – pure polysterene (PS);
2 – polystyrene with Na-MMT;
3 – intercalated PS with organomontmorillonite obtained in extruder at 185°C;
4 – intercalated / layered PS with organoclay obtained in extruder at 170°C in nitrogen atmosphere or in vacuum.

It is impossible to give an exact answer on the question about how the refractoriness of organoclay – based nanocomposites increases basing on only the upper experimental data but the obvious fact is that the increased thermal stability and refractoriness are due to the presence of the clays existing in the polymeric matrix as nanoparticles and playing the role of the heat isolators and elements preventing the flammable products of the decomposition to fly off.

There are still many problems unresolved in the field but indisputably polymeric nanocomposites will take the leading position in the chemistry of the advanced materials with high heat and flame resistance. Such materials can be used either as itself or in combination with other agents reducing the flammability of the substances.

The processes of the combustion are studied for the number of a polymeric nanocomposites based on the layered silicates such as nylon-6.6 with 5weight% of Cloisite 15A – montmorillonite being modified with the dimethyldialkylammonium (alkyls studied C_{18}, C_{16}, C_{14}), maleinated polypropylene and polyethylene, both (1.5%) with 10weight% Cloisite 15A. The general trend is two times reduction of the speed of the heat release. The decrease in the period of the flame induction is reported for all nanocomposites in comparison with the initial polymers [54].

Table 7. Flammability of several polymers and composites

Sample	Coke remnant, weight%	Max speed of heat release, kW/m^2	Average value of heat release, kW/m^2	Average heat of combustion, MJ/kg	Specific smoke release, m^2/kg	CO release, kg/kg
Nylon-6	1	1010	603	27	197	0.01
Nylon-6 + 2% of silicate (delaminated)	3	686	390	27	271	0.01
Nylon-6 + 5% of silicate (delaminated)	6	378	304	27	296	0.02
Nylon-12	0	1710	846	40	387	0.02
Nylon-12 + 2% of silicate (delaminated)	2	1060	719	40	435	0.02
Polystyrene	0	1562	803	29	1460	0.09
PS + 3% of silicate Na-MMT	3	1404	765	29	1840	0.09
PS + 3% of silicate C14-FH (intercalated)	4	1186	705	28	1790	0.09
PS + 3% of silicate 2C18-MMT (delaminated)	4	567	444	28	1730	0.08
Polypropylene	0	1525	536	39	704	0.02
PP + 2% of silicate (intercalated)	3	450	322	40	1028	0.02

In the paper "Polymeric nanocomposites based on the organomodified layered silicates: structure, manufacture and properties" by Mikitaev A.K., Bedanokov A.Yu. and Mikitaev M.A.

The influence of the nanocomposite structure on its flammability is reflected in the Table 7. One can see that the least flammability is observed in delaminated nanocomposite based on the polystyrene whereas the flammability of the intercalated composite is much higher [54].

The optical properties of the nanocomposites are of much interest too. The same materials could be either transparent or opaque depending on certain conditions. For example in Fig.19 we see transparency, plot (a), and turbidity, plot (c), of the material in dependence of the frequency of the current applied.

The effect in Fig. 19 is reversible and can be innumerately repeated. The transparent and opaque states exhibit the memory effect after the applied current switched off, plots (b) and (d) in Fig. 19. The study of the intercalated nanocomposites based on the smectite clays reveals that the optical and elecrooptical properties depend on the degree of intercalation [81].

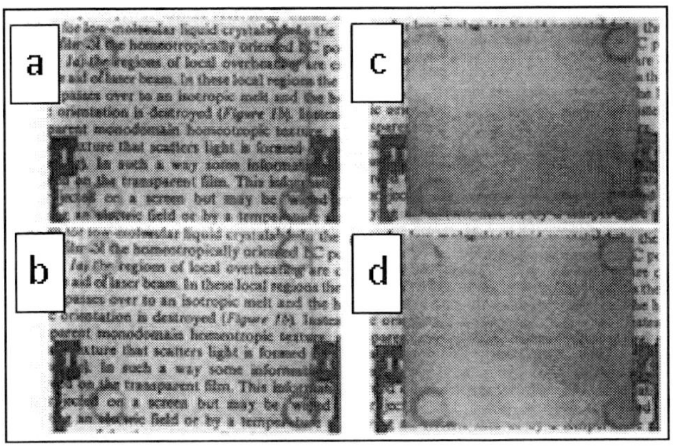

Figure 19. The optical properties of the clay – based nanocomposites in dependence of the applied electric current:
(a) – low frequency, switched on;
(b) – low frequency, switched off;
(c) – high frequency, switched on;
(d) – high frequency, switched off.

Conclusions

The quantity of the papers in the field of the nanocomposite polymeric materials has grown multiple times in recent years. The possibility to use almost all polymeric and polycondensated materials as a matrix is shown. The nanocomposites from various organoclays and polymers have been synthesized. Here is just a small part of the compounds for being the matrix referenced in literature: polyacrylate [83], polyamides [82,84,85], polybenzoxasene [86], polybutyleneterephtalate [11,82,87], polyimides [88], polycarbonate [89], polymethylmetacrylate [90], polypropylene [91,92], polystyrene [90], polysulphones [93], polyurethane [94], polybuthyleneterephtalate and polyethyleneterephtalate [10,65,68,79,95,99-107], polyethylene [96], epoxies [97].

The organomodified montmorillonite is of the special interest because it can be an element of the nanotechnology and it can also be a carrier of the nanostructure and of asymmetry of length and width in layered structures. The organic modification is being usually performed using the ion-inducing surface-active agents. The non-ionic hydrophobisation of the surface of the layered structures have been reported either. The general knowledge about the methods of the study is being formed and the understanding of the structure of the nanocomposite polymeric materials is becoming clear. Also scientists come closer to the realizing of the relations between the deformational and strength properties and the specifics of the nanocomposite structure. The growth of researches and their direction into the nanoarea forecasts the fast broadening of the industrial involvement to the novel and attractive branch of the materials science.

References

[1] Romanovsky B.V., Makshina E.V. // Sorosovskii obrazovatelniu zhurnal, 8(2) 2004: 50-55 [in Russian]

[2] Golovin Yu.I. // *Priroda*, 1 2004 [in Russian]

[3] Carter L.W., Hendrics J.G., Bolley D.S. *United States Patent №2,*531,396; 1950.

[4] Blumstain A. *Bull Chem Soc* 1961:899-905.

[5] Fujiwara S., Sakamoto T. *Japanese Application №* 109,998; 1976.

[6] Usuki A., Kojima Y., Kawasumi M., Okada A., Fukushima Y., Kurauchi T., Kamigatio O. *Journal of Appl polym science*. 1995;55:119.

[7] Usuki A., Koiwai A., Kojima Y., Kawasumi M., Okada A., Kurauchi T., Kamigaito O. *Journal of Appl polym science*. 1995;55:119

[8] Okada A., Usuki A. *Mater Sci Engng*. 1995;3:109.

[9] Okada A., Fukushima Y., Kawasumi M., Inagaki S., Usuki A., Sugiyama S., Kurauchi T., Kamigaito O. *United States Patent №4*,739,007; 1988.

[10] Mikitaev M.A., Lednev O.B., Kaladjian A.A., Beshtoev B.Z., Bedanokov A.Yu., Mikitaev A.K. // *Second International Conference* (Nalchik 2005)

[11] J.-H. Chang, Y.U. An, S.J. Kim, S. Im. *Polymer,* 2003;44:5655-5661.

[12] A.K.Mikitaev, A.Y.Bedanokov, O.B.Lednev, M.A.Mikitaev Polymer/silicate nanocomposites based on organomodified clays / Polymers, Polymer Blends, Polymer Composites and Filled Polymers. *Synthesis, Properties, Application*. Nova Science Publishers. New York 2006.

[13] D.M. Delozier, R.A. Orwoll, J.F. Cahoon, N.J. Johnston, J.G. Smith, J.W. Connell. *Polymer*, 2002;43:813-822.

[14] Kelly P., Akelah A., Moet A. *J. Mater. Sci*. 1994. V.29. P.2274-2280.

[15] J.-H. Chang, Y.U. An, D. Cho, E.P. Giannelis. *Polymer, 2003*;44:3715-3720.

[16] Yano K, Usuki A, Okada A. *J Polym Sci*, Part A: Polym Chem 1997;35:2289.

[17] Garcia-Martinez JM, Laguna O, Areso S, Collar EP. *J Polym Sci*, Part B: Polym Phys 2000;38:1564.

[18] Giannelis E. P., Krishnamoorti R., Manias E. *Advances in Polymer Science, Vol.138* Springer-Verlag Berlin Heidelberg 1999.

[19] D.M. Delozier, R.A. Orwoll, J.F. Cahoon, J.S. Ladislaw, J.G. Smith, J.W. Connell. *Polymer, 2003;44*:2231-2241.

[20] Y.-H. Shen. *Chemosphere, 2001*;44:989-995.

[21] E.P. Giannelis, *Adv. Mater*. 8 (1996) 29–35.

[22] H.R. Dennis, D.L. Hunter, D. Chang, S. Kim, J.L. White, J.W. Cho, D.R. Paul. *Polymer*, 2001;42:9513-9522.

[23] X. Kornmann, H. Lindberg, L.A. Berglund, *Polymer 42* (2001) 1303–1310.

[24] T.D. Fornes; D.R. Paul. Formation and properties of nylon 6 nanocomposites. *Polímeros vol.13 no.4 São Carlos* Oct./Dec. 2003.

[25] D. Voulgaris, D. Petridis. *Polymer,* 2002;43:2213-2218.

[26] H.-L. Tyan, Y.-C. Liu, K.-H. Wei. *Polymer*, 1999;40:4877-4886.

[27] Davis CH, Mathias LJ, Gilman JW, Schiraldi DA, Shields JR, Trulove P, Sutto TE, Delong HC. *J Polym Sci*, Part B: Polym Phys 2002;40:2661.

[28] Morgan AB, Gilman JW. *J Appl Polym Sci* 2003;87:1329.

[29] John N. Hay and Steve J. Shaw. Organic-inorganic hybridssthe best of both worlds? *Europhysics News* (2003) Vol.34 No.3.

[30] Chang JH, An YU, Sur GS. *J Polym Sci Part B*: Polym Phys 2003;41:94.

[31] Chang JH, Park DK, Ihn KJ. *J Appl Polym Sci* 2002;84:2294.

[32] Pinnavaia TJ. *Science* 1983;220:365.

[33] Messersmith PB, Giannelis EP. *Chem Mater* 1993;5:1064.

[34] Vaia RA, Ishii H, Giannelis EP. *Adv Mater* 1996;8:29.

[35] Gilman JW. *Appl Clay Sci* 1999;15:31.

[36] Fukushima Y, Okada A, Kawasumi M, Kurauchi T, Kamigaito O. Clay Miner 1988;23:27.

[37] Akelah A, Moet A. *J Mater Sci* 1996;31:3589.

[38] Vaia RA, Ishii H, Giannelis EP. *Adv Mater* 1996;8:29.

[39] Vaia RA, Jandt KD, Kramer EJ, Giannelis EP. *Macromolecules 1995*;28:8080.

[40] Greenland DG. *J Colloid Sci* 1963;18:647.

[41] Chang JH, Park KM. *Polym Engng Sci* 2001;41:2226.

[42] Greenland DG. *J Colloid Sci* 1963;18:647.

[43] Chang JH, Seo BS, Hwang DH. *Polymer 2002*;43:2969.

[44] Vaia RA, Jandt KD, Kramer EJ, Giannelis EP. *Macromolecules* 1995;28:8080.

[45] Fukushima Y, Okada A, Kawasumi M, Kurauchi T, Kamigaito O. Clay *Miner* 1988;23:27.

[46] Chvalun S.N. // *Priroda 2000;* 7 [in Russian]

[47] Brinker C.J., Scherer G.W. *Sol-Gel Science*. Boston, 1990.

[48] Mascia L, Tang T. *Polymer* 1998;39:3045.

[49] Tamaki R, Chujo Y. *Chem Mater* 1999;11:1719.

[50] Serge Bourbigot e.a. Investigation of Nanodispersion in Polystyrene–Montmorillonite Nanocomposites by Solid-State NMR. *Journal of Polymer Science: Part* B: Polymer Physics, Vol. 41, 3188–3213 (2003).

[51] O.B. Lednev, A.A. Kaladjian, M.A. Mikitaev, M.A. Tlenkopatchev. New polybutylene terephtalate and organoclay nanocomposite materials. // Abstracts of the International *Conference on Polymer materials* (México, 2005)

[52] Tretiakov A.O. // Oborudovanie I instrument dlia professionalov №02(37) 2003 [in Russian]

[53] Sergeev G.B. // Ros. Chem. J. (*The journal of the D.I.Mendeleev Russian chemical society*), 2002, Vol. XLVI, №5 [in Russian]

[54] Lomakin S.M., Zaikov G.E. // *Visokomol. Soed. B. 2005*, 47(1) 104-120 [in Russian]

[55] Xu H, Kuo SW, Lee JS, Chang FC. *Macromolecules* 2002;35:8788.

[56] Haddad TS, Lichtenhan JD. *Macromolecules* 1996;29:7302.
[57] Mather PT, Jeon HG, Romo-Uribe A, Haddad TS, Lichtenhan JD. *Macromolecules* 1996;29:7302.
[58] Hsu SLC, Chang KC. *Polymer* 2002;43:4097.
[59] Chang JH, Seo BS, Hwang DH. *Polymer* 2002;43:2969.
[60] Fornes TD, Yoon PJ, Hunter DL, Keskkula H, Paul DR. *Polymer 2002*;43:5915.
[61] Chang JH, Seo BS, Hwang DH. *Polymer 2002*;43:2969.
[62] Fornes TD, Yoon PJ, Hunter DL, Keskkula H, Paul DR. *Polymer 2002*;43:5915.
[63] Wen J, Wikes GL. *Chem Mater* 1996;8:1667.
[64] Zhu ZK, Yang Y, Yin J, Wang X, Ke Y, Qi Z. *J Appl Polym Sci* 1999;3:2063.
[65] Mikitaev M.A., Lednev O.B., BeshtoevB.Z., Bedanokov A.Yu, Mikitaev A.K. // Second International conference "*Polymeric composite materials and covers*" (Yaroslavl 2005, may) [in Russian]
[66] Fischer HR, Gielgens LH, Koster TPM. *Acta Polym* 1999;50:122.
[67] Petrovic XS, Javni L, Waddong A, Banhegyi GJ. *J Appl Polym Sci* 2000;76:133.
[68] Lednev O.B., Beshtoev B.Z., Bedanokov A.Yu, Alarhanova Z.Z., Mikitaev A.K. // *Second International Conference* (Nalchik 2005) [in Russian]
[69] J.-H. Chang, S.J. Kim, Y.L. Joo, S. Im. *Polymer, 2004*;45:919-926.
[70] Lan T, Pinnavaia TJ. *Chem Mater* 1994;6:2216.
[71] Masenelli-Varlot K, Reynaud E, Vigier G, Varlet J. J Polym Sci Part B: *Polym Phys* 2002;40:272.
[72] Yano K, Usuki A, Okada A. J Polym *Sci Part A: Polym Chem* 1997;35:2289.
[73] Shia D, Hui Y, Burnside SD, Giannelis EP. *Polym Engng Sci* 1987;27:887.
[74] Curtin WA. *J Am Ceram Soc* 1991;74:2837.
[75] Chawla KK. Composite materials science and engineering. New York: Springer; 1987.
[76] Burnside, S., D. and Giannelis, E. P., *Chem. Mater. 7*, 1995, 4597.
[77] Babrauskas, V., Peacock, R. D., *Fire Safety Journal*, 18, 1992, 225.
[78] Gilman, J., Kashiwagi, T., Lomakin, S., Giannelis, E., Manias, E., Lichtenhan, J., and Jones, P., in. *Fire Retardancy of Polymers: the Use of Intumescence. The Royal Society of Chemistry*, Cambridge, 1998, 203-221.
[79] Mikitaev A.K., Kaladjian A.A., Lednev O.B., Mikitaev M.A. // *Plastic masses 2004*; 12:45-50 [in Russian]
[80] Gilman, J., Morgan A. *10th Annual BCC Conference*, May 24-26, 1999.
[81] John N. Hay and Steve J. Shaw. Organic-inorganic hybrids: the best of both worlds? *Europhysics News* (2003) Vol. 34 No. 3.
[82] D.M. Delozier, R.A. Orwoll, J.F. Cahoon, N.J. Johnston, J.G. Smith, J.W. Connell. *Polymer*, 2002;43:813-822.
[83] Chen Z, Huang C, Liu S, Zhang Y, Gong K. *J Apply Polym Sci* 2000;75:796-801.
[84] Okado A, Kawasumi M, Kojima Y, Kurauchi T, Kamigato O. *Mater Res Soc Symp Proc* 1990;171:45.
[85] Leszek A. Utracki, Jorgen Lyngaae-Jorgensen. *Rheologica Acta, 2002*;41: 394-407.
[86] R. Wagener, T.J.G. Reisinger. *Polymer, 2003*;44:7513-7518.
[87] Li X, Kang T, Cho WJ, Lee JK, Ha CS. *Macromol Rapid Commun.*
[88] H.-L. Tyan, Y.-C. Liu, K.-H. Wei. *Polymer*, 1999;40:4877-4886.
[89] Vaia R, Huang X, Lewis S, Brittain W. Macromolecules 2000;33:2000-4.

[90] Okamoto M, Morita S, Taguchi H, Kim Y, Kotaka T, Tateyama H. *Polymer* 2000;41:3887-90.

[91] W.S. Chow, Z.A. Mohd Ishak, J. Karger-Kocsis, A.A. Apostolov, U.S. Ishiaku. *Polymer*, 2003;44:7427-7440.

[92] Antipov E.M., Guseva M.A., Gerasin V.A., Korolev Yu.M., Rebrov A.V., Fisher H.R., Razumovskaya I.V. // *Visokomol soed.* A. 2003; 45(11): 1885-1899 [in Russian].

[93] Sur G, Sun H, Lyu S, Mark J. *Polymer 2001*;42:9783-9.

[94] Wang Z, Pinnavaia T. *Chem Mater* 1998;10:3769-71.

[95] Bedanokov A.Yu., Beshtoev B.Z. / *Malij polimernij congress* (Moscow 2005) [in Russian]

[96] Antipov E.M., Guseva M.A., Gerasin V.A., Korolev Yu.M., Rebrov A.V., Fisher H.R., Razumovskaya I.V. // *Visokomol. Soed.* A. 2003; 45(11): 1874-1884 [in Russian].

[97] Lan T., Kaviartna P., Pinnavaia T. *Proceedings of the ACS PMSE* 1994;71:527-8.

[98] Kawasumi et al. Nematic liquid crystal/clay mineral composites. *Science & Engineering C, Vol 6,* pp 135-143, 1998.

[99] Lednev O.B., Kaladjian A.A., Mikitaev M.A. // *Second International Conference* (Nalchik 2005) [in Russian]

[100] Mikitaev A.K., Kaladjian A.A., Lednev O.B., Mikitaev M.A., Davidov E.M. // *Plastic masses* 2005;4:26-31[in Russian]

[101] Eid A., Mikitaev M.A., A.Y.Bedanokov, Mikitaev A.K. Recycled Polyethylene Terephthalate/Organo-Montmorillanite Nanocomposites, Formation And Properties // *The first Afro-Asian Conference on Advanced Materials Science and Technology* (AMSAT 06), Egypt, 2006.

[102] A.K. Mikitaev, A.Y.Bedanokov, O.B.Lednev, M.A.Mikitaev Polymer/silicate nanocomposites based on organomodified clays/ Polymers, Polymer Blends, Polymer Composites and Filled Polymers. *Synthesis, Properties, Application.* Nova Science Publishers. New York 2006.

[103] Malamatov A.H., Kozlov G.V., Mikitaev M.A. *Mechanismi uprochnenenia polimernih nanokompositov*, (Moscow, RUChT 2006) 240 p. [in Russian]

[104] Eid A. *Doctor Thesis*, (Moscow, RUChT 2006) 121 p. [in Russian]

[105] Lednev O.B. *Doctor Thesis* (Moscow, RUChT 2006) 128 p. [in Russian]

[106] Malamatov A.H. *Professor Thesis* (Nalchik, KBSU 2006) 296 p. [in Russian]

[107] Borisov V.A., Bedanokov A.Yu., Karmokov A.M., Mikitaev A.K., Mikitaev M.A., Turaev E.R. // *Plastic masses 2007*; 5 [in Russian]

In: Progress in Monomers, Oligomers, Polymers, Composites ... ISBN 978-1-60456-878-3
Editors: R. A. Pethrick, G.E. Zaikov et al.

Chapter 25

OPTICAL SPECTRA OF INDIVIDUAL MOLECULES AND PHTHALOCYANINES CHROMOGENES UV-BANDS

Yu.A. Mikheev, L.N. Guseva, Yu.A. Ershov* and G.E. Zaikov†
N.M. Emanuel Institute of Biochemical Physics,
Russian Academy of Sciences,Moscow, Russian Federation
N.E. Bauman State Technical University, Moscow, Russian Federation

ABSTRACT

It was shown, that amorphous copper phthalocyanin (CuPc) forms molecular solutions in heptane, water and in polymeric matrixes (polythene, cellulose triacetate, polycaproamide). Is established, that the spectrum of individual molecules CuPc essentially differs from spectra of blue-green pigments CuPc: in a spectrum of individual molecules CuPc are absent chromogenes bands. The intensive electronic-oscillatory UV-bands of an acene type are characteristic for a spectrum of individual molecules. These bands correspond to three electronic $\pi \rightarrow \pi^*$ transitions. In this spectrum there are also weak $n \rightarrow \pi^*$ bands of p-electrons transitions of nitrogen atoms of aromatic macrocircle. From the received results follows, that the accepted literature treatment of an phthalocyanines chromogenity origin is incorrect. According to this treatment a known spectral Q-band (λ = 670 nm), belongs to individual molecules. Actually Q-band determining pigments colors, belongs to supramolecular units of CuPc.

Key-words: phthalocyanin, optical spectra, pigments colors, supramolecular units, individual molecules, copper derivatives.

---

* Correspondence to: N.E. Bauman State Technical University, Vtoraya Baumanskaya str. 5, Moscow, 107005 Russian Federation, E-mail: Ershov_54@mail.ru

† Correspondence to: N.M. Emanuel Institute of Biochemical Physics, Russian Academy of Sciences, Kosygin str. 4, Moscow, 119334 Russian Federation. E-mail: Chembio@sky.chph.ras.ru

Phthalocyanines derivatives in the habitual kind are powders of blue-green colors. The large number phthalocyanines derivatives is synthesized. These derivatives contains atoms of various metals Me and various aromatic ligands in a phthalocyanin molecule [1]:

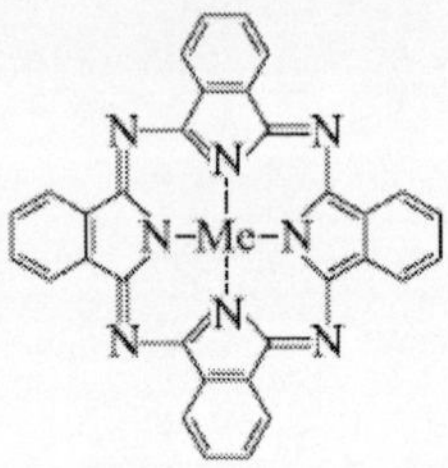

All of them have a characteristic light absorption band in the field of a spectrum 500-700 nm (20000 - 14300 cm^{-1}) named by a Q-band. In the literature until recently Q-band attribute to photoexcitation of individual molecules phthalocyanin on the ground that all of them have identical aromatic macrocircle, including atoms of nitrogen and containing 18 conjugated π - electrons [2]. However recently in the authors work [3] is established, that the Q-band belongs to phthalocyanine molecule dimers.

In [3] investigated copper-phthalocyanin CuPc samples, which apply to computer disks as the information carrier, and also in dying business. It was a high-dispersed powder, NIOPIC manufacture, Moscow, 1978.

Spectra of the investigated preparation in the concentrated sulfuric acid (fig.1, spectrum 1) and in dioxane (fig.1, the spectrum 2) solutions, corresponds to spectra known from the literature [3].

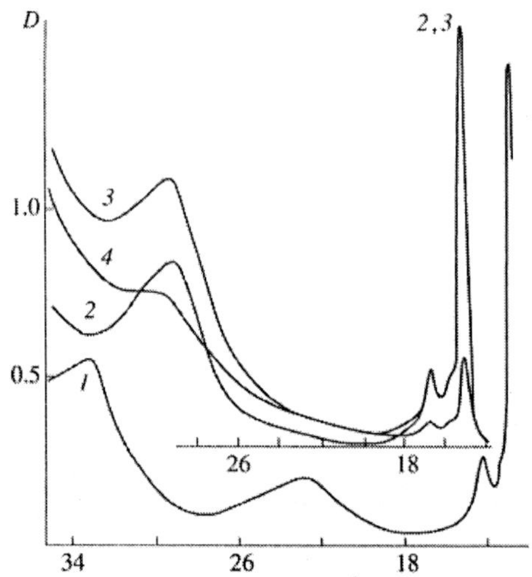

Figure 1. Spectra of (*1*) dioxane, (*2*) concentrated H_2SO_4, (*3*) yellow-green solution of the initial CuPc preparation in concentrated H_2SO_4, and (*4–6*) blue dioxane extracts from precipitates obtained by adding water to solutions of CuPc in (*4*) concentrated and (*5, 6*) 60% H_2SO_4. Spectrum *6* is what became of spectrum *5* after the test solution was left to stand for three days.

[fig.1] Spectra of yellow - green slns initial CuPc in conc. H2SO4

Technical copper-phthalocyanin extremely poorly is dissolved in the majority of the solvents. As the best solvents for it are dioxane [4] and concentrated sulfuric acid [5]. In the concentrated sulfuric acid the CuPc-pigment forms a solution of a pigment chromogene particles, which do not turn in sulphoacides. On it the technology of clearing of a crude product [5] is based. A crude pigment dissolve in the concentrated sulfuric acid. At dissolution of the received solution by water up to 70 % of an acid the pigment sulphate deposit drops out. Then the sulphate deposit carefully wash out by water. Thus occurs sulphate hydrolysis and the blue-green colour crystal of α - phase powder is formed. α-Phase is unstable and at heating transformes in β-phase.

Inability of chromogen particles of a pigment powder to react with a sulfuric acid essentially distinguishes this pigment from usual aril compounds. Such inertness has found an explanation in work [3]. In this work, with the help UV-spectroscopy, studied polymeric films sorption of CuPc-molecules from supersaturated water-dioxane solutions.

The water solutions prepared, mixing of an initial pigment dioxane suspension with water. Earlier in literature [4] was considered, that solution of chromogen copper-phthalocyanin molecules in the dioxane is formed. The results of experiences with sorption have appeared unexpected and have forced the authors [3] in detail to investigate composite structure of an initial preparation. Was established, that the preparation contains a small portion of an amorphous phase.

The experiments in the present work, have confirmed, what exactly this phase forms solutions containing individual molecules. According to the data [3] has appeared, that the individual molecules in a solution essentially differ on the spectral and chemical properties from chromogenes particles.

Is established, that Q-chromogenes are CuPc-molecules dimers. These dimers sediments in dioxane medium, though also slow. Sedimentation find out on decrease of a Q-band intensity. The spectrum 3, fig. 1 belongs to the freshly prepared filtered solution received at heating pure α-phase in dioxane. In three day the intensity of a Q-band is considerably reduced

(fig.1, curve 4), that testifies about chromogenes Q-particles sedimentation.

Same chromogenes with a Q-band disappear faster in unfiltered pigment suspention, which initial spectrum is submitted on fig.2, curve 1.

[fig.2] Spectra unfiltered dark blue suspention initial CuPc in dioxane

In 20 hours of a storage (curve 2) spectra has changed as a result of decrease of a light scattering level and simultaneous transformation chromogen bands of α-phase undergoing structural change. For this time a Q-band (λ = 670 нм, ν = 14 925 см-1) disappears completely, but completely is restored at heating. During a storage at T_{room} the Q-band again disappears. Such thermochromy of a solution is explained by adsorption of chromogenes Q-particles on CuPc microcrystals suspension.

At spectra fig.2 there are electron-oscillatory UV-bands at frequencies ν> 28 000 cm^{-1}. These bands do not change the intensity during suspension aging. The UV-bands are shown more evident in a spectrum of the filtered solution (fig.2, curve 3) and belong to individual copper-phthalocyanin molecules, which absorbed as well polymeric films, as against Q-particles and microcrystals.

The UV-spectra films with absorbed copper-phthalocyanin molecules are submitted on fig.3.

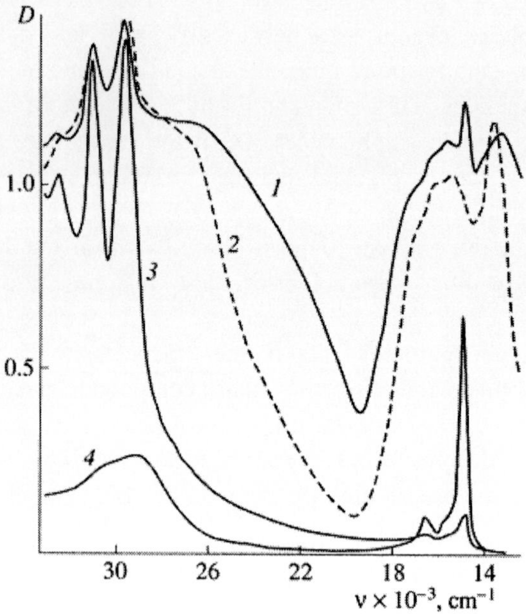

Figure 2. Spectra of (*1*) the water–acid layer over the precipitate formed after a solution of 0.3 g of the initial CuPc preparation in 5 ml of concentrated H_2SO_4 was allowed to standfor 2 h and then diluted with 150 ml of water; (*2–4*) the spectra of extracts (in (2) heptane, (3) ethanol, and (4) dioxane) from the precipitate formed after this dilution; (5, 6) the spectra of the extracts (in (5) ethanol and (6) dioxane) from the precipitate formed after the dilution with 150 ml of water of the same CuPc–concentrated H2SO4 solution but allowed to stand before the dilution for 7 days.

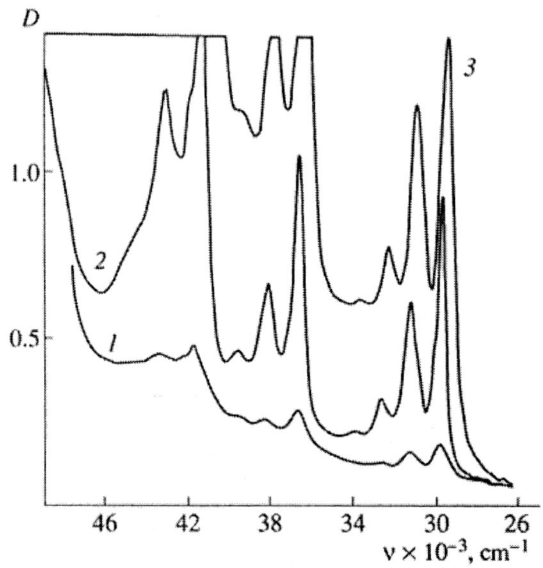

Figure 3. Spectra of (*1, 2*) aqueous extracts of sulfonated CuPc molecules from the precipitates obtained by adding water to 7-day-aged solutions of the initial CuPc preparation in concentrated $2SO_4$ and (3–5) sulfuric acid extracts ((3, 4) diluted in sulfuric acid and (5) diluted in water by a factor of 20) from a solution of the initial CuPc preparation in heptane.

[Fig.3] Spectra of molecules CuPc in water (1), films PE (2), pamide (3)

Here spectrum 1 concerns to freshly prepared filtered water-dioxane solution with a high degree supersaturation and spectra 2 and 3 accordingly to films of polythene and polycaproamide PC-4. Absorption carried out from water-dioxane solution, in which spectra of a UV-band are not visible. Therefore spectra 2, 3 testify to extremely effective film extraction of phthalocyanin molecules. The kind of spectra 2, 3 on fig.3 testifies to presence of the advanced system of electron-oscillatory bands CuPc-molecules in range ν > 30 000 см-1.

The spectrum of a solution of CuPc-molecules in heptane is even more informatical (fig.4, curve 1).

[Fig.4] Spectra slns of molecules CuPc after heating initial CuPc in гептане

Solution have prepared by heating of a large sample preparation (0.8 g) in heptane (volume 10 cm^{-3}) on water bath with the subsequent filtering through the dense filter.

The spectrum contains bands three electronic $\pi \rightarrow \pi^*$ transitions in ultra-violet range. The observable intervals between vibronic bands are 1400 - 1500 cm^{-1}, that is typical for aromatic substances of a antracene type. However at phthalocyanin this property was not marked earlier.

On a slope low-frequency $\pi \rightarrow \pi^*$ bands (fig.4, the curve 1) is visible system of weak bands. A such bands on the low intensity and narrow vibronic splitting correspond to $n \rightarrow \pi^*$ electron transition of the nitrogen atom which has been built-in directly in an aromatic circuit of conjugation. Such $n \rightarrow \pi^*$ bands are characteristic for asine aromatic molecules (pyrasine, pyrimidine) [6]. Averaging of known extinction coefficients of $n \rightarrow \pi^*$ bands of pyrasine and pyrimidine, and using the received average meaning for CuPc, it is possible to estimate extinction coefficient for longwave $\pi \rightarrow \pi^*$ band appropriate ~ 440 000 l/mole.cm.

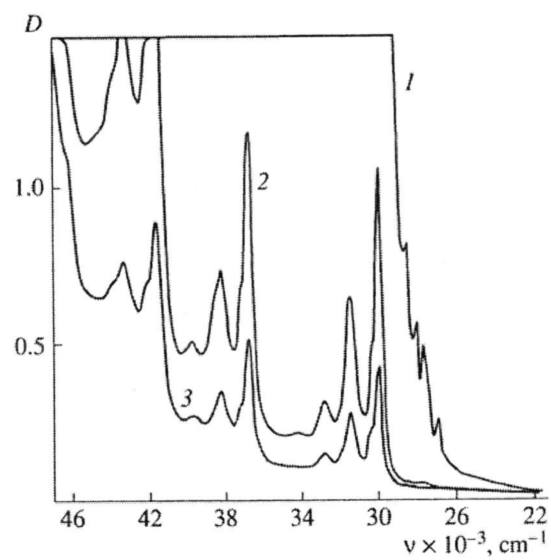

Figure 4. Spectra of (*1, 2*) unfiltered blue suspensions prepared by mixing 2.3 mg of initial CuPc preparation with 10 ml of dioxane, (*3*) a pale blue dioxane solution obtained by filtering a suspension of 20 mg of initial CuPc preparation in 10 ml of dioxane, and (*4*) a blue solution of CuPc in dioxane obtained by extraction and filtering of the .-phase precipitate produced by adding water to a solution of CuPc in concentrated H_2SO_4

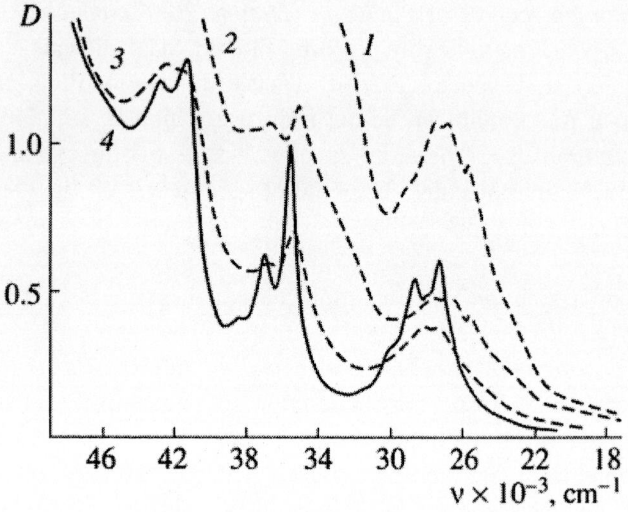

Figure 5. Spectra of (*1, 3*) 40-∝m-thick PE and (*2, 4*) 70-∝m PK-4 films (*3, 4*) with and (*1, 2*) without absorbed CuPc molecules. (*5*) Spectrum of a freshly prepared supersaturated aqueous solution of CuPc obtained from a suspension of 0.6 g of initial CuPc preparation in 10 ml of heated dioxane by filtration and mixing with 400 ml of water.

The character of the received spectrum allows to attribute it directly to aromatic macrocircle including atoms of nitrogen. This macrocircle is a basis structure of a copper phthalocyanin molecule, together with others phthalocyanines.

That in a heptane solution are just the individual molecules CuPc proves to be true by the fact sulphonating of them. Sulphonating carried out as follows.

CuPc heptane extract (fig.4, spectrum1) place in glass with the concentrated sulfuric acid and will carry out slow heptane evaporation. In 1.5 months in пробирке there is a dark - brown colour solution. The spectra of this solution strongly diluted by a sulfuric acid, are submitted on fig.5, curves 1, 2.

[Fig.5] Spectra of sulphonating solutions of molecules CuPc in conc. H2SO4,

If this dark-brown solution to dilute by water (up to the 4% sulfuric acid), is observed hypsochromic displacement of UV-bands of $\pi \rightarrow \pi^*$ transitions (curve 3). Such displacement is explained to that in the concentrated sulfuric acid sulphonated phthalocyanine molecules are as salts-sulphates with protonated nitrogen atoms.

Under action of water sulphates are hydrolyzed and the equilibrium is displaced to not salt form of sulphonated phthalocyanine molecules. The spectrum of a water solution pure not salt sulphonated form is given on fig.5, curve 4. The given solution is received passing a heptane stage extraction, washing out water a CuPc deposit, extracted by water from a initial CuPc solution in concntrated H_2SO_4.

It is possible to explain changes, noticed in spectra, as follows. Sulphoacids of copper phthalocyanine are formed as salts - sulphates already at dissolution of an initial preparation in the concentrated sulfuric acid, as in a preparation there is an impurity of an amorphous phase. At depositing of a pigment by water from a H_2SO_4 solution the small impurity of sulphonated molecules drops out in a deposit together with sulphate salts of chromogenic particles-dimers [3]. The washing of a deposit by water causes sulphate hydrolysis. In a deposit remains insoluble in water α-phase, and soluble sulphonated molecule are washed away. These one have found out at fraction washing of a deposit. On fig.5 the spectrum 4, registered for an intermediate water fraction is shown.

On fig.6 the spectra of fractions extracted with heptane (a curve 1), ethanol (curves 2, 4) and dioxane (curves 3, 5) of deposits are given which have not achieved complete sulphoacids loss at previous washing by water.

[Fig.6] Sp-ra sulphir CuPc in гептан 1), этанол (2), dioxane (3) extracts

The kind of spectra sulphoacids depends on a storage time of of an initial preparation solution in H_2SO_4. On fig.6 the spectra 1-3 correspond to two hours of a storage, spectra 4, 5 - seven day. The observable change is connected to growth of a degree sulphonating of molecules CuPc. That is molecules just sulphonated phthalocyanin, is unequivocally identified by occurrence of a deposit of dark blue crystals in the dendrites form after evaporation of the solvents.

Thus, the individual phthalocyanine molecules combine in themselves ability benzene substances to form sulphoacids with ability aromatic azenes to form sulphates.

As against individual molecules CuPc, chromogenic dimers with a Q-band in a spectrum form a steady solution in the concentrated sulfuric acid and do not give sulphoacids. Such dimers behaviour $(CuPc)_2$ is possible to explain to that the sulphonating reaction proceeds with change of of a benzene ring geometry at a stage of an intermediate product of reaction - σ-complex. In a dimers case formation σ-complex, apparently, is complicated. This is interfered by strong intermolecular interaction between flat phthalocyanin molecules, strongly stabilizing flat geometry of molecules in dimers (stacking).

The received result will not be coordinated by traditional representation, that chromogenity of phthalocyanine molecules is determined by photoexcitation individual aromatic macrocircles with atoms of nitrogen. Actually a chromogenic particles with a Q-bands are dimers $(CuPc)_2$.

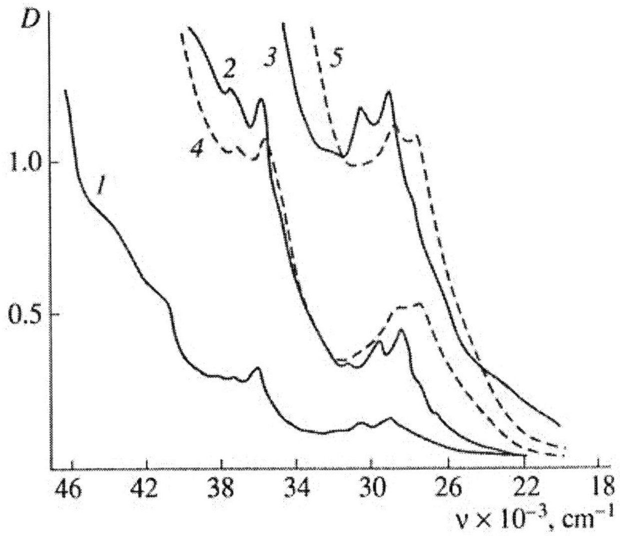

рис. 6

Figure 6. Spectra of (*1*) a solution of CuPc in heptane prepared by heating 0.8 g of initial preparation in 10 ml of heptane over a water bath with subsequent filtration, (*2, 3*) this solution diluted by a factors of (*2*) 60 and (*3*) 150, and (*4*) heptane.

According to results [3], confirmed in the present work, it is necessary anew to consider quantum-chemical models of phthalocyanine molecules, being based any more on idea of chromogenic molecules, and on the fact them noxchromogenity. Simultaneously there are tasks on finding - out of a nature chromogenicty of supramolecular dimers and study of photophysical and photochemical properties separately of molecules and separately chromogenes of phthalocyanines.

REFERENCES

[1] Electronic Spectra of Phthalocyanines and Related Compounds: A Catalogue, Ed. by E.A. *Luk'yanets (*MNPO NIOPIK-NIITEKhim, Cherkassy, 1989) [in Russian].

[2] J. Simon and J.-J. Andre, *Molecular Semiconductors* (Springer, Berlin, 1985; Mir, Moscow, 1988).

[3] Photoelectrical Properties and Solar Cells. Springer-Verlag, Berlin, 1985.

[4] Yu.A. Mikheev, L.N. Guseva, and Yu.A. Ershov, *Zh. Fiz. Khim. 2007*. V. 81, № 4. P.715. [Russ. J. Phys. Chem. 81,715 (2007)].

[5] A.N. Terenin, Photonics of Dye Molecules and Related Organic Compounds (*Nauka,* Leningrad, 1967) [in Russian].

[6] M.A. Chekalin, B.V. Passet, and B A. Ioffe, Technology of Organic Dyes and Intermediates (*Khimiya*, Leningrad, 1972) [in Russian].

[7] E. S. Stern and C. J. Timmons, Gillam and Stern's Introduction to Electronic Spectroscopy in Organic Chemistry, 2nd ed. (E. Arnold, London, 1970; Mir, Moscow, 1974).

In: Progress in Monomers, Oligomers, Polymers, Composites … ISBN 978-1-60456-878-3
Editors: R. A. Pethrick, G.E. Zaikov et al.

Chapter 26

KINETIC MODEL OF COPOLYMERIZATION AT LINEAR CHAIN TERMINATION

***Yu.G. Medvedevskikh**, *G.I. Khovanets'*, *I.Yu. Yevchuk* and *G.E. Zaikov*†**

Department of Physical Chemistry of Combustible Minerals
L. M. Lytvynenko Institute of Physico-organic Chemistry
and Coal Chemistry NAS of Ukraine
N. M. Emanuel Institute of Biochemical Physics RAS
Moscow, Russian Federation

ABSTRACT

Numerous experimental data on kinetics of photoinitiated stationary and nonstationary polymerization of mono- and di(meth)acrylates to high conversions show the possibility of the change of chain termination mechanism in reaction zones, where translation mobility of radicals (and, respectively, the rate of bimolecular chain termination) abruptly decreases. In this case linear from the formal stoichiometric point of view chain termination mechanism, which appears as an act of self-burial of active radicals at their propagation in polymer matrix, becomes determinative. Suggested earlier kinetic models with taking into consideration possible role of linear chain termination are in satisfactory agreement with all experimental data on homopolymerization, but need to be extended on copolymerization processes. In present article we propose kinetic model of copolymerization at linear chain termination, which appears as an act of self-burial of active radical in reaction of its propagation.

Key words: kinetic model, copolymerization, linear chain termination

---

* Correspondence to: Department of Physical Chemistry of Combustible Minerals, L. M. Lytvynenko Institute of Physico-organic Chemistry and Coal Chemistry NAS of Ukraine 79060 Lviv Naukova St., 3a, e-mail: vfh@org.lviv.net

† Correspondence to: N. M. Emanuel Institute of Biochemical Physics RAS, 119334 Moscow, Kosygin St. 4, e-mail: chembio@sky.chph.ras.ru

INTRODUCTION

Process of polymerization to high conversions is accompanied by the change of phase state of polymerizing system. It affects all factors, which determine kinetics of polymerization. But initiation of reaction and chain propagation, changing kinetic parameters, retain the essence of their elementary acts. Conversely, chain termination can change its mechanism. It is related to the fact, that at abrupt increasing of diffusion braking of bimolecular chain termination, linear (i. e. monomolecular on radicals) mechanism of chain termination goes into effect, which is the act of chain propagation leading to the capture of macroradical by polymer matrix [1] or to its freezing [2, 3].

Therefore, if the rate of bimolecular chain termination, which is accompanied by active radicals decay, becomes smaller than the rate of "self-burial" of active radicals at their propagation in polymeric matrix, what is possible at abrupt decreasing of macroradicals translation mobility, mechanism of chain termination from formal stoichiometric point of view becomes linear. It affects increasing of the order of reaction on initiator at high degrees of polymerization and achieving of high concentration of radicals till the end of the process [4, 5].

According to the concept of different reaction zones [6, 7], which is the development of the concept of microheterogeneity of polymerizing system [8], process of polymerization of bifunctional monomers is concentrated in two reaction zones already at early stages: in monomer phase and at interface liquid monomer – micrograins of polymer matrix at practically limited conversion on monomer. Kinetic model of photoinitiated polymerization of di(meth)acrylates is in good agreement with all experimental data at assumption, that at the stage of autoacceleration the main contribution into the summary kinetics of polymerization is made by the process at the interface layer, where because of the abrupt decreasing of segmental and, respectively, translation macroradicals mobility the chain termination is linear [6, 7, 9].

Investigations of kinetics of nonstationary process of postpolymerization of di(meth)acrylates, i. e. dark polymerization after *UV* irradiation switching off, confirmed this assumption and allowed to ascertain, that linear chain termination obeys the Kolraush stretched exponent law, and to obtain the numerical estimations of the chain termination rate constant k_t over the range $(1\div5)\cdot10^{-5}$ m^3/mol·s. Theoretical derivation [10] of the stretched exponent law was suggested on the base on the supposition that linear chain termination is the act of chain propagation leading to the trap.

At photoinitiated polymerization of monofunctional (meth)acrylates in bulk theoretical describing of the process to high conversions [9] satisfactorily agrees with experimental data at assumption that in monomer-polymer phase and at interface between it and polymer-monomer matrix the chain termination is bimolecular, but in polymer-monomer matrix it is linear [9].

Studing of dark postpolymerization kinetics of (meth)acrylates at conversions ≥ 0.7, which correspond to maximum of polymerization rate at the stage of autoacceleration, confirms this assumption [9, 11]. At that, similar to di(meth)acrylates, the linear chain termination obeys the stretched exponent law, and the rate constant is of the same order of 10^{-5} m^3/mol·s [9, 11].

Hence, kinetic models of homopolymerization to high conversions both bi- and mono(meth)acrylates are in satisfactory agreement with experimental data at assumption that at least at the stage of autoacceleration of these processes the linear chain termination as an act of self-burial of active radicals in reactions of chain propagation is of a great significance. Since considerable part of technologically valuable materials are copolymers, properties of which depend on their syntheses conditions in many respects, it was necessary to derive the kinetic model of copolymerization at linear chain termination.

KINETIC MODEL DERIVATION

Let us use the method of routes, developed both for catalytic and radical-chain processes [11, 12, 13]. According to it, the rate *r(v, w, u)* on *i*-route in the basis *w* of the group *w* of the graph $G = \{R, S\}$, what represents topology of kinetic scheme on multiplicity of radicals $R = \{R_i\}$ and elementary reactions $S = \{S_i\}$, which join radicals, is described by expression

$$r(v,w,u) = L(\sum_i M_i)^{-1} M(v)\Delta(v,w,u). \tag{1}$$

Here: L is total radical concentration; M_i is additional minor of a single group $v = \{R_i\}$: $M_i = (-1)^{n-1} \left| B_i \right|$, where $\left| B_i \right|$ is matrix determinant, which is obtained from initial matrix B of the n order by deleting of i-line and i-column; $M(v)$ is additional minor of the group v of power l, i. e. which consists of l tops of the graph $G = \{R, S\}$: $M(v) = (-1)^{n-l} \left| B(v) \right|$, where $\left| B(v) \right|$ is determinant of matrix $B(v)$, which is obtained from initial matrix B by deleting of those lines and columns, determinants of which coincide with determinants of the tops of the graph $G = \{R, S\}$, represented in given group; $\Delta(v, w, u)$ is Δ-factor of the route, which is determined by product of frequencies of elementary reactions, which form cyclic sequence of the route in given basis w of the group v.

According to the assumption, that linear chain termination is the act of self-burial of active radical R_i in reaction of its propagation, as a result it transforms into inactive radical R_z, kinetic scheme of propagation and linear chain termination at copolymerization of monomers A_1 and A_2 becomes as follows

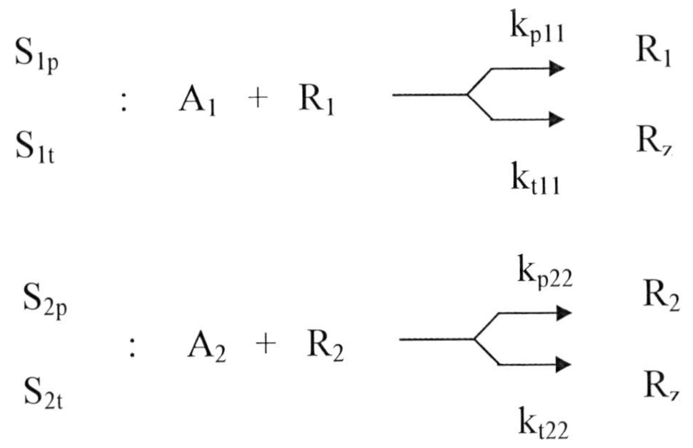

(2)

$$S_{3p},\ S_{3t}:\quad A_1 + R_2 \;\begin{cases} \xrightarrow{k_{p12}} R_1 \\ \xrightarrow{k_{t12}} R_z \end{cases}$$

$$S_{4p},\ S_{4t}:\quad A_2 + R_1 \;\begin{cases} \xrightarrow{k_{p21}} R_2 \\ \xrightarrow{k_{t21}} R_z \end{cases}$$

The graph $G = \{R, S\}$, which presents topology of kinetic scheme (2) with taking into account only elementary reactions of chain propagation, is shown on Fig. 1.

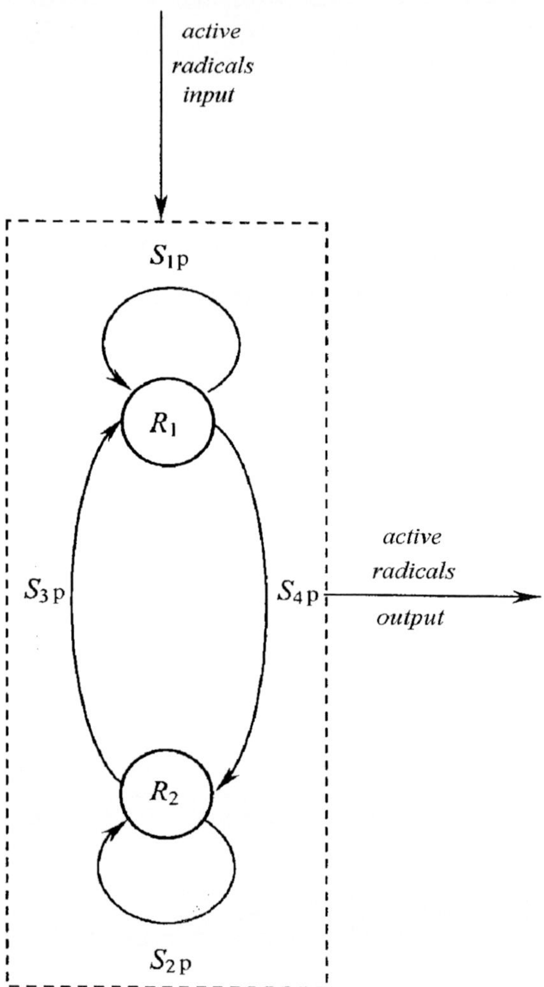

Figure 1. The graph $G = \{R, S\}$, which presents topology of kinetic scheme of chain propagation at copolymerization

On kinetic matrix determination $B = (b_{ij})$ its diagonal elements obey the relation $b_{ij} = -\sum_{i\neq j} b_{ij}$, and nondiagonal ones b_{ij} are the intensities of transformation from the state R_j into the state R_i. At present case the matrix order is equal $n = 2$, and its elements are determined only by the intensities of elementary reactions S_{3p} and S_{4p}, which affect relative concentrations of radicals R_1 and R_2:

$$B = \begin{pmatrix} -k_{p12}[A_2] & k_{p12}[A_1] \\ k_{p21}[A_2] & -k_{p12}[A_1] \end{pmatrix}. \tag{3}$$

At long chains approximation the condition of stationarity on active radicals assumes that their input v_{in} and output v_t rates are equal: $v_{in} = v_t$. This condition according to (2) is defined by expression

$$v_t = \sum_{i,j} k_{tij}[A_i][R_j] = L\sum_{i,j} k_{tij}[A_i]x_j, \tag{4}$$

in which $x_i = [R_i]/\sum_i [R_i] = L_i / L$ is relative radicals concentration of radicals of given type. It is determined from the condition $Bx = 0$, the solution of which is the expression

$$x_i = M_i / \sum_i M_i. \tag{5}$$

Using (5) in the condition of stationarity $v_{in} = v_t$. and taking into account (4), we obtain

$$L = v_{in} \sum_i M_i / \sum_{i,j} k_{tij}[A_i]M_j. \tag{6}$$

Thus, kinetic model of the route (1) becomes as follows

$$r(v,w,u) = v_{in} M(v)\Delta(v,w,u) / \sum_{i,j} k_{tij}[A_i]M_j. \tag{7}$$

According to kinetic matrix B on (3), additional single minors M_i are determined by expressions:

$$M_1 = k_{p12}[A_1],\ M_2 = k_{p21}[A_2]. \tag{8}$$

In the graph $G = \{R, S\}$ we have 3 routes, two of them are represented on one elementary reaction and form cycles at the tops R_1 and R_2, the third route is represented by the sequence of elementary reactions S_{3p} and S_{4p}. Thus, according to abovementioned algorithm, we have

$$u_1 = S_1, \quad \Delta_1 = k_{p11}[A_1], \quad M(1) = M_1 = k_{p12}[A_1],$$

$$u_2 = S_2, \quad \Delta_2 = k_{p22}[A_2], \quad M(2) = M_2 = k_{p21}[A_2], \tag{9}$$

$$u_3 = S_3 + S_4, \Delta_3 = k_{p12}k_{p21}[A_1][A_2], M(3) = M(1,2) = 1.$$

From here the expression for the rates on the routes follows:

$$r_1 = k_{p11}k_{p12}[A_1]^2 v_{in} / \sum_{i,j} k_{tij}[A_i]M_j,$$

$$r_2 = k_{p22}k_{p21}[A_2]^2 v_{in} / \sum_{i,j} k_{tij}[A_i]M_j, \tag{10}$$

$$r_3 = k_{p12}k_{p21}[A_1][A_2]v_{in} / \sum_{i,j} k_{tij}[A_i]M_j,$$

where

$$\sum_{i,j} k_{tij}[A_i]M_j = k_{t11}k_{p12}[A_1]^2 + (k_{t12}k_{p21} + k_{t21}k_{p12})[A_1][A_2] + k_{t22}k_{p21}[A_2]^2 \tag{11}$$

Taking into account the stoichiometry of the routes, the rates of monomers A_1 and A_2 consumption are equal

$$-d[A_1]/dt = r_1 + r_3,$$

$$-d[A_2]/dt = r_2 + r_3. \tag{12}$$

Respectively, the total rate of copolymerization is equal

$$-d([A_1] + [A_2])/dt = r_1 + r_2 + 2r_3. \tag{13}$$

Instantaneous composition of copolymer is defined by

$$d[A_1]/d[A_2] = (r_1 + r_3)/(r_2 + r_3). \tag{14}$$

CONCLUSION

Derivation of kinetic model of copolymerization at linear chain termination, which is considered as the act of self-burial of active radical in the reaction of chain propagation, has been suggested. The obtained equations will be used at the analysis of kinetics of photoinitiated copolymerization of mono- and di(meth)acrylates to high conversions.

REFERENCES

[1] Dyachkov A. I., Efimov A. L., Efimov L. I. et al. About chain termination mechanism at radical polymerization of methymethacrylate to high conversions. *Vysokomol. soyed.* V. 25(A), 10, 1983, P. 2176-2181.

[2] Zhu S., Tian Y., Hamieles A. E. Radical trapping and termination in free-radical polymerization of methylmethacrylate. *Macromolecules*, 10, 1990, P. 1144-1150.

[3] Kurdikar D. L., Peppas N. A. A kinetic study of diacrylate photopolymerizations. *Polymer*. V. 35, 5, 1994. P. 1004-1011.

[4] Guzeeva E. V., Efimov A. L., Dyachkov A. I. et al. Chain propagation limitation mechanism in radical polymerization of styrene to high degrees of conversion. *Vysokomol. soyed.* V. 28(A), 8, 1986, P. 587-588.

[5] Brun E. B., Ivanov V. A., Kaminsky V. A. Kinetics of postpolymerization at high degrees of conversion. *Proceedings of AS of USSR*. V. 291, 3, 1986, P. 618-622.

[6] Medvedevskikh Yu. G., Zagladko E. A., Turovsky A. A., Zaikov G. E. Kinetics of the three-dimentional thermopolymerization. *Intern. J. Polym. Mater*. 43, 1999. P. 157-168.

[7] Medvedevskikh Yu., Zahladko E., Turovsky A., Zaikov G. Kinetic model of stationary and non–stationary process of three–dimensional block polymerization. *Russian Polymer News*. V. 4, 3, 1999. P. 33–39.

[8] Berlin A. A., Korolyov G. V., Kefeli T. Ya. et al. Acrylic oligomers and materials on their base. *M.: Khimiya*, 1983. 232P.

[9] Medvedevskikh Yu., Bratus A., Hafijchuk G, Zaichenko A., Kytsya A., Turovsky A., Zaikov G. Kinetic model of block photopolymerization of glycidyl methacrylate to high conversion. *Journal of Applied Polymer Science*. V. 86, 2002. P. 3556–3569.

[10] Medvedevskikh Yu. G., Kytsya A. R., Bazylyak L. I., Turovsky A. A., Zaikov G. E. Kinetic models of stationary and non–stationary photoinitiated polymerization of dimethacrylates. *Oxidation Communications*. V. 28, 1, 2005. P. 173–207.

[11] Medvedevskikh Yu. G., Kytsya A. R., Bazylyak L. I., Turovsky A. A. Zaikov G. E. Stationary and nonstationary kinetics of the photoinitiated polymerization. 2004. *VSP*. Utrecht. Boston. P. 313.

[12] Medvedevskik Yu. G., Kucher R. V. Routes and kinetic models of linear catalytic reactions i complicated systems. *Proceedings of AS of Ukrainian SSR* (B), 1, 1984. P. 42-46.

[13] Berlin A. A., Medvedevskikh Yu. G., Kucher R. V. Kinetic model of the routes of unbranched radical-chain processes. *Proceedings of AS of Ukrainian* SSR (B), 6, 1986. P. 32-36.

In: Progress in Monomers, Oligomers, Polymers, Composites … ISBN 978-1-60456-878-3
Editors: R. A. Pethrick, G.E. Zaikov et al.

Chapter 27

EXTRUSION HEADS USED IN MANUFACTURING POLYMERIC CASINGS FOR PREVENTING PIPES FROM CORROSION

A.K. Panov, T.G. Beloborodova, T.A. Anasova and G.E. Zaikov
Academy of Science of Republic of Bashkortostan,
Sterlitamak Branch, Sterlitamak, Russian Federation
N.M. Emanuel Institute of Biochemical Physics, RAN, Moscow, Russian Federation

ABSTRACT

The contributors gave information about extrusion heads used in manufacturing polymeric casings for preventing pipes from corrosion. It was shown the high efficiency of this new design.

Key-words: polymers, pipes, corrosion, extrusion heads, manufacturing.

INTRODUCTION

On the basis of executed theoretical and experimental investigations highly productive constructions of extrusion heads used in manufacturing polymeric casings for preventing pipes from corrosion were worked out. Constructions of extrusion heads used in manufacturing of outer and inner casing for pipes are represented in the following paragraphs. The order of calculations of forming channels of extrusion heads are represented here too.

1. Construction of Extrusion Heads Used in Manufacturing Casing for Outer Pipe Framing

Polymeric materials due to their qualities find their more and more usage as construction materials, so allowing producing different articles including polymeric casings for pipes. Polymeric casing looks like a hollow body with a cylindrical outer surface and there are longitudinal lugs in the inner surface. Longitudinal lugs serve as a support for a pipe which is inside the casing. Thus, according to **Fig.1**, polymeric casing (1) frames pipe (2) and because of longitudinal lugs make channels (4) between outer surface of pipe and inner surface of casing (1).

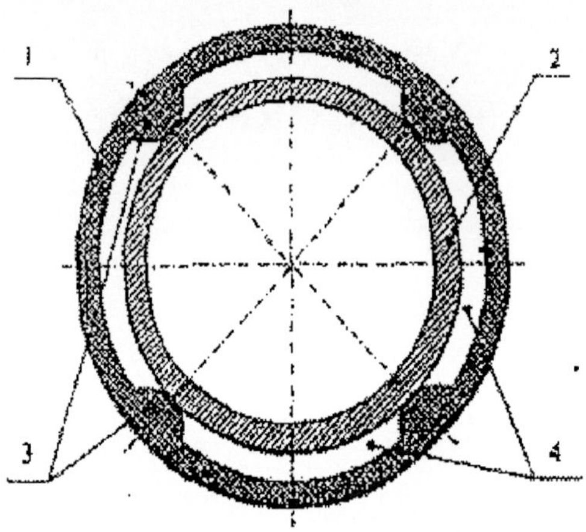

Figure 1. Pipe cross-section in casing.

Mentioned above construction of casing allows not only to isolate pipe from environment damages, saving it from corrosion and lowering heat exchange through the walls, but it also prevents overflow of transported product when hermetidy of pipe is damaged. And thus environment soiling is excluded.

We offer to use extrusion head used for manufacturing polymeric casing which is comprised of disconnected body, parts of which are tied by means of connecting bolts; mandrel-keeper, cone, grating matrix and disconnected mandrel, which form shaping channel. At the same time mandrel is executed with longitude slots, surfaces of which are formed by a curve of the second order and with a smooth section in the place of connecting of longitude slot with a cylindrical part of mandrel surface. There is an opportunity of cooling of inner surface of receiving casing [1, 2].

Extrusion head used for manufacturing polymeric casings, according to figures 2 and 3, is comprised of body (1),inside which cone (3) is located with the help of mandrel-keeper (3). In mandrel-keeper there are some slots, in which there is a pipe for water outlet (4) and a pipe for water supply (5). Fixing of mandrel-keeper (3) in the head is carried out with the help of the part of body (6), where regulating bolts (7) are established. They serve for radial removing of matrix (8). Longitude slots (10) are in mandrel (9) for shaping support lugs in polymeric casing, and at the same time in the place of connecting longitude slot (10) with cylindrical part of the surface of mandrel (9), a smooth section is made along the arch of a circle. Heating of extrusion head is made by electric heaters (11). The inner surface of cavity for melt current form parts of mandrel (12) and (13). In the cavity for melt current there is a cylindrical grating (14). A pipe for water feeding into calibrating mechanism (15) is placed inside of bar (16). Bar (16) is fixed in the base of bar (17), which is heat isolated with the help heat steady packing (18).

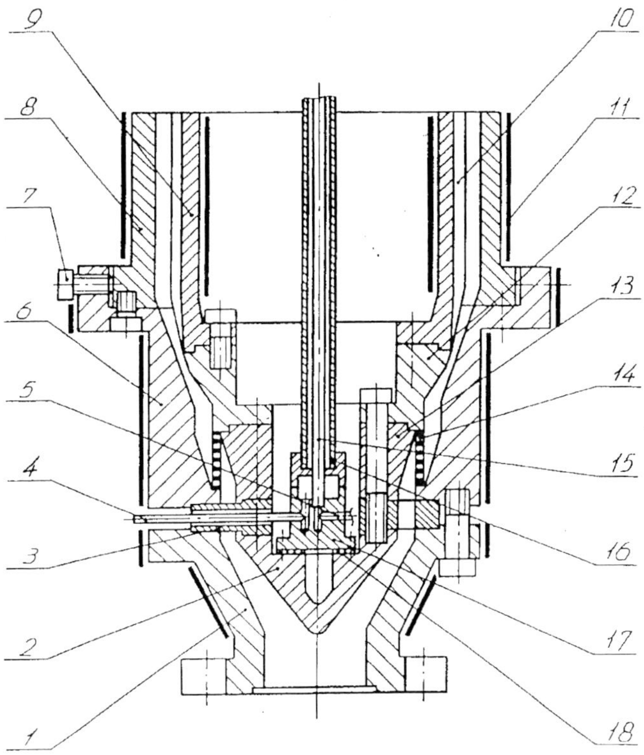

Figure 2. longitude section of extrusion head used for manufacturing outer polymeric casings.

Shaping channel is formed by matrix (8) and mandrel (9). Cross section of shaping channel repeats the shape of cross section of polymeric casing pipe framing. Surfaces of longitude slots (10), made in mandrel (9) are formed by the second order curve, for example by a curve $Y=kx^2$, where k is the coefficient of pressing of parabola branches, mm^{-1}, which is chosen according to the equation condition of pressure overfall in the ring part and in the slot part of shaping channel (e.g. for extrusion head, shaping channel length – 150mm, outer diameter of shaping channel – 96 mm; inner diameter of shaping channel – 76 mm and longitude slot depth – 8 mm; coefficient $k=0{,}024mm^{-1}$). Longitude slots (10) form support lugs in the inner surface of polymeric casing, providing equation of pressure overfall along all the section of shaping channel of extrusion head.

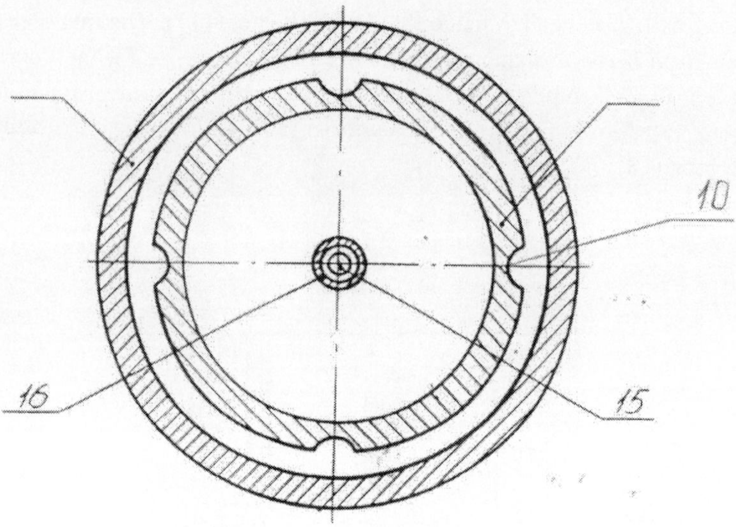

Figure 3. cross-section of extrusion head used for manufacturing polymeric casings of outer pipe framing.

2. Description of Construction Extrusion Head Used for Manufacturing Polymeric Casings of Inner Pipe Framing

The developed design of extrusion head used for manufacturing polymeric casings of inner pipe framing provides hightened strength characteristics, increases hardness of polymeric casing of inner pipe framing because of support lugs on its outer surface are shaped. Increasing of quality and dimensional precision of polymeric casing of inner pipe framing is achieved thanks to the fact that making longitudinal slots in the inner surface of matrix makes it easier to calibrate casing. Disposition of longitudinal slots on the inner surface of matrix allows to shape support lugs on the outer surface of polymeric casing of inner pipe framing and this increases hardness of polymeric casing and also its strength characteristics.

Such a design makes the process of calibrating of polymeric casing because it is done on the outer diameter with the help of calibrating nozzle, connected with cooling bath. As a result it is not necessary to use parts, providing inner cooling of casing and the quality of outer of outer surface of polymeric casing increases.

Figure 4 demonstrates longitudinal section of extrusion head used for manufacturing polymeric casing of pipe inner framing. Figure **2** demonstrates a cross-section of a moulding channel of extrusion head used for manufacturing polymeric casing of pipe inner framing.

According to Figure 4 an extrusion head used for manufacturing polymeric casing of pipe inner framing consists of case (1) within which cone (2) is set by means of mandrel-keeper (3). Fixation of mandrel-keeper (3) in the head is provided by means of case (4) detail, within which regulating bolts (5) for a radial displacement of matrix (6) are set. To form support lugs in extruded polymeric casing in matrix (6), longitudinal slots are made; simultaneously there is a smooth section of matrix along the circumference's arch in a part where a longitudinal slot surface joins a cylindrical part of matrix surface. The hollow inner surface for polymers melt flowing is formed by mandrel (8) and its details (9) and (10). For melt flowing there is a cylindrical grating (11) in a hollow. Heating of extrusion head is provided by electrical heaters (12).

Extrusion head used for manufacturing polymeric casing of pipe inner framing works as follows. Having gone through the filter polymer's melt is fed to a round cross channel situated in case (1), according to Figure 1 by means of cone 2 polymer's stream is cut and gets the ring shape. While passing through mandrel-keeper the melt is cut by mandrel-keeper's ribs. To provide welding ability of polymer's stream there is a cylindrical grating in the shape of a glass, set behind mandrel-keeper.

The hollow of feeding channel is farmed by a case detail (4) and mandrel's details (9, 10). Just in front of mandrel (8) melt stream passes the section of narrow spot, this allows to improve the welding ability of polymer's melt stream.

According to Figure 5 a moulding channel is formed by matrix (6) and mandrel (8). The shape of a moulding channel cross-section coincides with a shape of cross-section of polymeric casing of pipe inner framing.

The longitudinal surfaces (7) in matrix (6) are formed by the second order curve. The longitudinal slots form support lugs on the outer surface of polymeric casing of pipe inner framing; providing equality of pressure overfall all section of extrusion head moulding channel long. A smooth matrix cutoff along the circumference's arch in a place where longitudinal slots' surfaces join a cylindrical part of a matrix surface eliminates cutting off a profile of polymeric casing of pipe inner framing diminishing shear strength in a polymer melt stream in longitudinal slots.

Extrusion head has outer and inner heating, provided by means of electrical heaters (12). Openings for thermo pairs setting within them are provided for head details, their number depends on the number of independently regulated parts of heating. The outer surface framing casing is formed by matrix (8), which position is changeable depending on its displacement to radial direction by means of regulating bolts (7). To supply a calibrating device with cooling water and to draw it aside the following details are used: a tube for water pipe-bend (4), supply tube (5), a tube for water supply into calibrating device (15), bar (16) and bar base (17). By means of these details before hand cooling of polymeric casing inner surface becomes possible.

An extrusion head is heated by electrical heaters (11). Openings for setting thermo pairs within them are provided for head details in accordance with the number of independently regulated heating parts. Inner extrusion head filled with air and a thermo stable pad (18) as well, essentially reduce an unwilling heat exchange between a hot head and the details providing cooling water supply and pipe-bend.

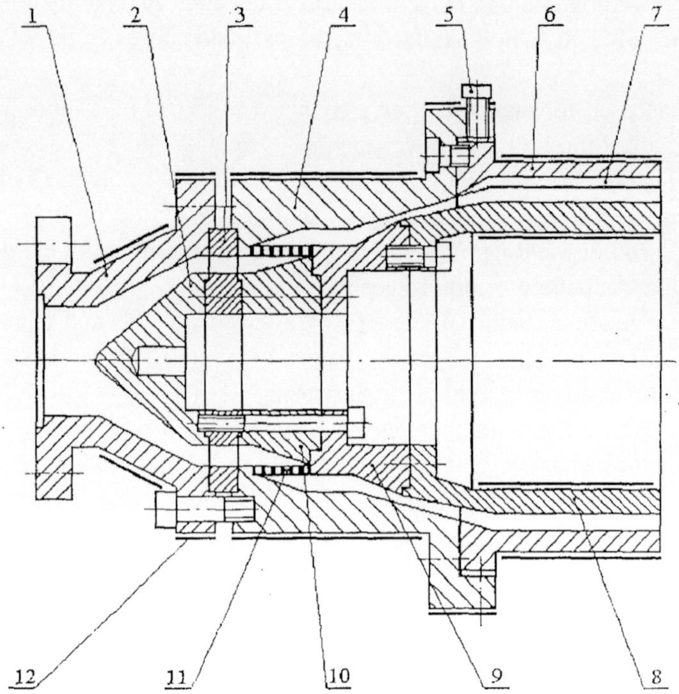

Figure 4. Extrusion head for manufacturing polymeric casing of pipe inner framing

Thus, the design of extrusion head for manufacturing outer polymeric casing is analogous to the design of extrusion head for manufacturing polymeric pipes. The difference is in the structure of shaping channel, i.e. the cross-section of polymeric casing (Figure 4). Correspondingly, in a shaping channel there is a section of longitudinal slots parallel to ring's section (like in extrusion heads for pipe manufacturing), which forms longitudinal lugs on the outer casing surface.

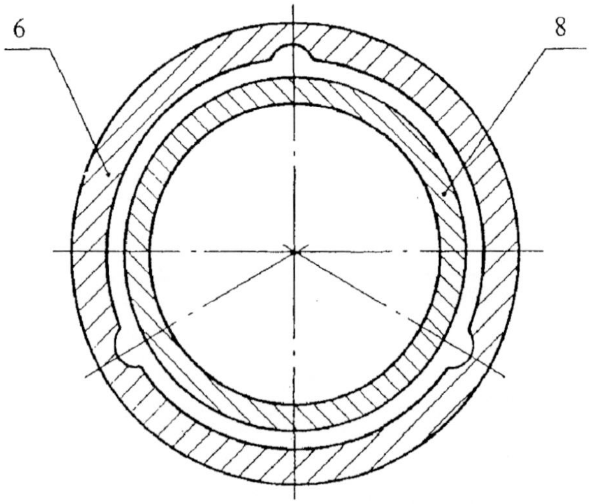

Figure 5. Cross-section of extrusion head for manufacturing pipe inner framing.

3. Description of Extrusion Head of Calculation Method Used for Manufacturing Polymeric Casing of Pipe Framing

Designing the extrusion heads construction required selection of polymeric material for casing manufacturing as well as calculation of its dimensions taking into account the outer diameter of steel pipe-line. Polymeric casing is designed for its framing.

Having taken into consideration the most frequently used polymeric materials, such as high density polyethylene (HDP), low density polyethylene (LDP), polypropylene (LP), polyvinylchloride (PVC) as material for manufacturing polymeric casing of pipe framing, polyvinylchloride was chosen. The selection is explained by PVS's advantages to other large capacity plastics. They are: the raw material availability, the durable qualities, taking care of natural resources, solving ecological problems.

In our work we use PVC composition with mixing of the following components (%): PVC- 77,0-91,0; calcium stearat – 2,5; plumber silicate – 1,2; dioetilphtalate – 3,0-20,0.

To ascertain polymeric casing dimensions, thickness of its wall S and height of support lug h [3] were calculated. The calculation of height h was carried out on permissible contact stress $[\sigma_k]$ which is equal to 3-8 MPa for a selected polymeric material.

It was supposed that a cross-section of support lug is semicircle with radius h and that steel pipe with radius R is based on two support lugs.

A final formula for calculating the height of support lug is as follows:

$$h = \left(\frac{[\sigma_k]^2 \cdot \pi \cdot (1 - \mu^2) \cdot (E_1 + E_2)}{g_E \cdot E_1 \cdot E_2} - \frac{1}{R} \right)^{-1}, \quad (1)$$

where E_1, E_2 - are elasticing modules of casing and pipe-line materials;

q_k –force given for a contact length unit.

The calculation of wall thickness was carried out under the following conditions: stability of polymeric casing shape while pipe-line laying in the ground, durability of casing material under pressure which is equal to a working one in a pipe-line being framed.

Carrying out the calculation of polymeric casing concerning durability of shape, it was supposed that outer framing surface is under inner pressure P, which is equal to the pressure of ground on the casing surface:

$$P = G_Г / S_{ОБ}, \quad (2)$$

where $G_Г$ - gravity force of the ground over the casing;

$S_{ОБ}$ – casing surface influenced by gravity force G_R.

Critical outer pressure of casing p_{kp} was calculated with a formula:

$$p_{кр} = \frac{E \cdot S^3}{4 \cdot (1-\mu^2) \cdot R^3}, \tag{3}$$

Where E is elasticity module of casing material; R is inner radius of a cylindrical part of casing.

It is possible to calculate the minimum thickness of casing wall S for condition of shape stability taking into account the condition $p \leq p_{кр}$ and introducing the coefficient of stability reserve:

$$S = R\sqrt{\frac{4m\rho(1-\mu^2)}{E}}. \tag{4}$$

To provide casing material durability under working P in a framed pipe-line, the pipe framing casing is to bear a working pressure of substance pumped over the framed pipe-line. So the calculation of wall thickness is carried out proceeding from permissible strain for a selected material σ_g, at the same time taking into account the maximum pressure P for a selected material.

Thus, we conclude that the suggested construction of an extrusion head for manufacturing polymeric casing allows producing pipe-line casings with dimensions and quality required.

REFERENCES

[1] *Patent RF № 2134640, 6B29C47/20* Extrusion head used for manufacturing polymeric casing pipe framing / A.K. Panov, A.N. Mehlis(RF). Applicated 24.02.98. Published 20.08.99.

[2] A.K. Panov, A.N. Mehlis, A.A. Panov Elaboration of extrusion head construction used for manufacturing polymeric casing // *Technique on the bourderline of the 21 century: Collection of scientific works. Ufa:* Gigem. 1999. p. 48-57.

[3] A.K. Panov, A.N. Mehlis. Method of dimension calculation of polymeric casing pipe framing // Cybernetics method of chemical-technological processing: *Collection of reports at scientific conference. Ufa*: USOTU. 1999. V.2. Book 1, p.115

In: Progress in Monomers, Oligomers, Polymers, Composites … ISBN 978-1-60456-878-3
Editors: R. A. Pethrick, G.E. Zaikov et al.

Chapter 28

PLANT USED FOR MANUFACTURING BUILDING TILE ON THE BASIS OF POLYMER MATERIALS

A.A. Panov, T.G. Beloborodova,T.A.Anasova and G.E. Zaikov
Academy of Science of Republic of Bashkortostan, Sterlitamak Branch, Sterlitamak, , Russian Federation
N.M. Emanuel Institute of Biochemical Physics, RAN, Moscow, Russian Federation

ABSTRACT

The contributors gave information about plant used for manufacturing building tile on the basis of polymer materials. Given plant allows to produce qualitative building tile

Key-words: plant, tile, building, manufacturing, polymer materials.

INTRODUCTION

Elaborated press-form used for manufacturing building tile on the basis of polymer materials consists of 4 chambers, square section matrix, punch, high pressure cylinder, switch-board chamber, rod with a disk, piston with pointed pivot and chamber-bunker. Given plant allows to produce qualitative building tile.

DESCRIPTION OF A PLANT USED FOR MANUFACTURING BUILDING TILE ON THE BASIS OF POLYMERIC MATERIALS

There is a known construction of plant for making polymer articles. Press-form provides reciprocal centring of punch and matrix: conjugation of punch with loading chamber fit of forming signs, which have elements both into matrix and punch, fitting size of punch matrix into their iron rings are carried out the seventh qualitate. The press-form has a system of heating, consisting of cylindrical pivots of electric heaters, which contact directly with matrix and punch (1).

The shortcoming of given article is that the absence of cooling system of pressed article does not allow to push out ready article after pressing. That is why the productivity of the plant and quality of ready article are significantly reduced. It should also be noted the complicated construction of press-form.

The article for getting on the basis of polyvinylchloride with steam generator is the most similar technically and by achieved effect to the previous one. It also contains high pressure cylinders for preparation of melts, press-form with system of heating, consisting of upper and lower matrix between which intermediate shaping calibrating removable plug is placed. A movable disk, which is placed on the rod, is installed into plug. (2)

The shortcoming of known press-form is complication of conducting of the process of pressing. This happens because of imperfection of pressing group and pushing out pressed article.

Absence of cooling system of pressed article reduces productivity and quality of manufactured polymer tile.

Elaborated construction allows to solve problems of simplification of the construction of press-form and raising of quality of manufactured polymer tile.

Mentioned tasks are solved with the help of the fact that press-form for polymer materials consists of matrix, punch, mechanism of reflexive-receiving punch movement and the system of heating. And for the purpose of using polymer materials for manufacturing building tile it is noted for the matrix is cross-section by lid with cooling system of punch, mounted on the rod with possible reflexive-receiving movement when it is fed with compressed air into distributing chamber for compressing and pushing out of pressed article.

Technical result, which is provided by plant for making building tile on the basis of polymer materials, is manifested in simplification of construction and raising of quality of manufactured tile, owing to the choice of optimal geometry of matrix with square cross-section with cooling system. There is also mechanism of reflexive-receiving punch movement, fixed on the rod, because of compressed air, fed into distributing chamber. In the process of pressing tile is manufactured without any signs of warping and high size precision and quality are also provided.

Arrangement of a given construction is explained by draughts: picture 1 shows plant with all necessary devices for conducting the process of pressing; figure 2 shows cross-section of A-A matrix with cooling chamber.

Plant for making building tile on the basis of polymer materials, as shown in picture 1 contains 4 chambers : matrix of square cross-section 1 for pressing with lid 2 and housing 3 with pipe socket 4; with lids 6 and 7, punch 8, rod with disk 9 and pipe sockets with valves 10; high pressure cylinder 11 for melting polymer with lid 12 and piston 13 with graduated core14, electric heating 15 and insulation 16 and pipe sockets with valves 17 and lock tap 18; chamber-bunker 19 with pipe socket 20and lock tap 21. Figure 2 shows matrix pressing 1in section with housing 3 and vacant space 22 for coolant and also compressed tile 23. Plant used for manufacturing building tile on the basis of polymer materials works in the following way.

Polymer is loaded into camera-bunker 19, then granulated polymer goes to high pressure cylinder 11 by means of compressed air, which goes through pipe socket 20. After cylinder is filled up in high pressure cylinder, valve 21 is locked, electric heating system 15 is turned on and granulated polymer becomes a melt. Then compressed air is given to cylinder 11 through upper pipe socket 17 and melt is pressed through into chamber-matrix 1 with the help of piston 13. The dose of melt is given according to points of the core 14, originally dosed on the polymer melt. After filling camera 1 with necessary dose of melt is locked by lock tap 18, then compressed air goes through pipe-socket 10, which moves disk with rod 9 and punch 8, and that is how tile 23 is pressed. After pressing the process of cooling is carried out by means of compressed air, which is given through pipe-socket 4 into vacant space for coolant 22.

The lid of camera 2 after cooling opens and pressed tile 23 is pressed through outside by piston. Further all operations are repeated.

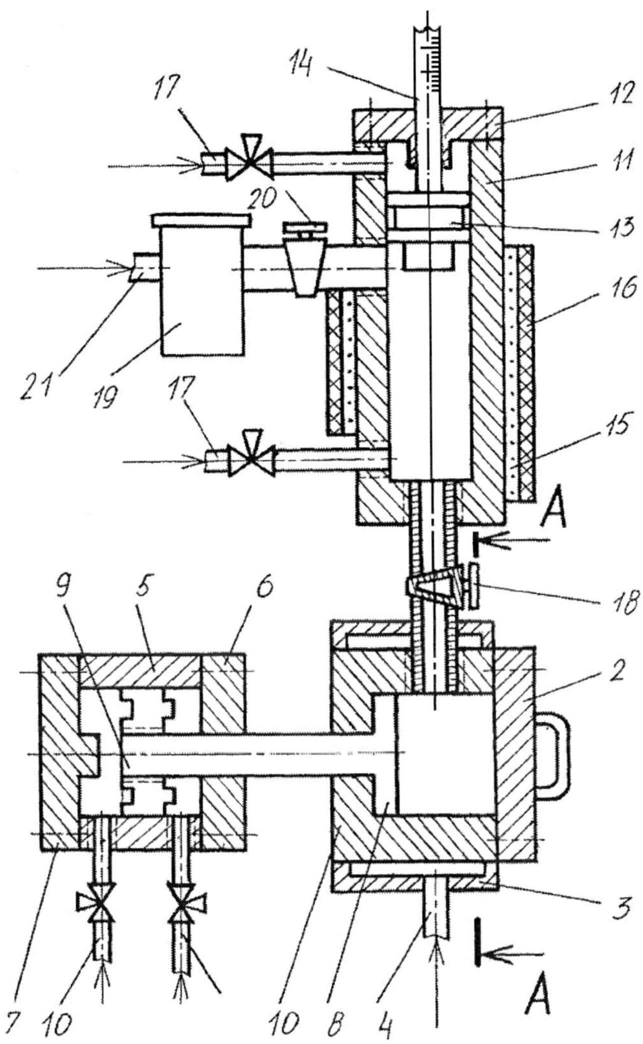

Figure 1. Plant for conducting the process of rolling.

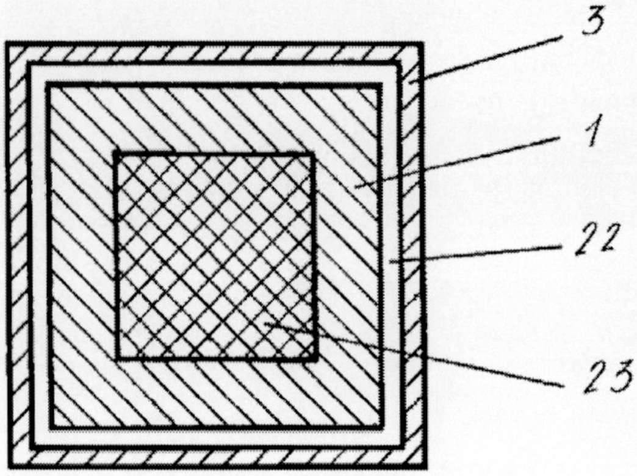

Figure 2. Cross-section of matrix with cooling camera.

On the elaborated construction of the plant they got tile from polypropylene 01003 TC 211-020-0020-3521-96, size 150*150 mm, 4 mm thick. Optimal temperature for pressing is 468-478 K and pressure 3.5 / 4.0 MPA.

This offered plant used for manufacturing tile on the basis of polymer materials allows to raise quality of pressed building tile.

REFERENCES

[1] Kazankov Y.V., Makarov M.S., Mironov B.A. Calculation and construction of shaping instrument M., *Rotoprint MICh.M*, 1992, p.p. 4-15

[2] A.c. *USSR №1391932 Press-form for getting on the basis of polyvinylchloride* A.K Panov, V.P Churov, T.F. Iljina, T.V. Grigorjeva (USSR), published 10.04.88, issue 16.

In: Progress in Monomers, Oligomers, Polymers, Composites … ISBN 978-1-60456-878-3
Editors: R. A. Pethrick, G.E. Zaikov et al.

Chapter 29

CONSTRUCTIONS OF MULTI-JET EXTRUSION HEADS FOR MANUFACTURING PROFILE ARTICLES

A.A. Panov, T.G. Beloborodova, A.K. Panov and G.E. Zaikov
Academy of Science of Republic of Bashkortostan, Sterlitamak Branch,
Sterlitamak, Russian Federation
N.M. Emanuel Institute of Biochemical Physics, RAN,
Moscow, Russian Federation

ABSTRACT

The contributors of this paper give information about constructions of multi-jet extrusion heads for manufacturing profile articles. It was shown high efficiency of these constructions.

Key-words: multi-jet, profile articles, extrusion, constructions, heads.

INTRODUCTION

The increase in production of plastic consumer items requires the improvement of the existing equipment and the elaboration of new highly productive equipment. In particular, multi-jet polymeric articles in the form of a frame plank and profile lath. The analysis of the articles has shown that generally they have just one shaping channel for extrusion of a produced profile. Consequently the productivity is not very high.

MULTI-JET EXTRUSION HEAD FOR MANUFACTURING POLYMERIC COINSTRUCTION FRAME PLANK

The elaborated construction multi-jet head for manufacturing polymeric construction frame plank thanks to the selection of the optimum geometry of the cross-section of the moulding channels makes it possible to provide the best quality and high productivity during the process of manufacturing. The optimum geometry of the cross-section of the moulding channels is achieved by a coordinated position of the upper and lower sections of a head case; in the upper section there are hollows in the kind of turnings. Sinusoidal curved lines at the incidence angle of 45° form the lateral sides, by the amplitude change; the sinusoidal curved line period and the incidence angle are constructively selected within 45° to the plank base.

The form lateral surfaces of a hollow is described by the equation

$$y = A\sin\left[\frac{2\pi x}{T}\right], \tag{1}$$

where

y – an ordinate of sinusoid's points , mm;

x – an abscissa of sinusoid, mm;

A – an amplitude of sinusoid, mm;

T – a period of sinusoid, mm.

By selection the amplitude's values, sinusoidal curved line's period, the dimensions of rectangular hollows in an upper and lower section as well as the height of moulding channels the uniform even speed distribution in all the points of the cross section of the extruded item is achieved. Thus it becomes possible to diminish residual displacement strains, to eliminate uneven yield of extrudate and warping of the manufactured article. This provides rather high dimensional exactness and quality of an extruded construction frame profile plank.

The structure of an extrusive head is explained by means of figures:

Figure 1 is the cross section of multi-jet extrusion head for manufacturing polymeric construction frame plank;

Figure 2 is a longitudinal section of extrusion head along A $\rightarrow$ A;

Figure 3 is the cross section of a finished frame plank.

Multi-jet extrusion head for manufacturing polymeric construction frame plank contains a split case consisting of two sections: an upper one (1) and a lower one (2), where moulding channels are situated (3) in the form of the rectangular turning (4). The hollows are made in the upper section and the lateral sides of channels are formed by sinusoidal curved lines (5) at an incidence angle of 45'. Rectangular turnings are envised in the lower section (6). The upper (1) and the lower (2) sections are connected to each other with bolts (7) and are fixed to the joint flange (8) with bolts (9).Case sections (1) and (2) are provided with the cooling system consisting of longitudinal channels (10) for water flowing and nozzle (11) for providing water supply and medium withdrawal. Moulding channels (3) communicate with a feeding channel (12).

An extrusion head is joined to extruder by the joint flange (8).

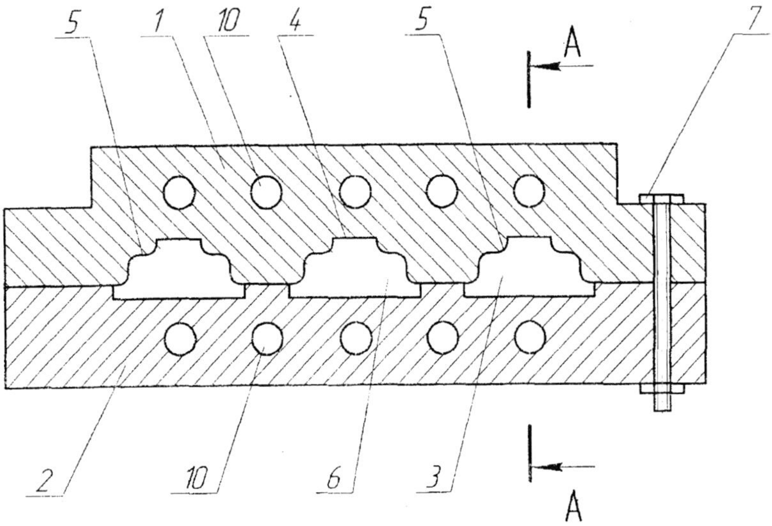

Figure 1. is cross section of multi-let extrusion head for manufacturing polymeric construction frame plank;

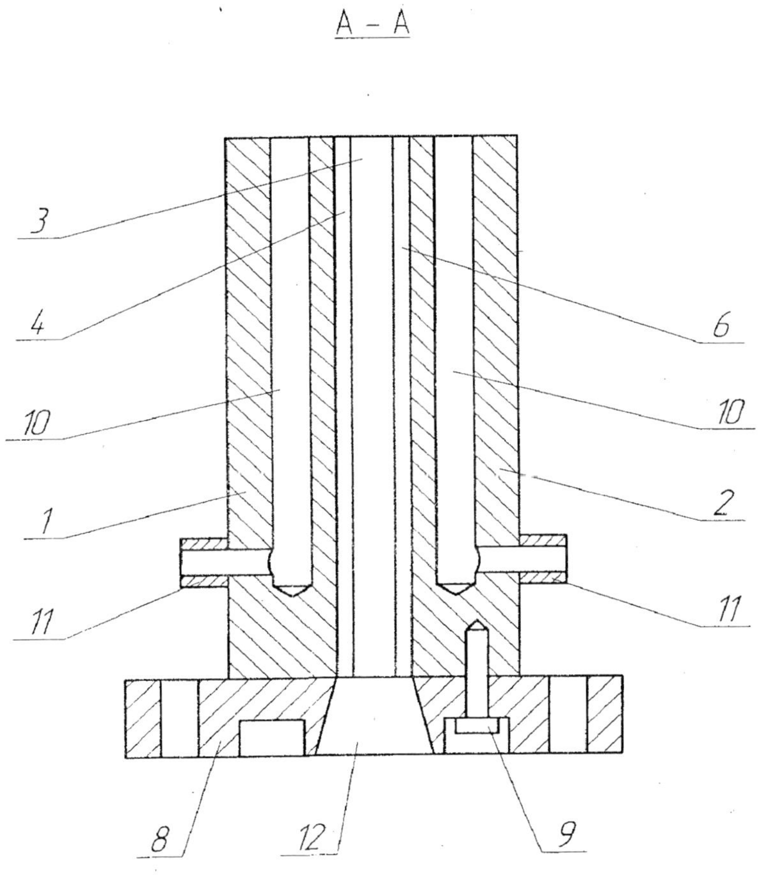

Figure 2. is a longitudinal section of extrusion head along A-A

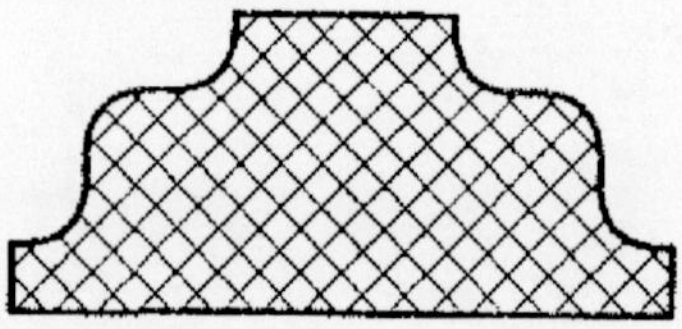

Figure 3. is the cross section of frame plank.

Multi-jet extrusion head for manufacturing polymeric construction plank works as follows. Polymeric melting supercharged by extruder's shneck gets into supply channel (12), then it goes into moulding channels (3) where it takes the shape of construction frame plank as shown in Figure 3.

The outer surface of a construction frame profile plank is formed by inner surfaces of moulding channels which are formed by the upper and lower head sections. To make dimensional exactness and quality of the articles higher , sections (1) and (2) of the case are made of steel 45 with a high of surface purity (10- 12), matching the mentioned details allows to produce whole moulding channels without any clearance during assembly. While pressing through meeting streams of polymer through moulding channels no traces assembling details are left on the construction frame plank surface. The construction frame plank is produced with a glossy surface and a high class of purity. It is investigated that in sections (1), (2) there is a cooling system of extruded construction frame plank by supply of medium through nozzle (11) into the longitudinal cooling channels (10). Construction frame plank made of propylene of trade mark 01003 TY 2211-020-0020-3521-96 with its base 40 mm wide; 15mm high; the top of a lug 12mm, the lateral inclined sinusoidal side 21 mm long. Optimum extrusion condition at temperature in within the limits 468-478 K and extrusion pressure within 4, 0 -4, 5 M Pa.

Two-Channel Extrusion Head for Manufacturing Polymeric Construction Assembly Profile Rank

Elaborated two-channel extrusion head for manufacturing polymeric construction assembly profile rank allows to manufacture profile ranks of high quality widely used in construction and assembly work with application of existing extrusion equipment . The selection of optimum geometry of cross section of shaping channels is achieved by coordinated position of upper and lower head case sections, the upper section has hollows in the form of turnings, the lateral sides of which are formed by sinusoidal curved lines at the incidence angle of 45', by amplitude change, and sinusoidal curved line's period.

The selection of curvature radius of the hollow of the upper section and the lug of the lower section is done constructively and it depends on the thickness and the width of the rank.

The surface shape of the edges of the hollow of the upper section is a sinusoidal curved line with changing amplitude and period, the surface of which is described with an equation of sinusoid

$$y = A\sin\left[\frac{2\pi x}{T}\right],$$

where

y- an ordinate of sinusoid's points, mm;

A- fading amplitude of sinusoid , mm;

T- changing period of sinusoid, mm.

An even distribution of speeds in all the points of cross section of an extruded article is achieved by selecting amplitude's values, sinusoidal curved line's periods of which depends on the thickness of a manufacturing rank. The uneven yield of extrudate and the warping are eliminated. Thus a high degree of a dimensional exactness and quality of extruded profile lath are provided. The structure of a two-channel extrusion head for manufacturing polymeric construction and assembly lath is given in figures 1-3: Figure 1 is a cross section of a polymeric construction and assembly lath ; Figure 2 is a longitudinal section along A- A; Figure 3 is a cross section of a finished profile lath.

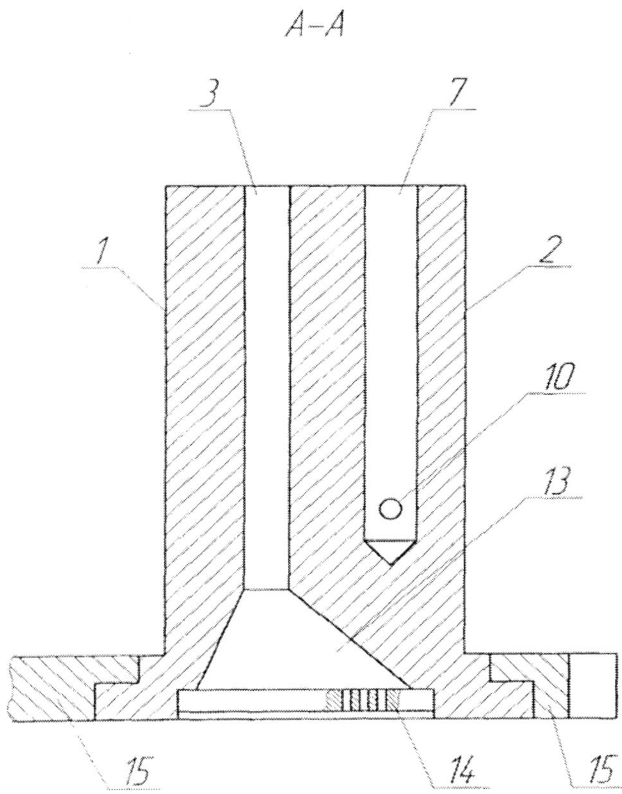

Figure 4. a cross section of two-channel extrusion head for manufacturing polymeric construction - assembly lath

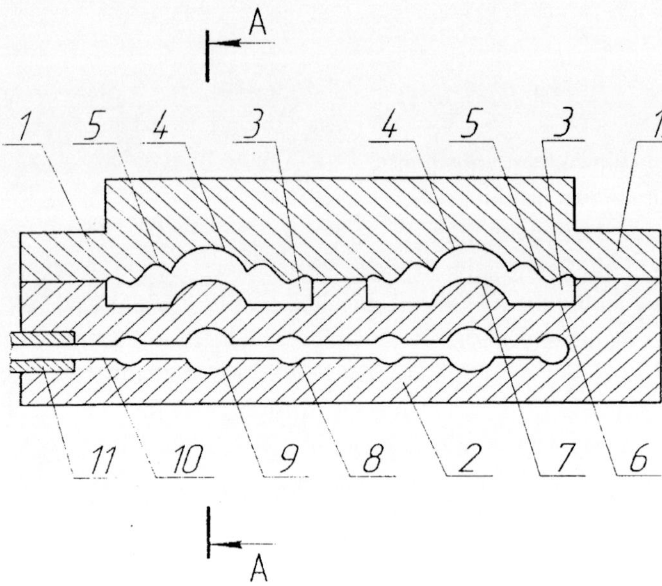

Figure 5. a longitudinal section along A→A

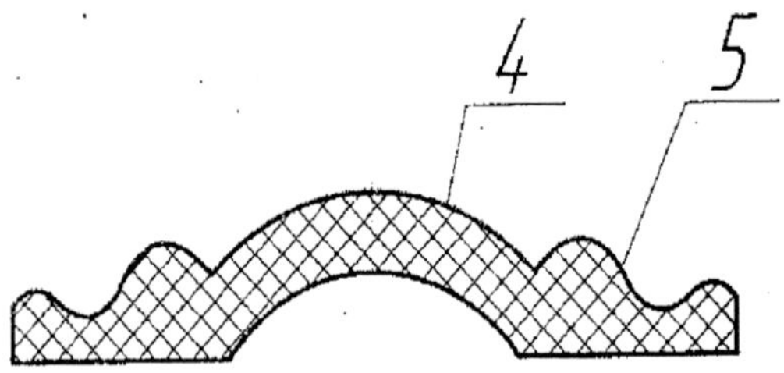

Figure 6. is a cross section of a finished profile rank.

Two-channel extrusion head for manufacturing polymeric construction - assembly lath according to figures 1 and 2 includes a sectional case consisting of two sections : the upper one (1) and the lower one (2), in which moulding channels (3) are located, made in the sections. In the upper sections there are profile turnings, their surface in the middle has a shape of oval curved lines (4) and the surface of hollows' edges has a shape of a sinusoidal curved line with a fading amplitude (5), the lower section contains hollows (6), in the central part of them there are oval lugs (7) symmetrical to the curved hollows' surface (4) of the upper section (1). Moulding channels (3) are provided with a cooling system consisting of cooling channels (9) and (8), channels for water flowing (10) and nozzle (11) through which the supply and pipe-bend of medium is provided.

The upper section (1) and the lower section (2) are joined by bolts (12). The moulding channels (3) communicate with a supply channel (13) with a filter grating in the entrance. The joining of an extrusion head to an extruder is made with a joining flange (15). A two-channel extrusion head for manufacturing polymeric construction and assembly profile lath functions as follows.

Polymeric melt supercharged by extruder's shneck gets into a supply channel (13), and then to the moulding channels (13) formed by an upper section (1) and a lower section (2). Moulding channels (3) give a shape of a profile lath to the melt according to **figure 3**.The upper section (1) and a lower section (2) are made of steel 45 with a high degree of purity (10-12) of the surface processing. It makes possible to assemble the whole without any clearance moulding channels (3), when pressing polymer's stream through it there are no traces of joining details. The profile lath is produced through the moulding channels (3) with a glossy surface with a high purity class.

Cooling of extruded profile lath is envisaged in the lower section of a case (2). It is carried out by supple and pipe-bend of a medium through nozzle (11) and cooling channels (8, 9) using flowing water channels (10).

The elaborated construction extrusion head for manufacturing of polymeric construction and assembly lath allows increasing the quality of the articles' output.

In: Progress in Monomers, Oligomers, Polymers, Composites … ISBN 978-1-60456-878-3
Editors: R. A. Pethrick, G.E. Zaikov et al.

Chapter 30

THE NANOADHESION INFLUENCE ON STRUCTURE AND PROPERTIES OF PARTICULATE-FILLED POLYMER NANOCOMPOSITES

G.V. Kozlov[1,], A.I. Burya[1], Yu.G. Yanovskii[2] and G.E Zaikov[3,†]*

[1]State Agrarian University, Dnepropetrovsk, Ukraine
[2]Institute of Applied Mechanics of Russian Academy of Sciences, Moscow, Russian Federation
[3]N.M. Emanuel Institute of Biochemical Physics, Russian Academy of Sciences, Moscow, Russian Federation

ABSTRACT

The results of the present paper have shown the existence in polymer particulate-filled nanocomposites nanoadhesion effect, which consists in enhanced level of polymeric matrix-nanofiller interaction. This effect has purely dimensional origin and was realized for nanofiller particles with diameter less than 80 nm. The nanoadhesion effect influences strongly on macroscopic properties of those nanocomposites, in which it was realized. So, essential thermal stability enhancement and considerable thermal expansion coefficient reduction of nanocomposites on the base of phenylone at introduction 1 mass. % aerosil only are due to the nanoadhesion effect. Appreciation of nanoadhesion effect allows qualitatively and quantitatively to describe exactly nanocomposites reinforcement degree.

Keywords: Polymer nanocomposite, interfacial adhesion, thermal expansion, nanoadhesion, structure, thermal stability.

* Correspondence to: State Agrarian University, 49027, Ukraine, Dnepropetrovsk, Voroshilov st., 25 , I_dolbin@mail.ru

† Correspondence to: N.M. Emanuel Institute of Biochemical Physics, Russian Academy of Sciences, 119334, Russia, Moscow, Kosygin st., 4, Chembio@sky.chph.ras.ru

INTRODUCTION

At present polymer nanocomposites are some of the most perspective engineering materials [1]. Moreover the main attention is paid to nanocomposites filled by layered silicates (organoclays) [2] or nanotubes [3], whereas particulate-filled nanocomposites were studied to a far smaller extent. Nevertheless, possibilities of the last class of polymer nanocomposites are far from being exhausted [4].

As it is well known [5], for multiphase materials the interaction level between phases plays a defining role in properties formation of these materials. The said above relates completely to polymer composites too (nanocomposites). So, the authors [6] have demonstrated that the filler contents increase in polymer composites results in good interfacial adhesion case of polymeric matrix-filler to modulus elasticity enhancement, but interfacial adhesion level determination by experimental methods in polymer composites case comes up on definite difficulties. So, the adhesional strength value determined by standard methods can differ sufficiently from the same for composites in virtue of the existence of various kinds of nonaccounted effects in real composites: filler particles aggregation [7], polymeric matrix structure change due to filler introduction [8] and a number of other factors. The indicated above importance of the interfacial phenomena consists in their influence practically on all polymer composites properties. Therefore the most suitable from practical point of view method of the interfacial adhesion level determination in bulk composites is the estimation of this factor according to composite macroscopic properties, for which a number of techniques (for example, [9]) was elaborated. The purpose of the present paper is to determine the interfacial adhesion level and the factors, influencing on it, and the influence of the indicated level on thermal and mechanical properties of particulate-filled nanocomposites on the base of phenylone as well.

EXPERIMENTAL

As polymer binding of the studied nanocomposites the linear heterochain copolymer phenylone S-2 of the following chemical constitution was used:

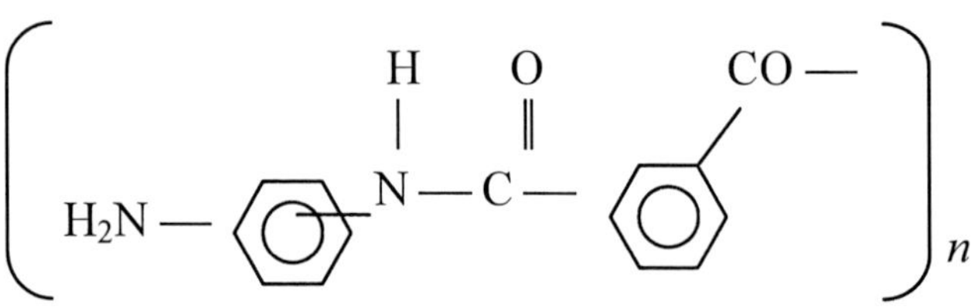

As particulate nanofiller the aerosil (A) with particles average diameter 15 nm and specific surface 170 m^2/g and oxinitride silicium-yttrium (OSY) Si_3N_4-Y_2O_3 with average diameter of particles 64 nm an specific surface 43 m^2/g were used. The contents of the first from the indicated nanofillers makes up to 0.3, 0.5 and 1.0 mass. %, the second one is varied within the range 0.2-10 mass. %.

The nanofiller introduction in polymeric matrix was realized in rotating electromagnetic field with the aid of nonequiaxed ferromagnetic particles with the ratio of length to diameter 4-5; the ferromagnetic particles volume, loading in an apparatus reactor, was within the range 0.04-0.05 from electromagnetic field action volume. The electromagnetic induction value of this field was within the range 0.08-0.12 Tl. At the indicated parameters it was found out experimentally that the optimal processing duration of nanocomposites in electromagnetic field was 270-300 s. Nanocomposites samples preparation was realized by compressional pressing method at temperature 537-616 K, pressure 40-100 MPa and exposure at the indicated parameters at a rate of 1 min/mm per product thickness.

The processing of powder-like compositions phenylone/A and phenylone/OSY in rotating electromagnetic field allows to inhibite (or decrease) nanofiller particles aggregation. For checking of the achievement of this effect the phenylone/A nanocomposite samples were processed by simple mechanical blending. Further the samples of nanocomposites, filled by aerosil and obtained with rotating electromagnetic field application, will be designated as phenylone/A-MF and obtained by mechanical blending – as phenylone/A. Nanocomposites filled by oxinitride silicium-yttrium were designated as phenylone/OSY.

The linear coefficient of thermal expansion of nanocomposites was determined on dilatometer DKV-SAM, intended for automatic registration of dilatometric curves of different solids, within the temperature range 298-498 K. For the tests the prismatic (6×4 mm) samples with the length 50±3 mm were used. The nonparallelism of sample grinded butt-ends consists of not more than ±0.02 mm. The number of samples for tests for each nanocomposite was no less than three.

For the surface structure study the three-dimensionally reflecting surface structure analyzer Zygo NewView 5022 was used.

The nanocomposites study by method of thermogravimetric analysis (TGA) was made on derivatograph Q-1500 of Paulik-Erdei system of the firm MOM (Hungary). The studies were carried out within the temperature range 298-873 K with the temperature rise rate 10 K/min, mass of a sample was equal to 100 mg. The curves recording was made with maximal sensitivity.

The mechanical tests on compression were made on testing machine FP-100 at temperature 293 K and strain rate 10^{-3} s^{-1}. The testing samples have diameter 10±0.5 mm and height 15±0.5 mm. For each nanocomposite five samples were tested.

Results and Discussion

In Fig. 1 the curves 1-3 represent three main types of the dependences of linear coefficient of thermal expansion of the studied nanocomposites α_n on nanofiller volume contents φ_n. The straight line 1 illustrates the case, when between two phases of nanocomposites adhesion is absent and at $\alpha_m > \alpha_n$ (α_m, α_n are thermal expansion coefficients of polymeric matrix and nanofiller, respectively) and the absence in interfacial layers residual compression strain polymeric matrix at heating will be expanded irrespective of filler particles and in this case $\alpha_n = \alpha_m$ [10]. The straight line 2 corresponds to a simple rule of mixtures [10]:

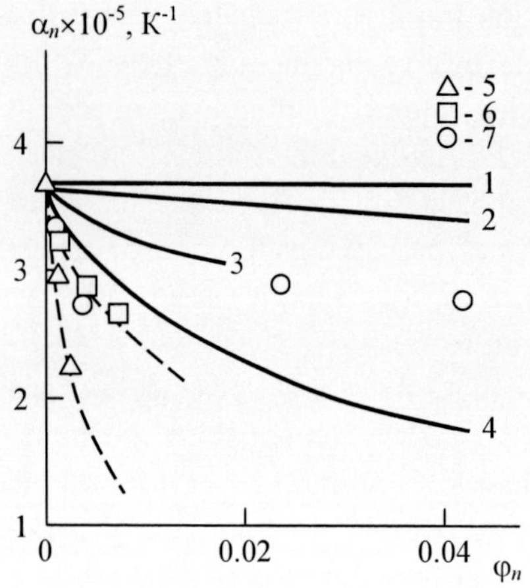

Figure 1. The dependences of linear coefficient of thermal expansion α_n on nanofiller volume contents φ_n. 1 – adhesion absence on interfacial boundary; 2 – mixtures rule; 3 – the Terner equation; 5-7 – experimental data for nanocomposites phenylone/A-MF (1, 2, 3, 5), phenylone/A (1, 2, 3, 6) and phenylone/OSY (1, 2, 4, 7).

$$\alpha_n^m = \alpha_m(1-\varphi_n) + \alpha_n\varphi_n, \tag{1}$$

where α_n^m is the thermal expansion coefficient of mixture, φ_n is the volume fraction of filler.

The rule (1) is correct for the ideal case only, when each phase expands independently one from another. And lastly, the straight line 3 corresponds to Terner equation [10]:

$$\alpha_n^T = \frac{\alpha_m(1-\varphi_n)K_m + \alpha_n\varphi_n K_n}{(1-\varphi_n)K_m + \varphi_n K_n}, \tag{2}$$

where K_m and K_n are a bulk elasticity modulus of polymeric matrix and nonofiller, respectively.

At the equation (2) derivation the strains equality method was used for calculation of mixtures thermal expansion coefficient depending on density, elasticity modulus, thermal expansion coefficient and mass relationship of composing components. If the made assumptions are correct, then the formula (2) will also be applicable for polymer nanocomposites.

Besides, in Fig. 1 values α_n determined experimentally are shown by points for two series of the studied nanocomposites including phenylone, for which α_n=α_m. Let's note, that in Terner equation Young modulus E magnitudes were used instead of the bulk elasticity modulus K. As it was known [10], such replacement allows to obtain a lower boundary of α_n value for polymer composites. From the data of Fig. 1 it follows that α_n values for both nanocomposites series, filled by aerosil, are disposed substantially lower than the curve calculated according to Terner equation, i.e. they are substantially smaller than the lower limiting values α_n for polymer composites. Besides α_n values for nanocomposites phenylone/A-MF are significantly smaller than the corresponding values for nanocomposites phenylone/A. In other words, inhibition (or weakening) of aerosil particles aggregation process substantially decreases α_n value. In case of nanocomposites phenylone/OSY the values α_n at small φ_n are somewhat smaller than the ones calculated according to Terner equation, although they are close to them, and at φ_n>0.05 the values α_n for them are disposed between the curves, calculated according to the mixtures rule and Terner equation, i.e. they have α_n values intermediate on absolute value and typical for polymer composites [10].

Let's consider physical significance of the described above α_n changes for nanocomposites, studied in this paper. It is possible to estimate polymeric matrix-filler interaction level (interfacial adhesion level) with the aid of the parameter b, determined according to the equation [10]:

$$\alpha_n = \alpha_n^m - b\left(\alpha_n^m - \alpha_n^T\right). \tag{3}$$

The larger the value b the higher interfacial adhesion level. For a large enough number of polymer composites with different matrixes and fillers b value varies from -0.19 to 1.39 [10]. Calculation according to the equation (3) for the studied nanocomposites shows that for phenylone/A-MF $b\approx12$, phenylone/A $b\approx6$ and for phenylone/OSY value b varies within the range ~ 5.40-0.35, decreasing at φ_n increase (Fig. 2). Therefore, polymeric matrix-filler interaction level in the studied nanocomposites can exceed substantially (on one order of magnitude) this value for microcomposites. Following to the authors [11] we called this effect nanoadhesion. Let's remind that in paper [11] the adhesion level for two-layered films of polystyrene on micro- and nanolevel was measured. It turns out, that in the last case adhesional strength is substantially higher in comparison with the first one. The authors explained [11] the indicated effect of polymers nanoadhesion by different degree of interfacial layer formation completion between two films of polystyrene.

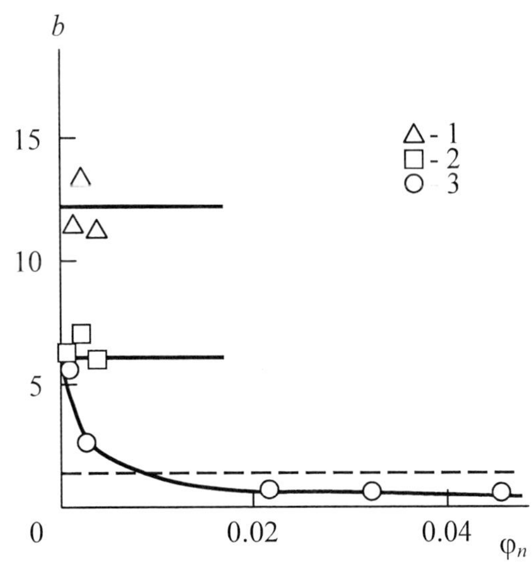

Figure 2. The dependences of parameter b on nanofiller volume contents φ_n for nanocomposites phenylone/A-MF (1), phenylone/A (2) and phenylone/OSY (3). The horizontal shaded line indicates upper limiting value b for microcomposites.

In the nanocomposites case the nanoadhesion cause is a small contact area of polymeric mztrix-nanofiller. From the comparison of aerosil and OSY particles size and Fig. 2 data one can make a conclusion that two factors as minimum influence on value b: nanofiller particles size and their aggregation level, i.e. in the long run nanofiller aggregates size as well. So, value b for aerosil is higher than for oxinitride silicium-yttrium (particles size are 15 and 64 nm, respectively) and for nanocomposites phenylone/A-MF value b is substantially larger (twice) than for phenylone/A as that was mentioned above, processing of phenylone and aerosil powder-like blend in rotating electromagnetic field reduces aerosil particles aggregation level. Besides, the effect of b decrease at φ_n increase for nanocomposites phenylone/OSY (Fig. 2) is also due to enhancement of oxinitride silicium-yttrium particles aggregation level at its contents rise.

In Fig. 3 the micrographs of the surface of phenylone/A (Fig. 3a, b, c) and phenylone/A-MF (Fig. 3d, e, f) with aerosil content 1 mass. % samples are shown, from which one can see the aggregation process inhibition for the last nanocomposites series. Fig. 2d is particularly outstanding in this sense, where one can see "cloud" of aerosil particles, which could not unite in an aggregate owing to the application of processing in electromagnetic field. And on the contrary, Fig. 2a, b, c clearly demonstrate substantial aggregation of aerosil particles in large enough aggregates. The obtained data assume that the studied nanoadhesion effect is purely dimensional effect or true nanoeffect [12]. It is well known that in adhesional strength measurements of different material pairs it is impossible to achieve the full contact of their surfaces. That is why the indicated contact is realized according to some hillocks only, that decreases contact area and, correspondingly, adhesional strength [13]. At very small (nanometer scale) contact area sizes the existence of large hillocks and depressions on the contact surface is scarcely probable that results to sharp adhesion level enhancement. This condition can be expressed more specifically: the nanoadhesion is achieved at the same scale of its participant sizes – nanoparticle and macromolecular coil.

In Fig. 4 the dependence of parameter b on nanofiller particles surface average area S_p for nanocomposites phenylone/aerosil and phenylone/OSY is shown. As one can see, the decrease b or adhesion level reduction at the increase S_p (nanofiller particles size) is observed and at $S_p \approx 2\times10^4$ nm^2, that corresponds to particles diameter ~ 80 nm, this dependence tries to attain $b \approx 1$, i.e. to perfect microadhesion level ($\alpha_n = \alpha_n^T$, that follows from the equation (3)).

The nanoadhesion effect influences strongly on polymer nanocomposites macroscopic properties. As follows from the data of Fig. 1, the introduction 1 mass. % of aerosil in phenylone results to α_n decrease in about 3 times (for nanocomposites phenylone/A-MF). In case of microcomposites such α_n decrease requires the introduction of filler with volume contents φ_f=0.3-0.7 [10], i.e. practically on two orders more. Further as an example let's consider nanoadhesion influence on thermal and mechanical properties of the studied nanocomposites.

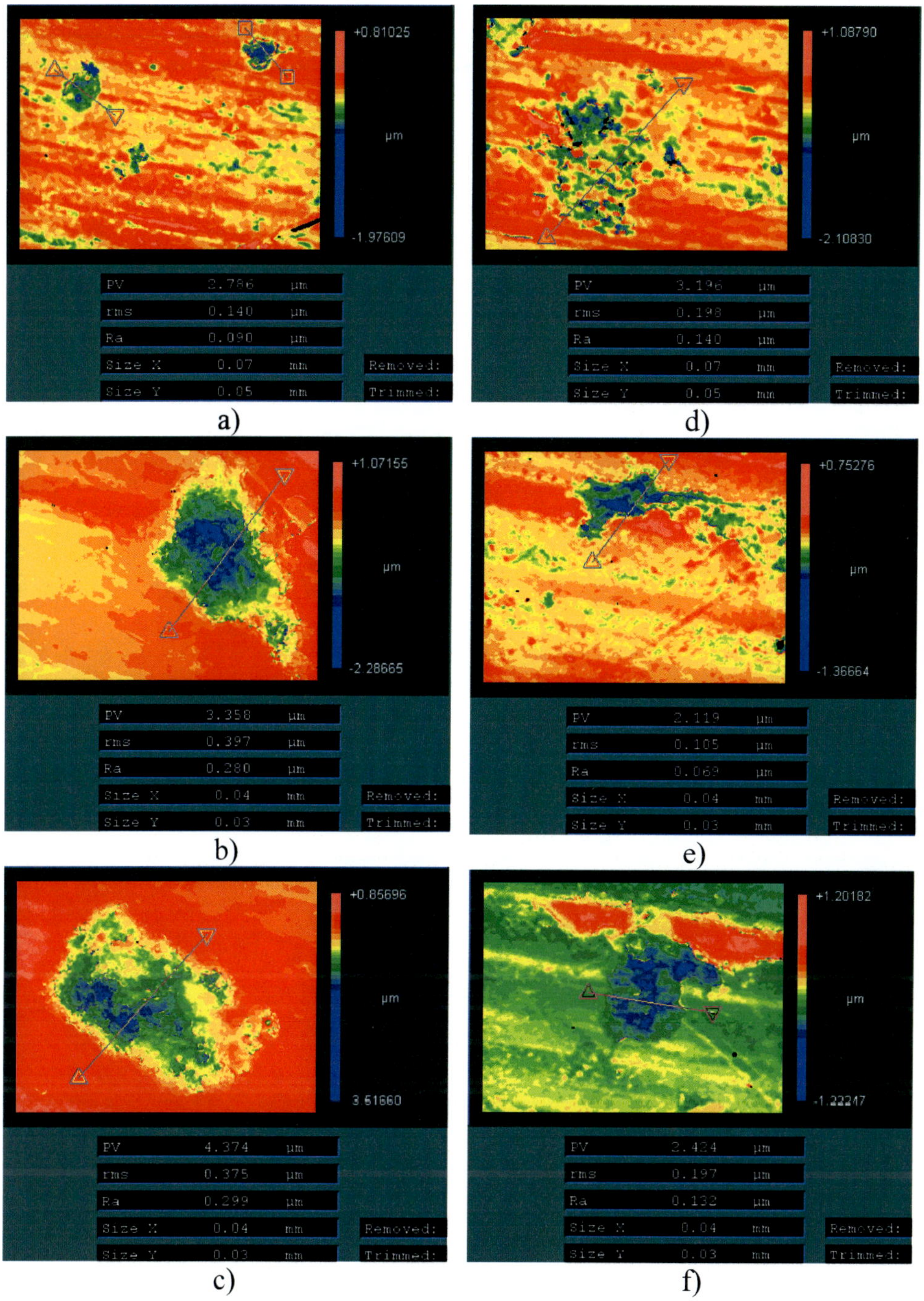

Figure 3. Electron micrographs of surfaces of nanocomposites phenylone/A (a, b, c) and phenylone/A-MF (d, e, f).

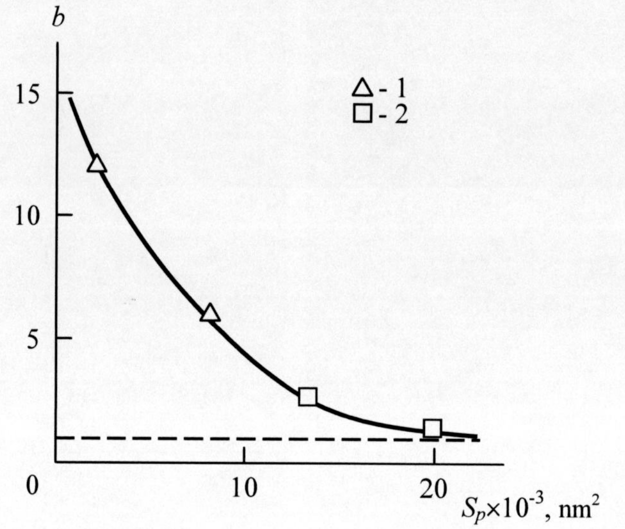

Figure 4. The dependence of parameter b on average area of nanofiller particles surface S_p for nanocomposites filled by aerosil (1) an oxinitride silicium-yttrium (2). The horizontal shaded line indicates perfect microadhesion level (b=1).

At present a large number of kinetic equations for the description of thermogravimetric analysis (TGA) data is known [14, 15]. However, all the indicated models do not take into consideration the polymer melt structure and the character of diffusive processes proceeding in thermooxidative degradation course. The last are important in the respect of controlling of oxidizer (for example, oxygen) access to reactive centers of polymeric macromolecules [16]. As a rule, the temperature of thermal degradation beginning in TGA tests (for example, temperature of 20 %-th sample mass loss $T_{20\%}$) is higher than temperature of the so-called transition "liquid 1 – liquid 2" T_{ll}, which can be estimated as follows [17]:

$$T_{ll} \approx (1.20 \pm 0.05)T_g, \qquad (4)$$

where T_g is glass transition temperature of polymer.

At T_{ll} there occurs a transition of the polymer melt from "liquid with fixed structure" (where residual structural ordering is observed [17]) to true liquid state or "structureless liquid". Nevertheless, melt "structureless" at $T>T_{ll}$ relates to the absence of the supermolecular structure, but the structure of a macromolecular coil in melt remains an important structural factor (in essence, unique at $T>T_{ll}$) [16]. Therefore structural analysis of thermal stability of polymer nanocomposites on the base of phenylone, filled by aerosil, will be given fur ther.

In Figs. 5 and 6 the TGA curves for nanocomposites phenylone/A and phenylone/A-MF are shown, respectively. As the plots of these figures show, on the first stage for the studied materials within the temperatures range 293-383 K sample mass gradual decrease on 1.5-2.0 is observed, connected with moisture loss. Then, up to T=573 K sample mass remains practically nonchanged. The intensive thermooxidative degradation of phenylone and nanocomposites on its base, accompanied by a considerable mass loss, begins in the air above 663 K.

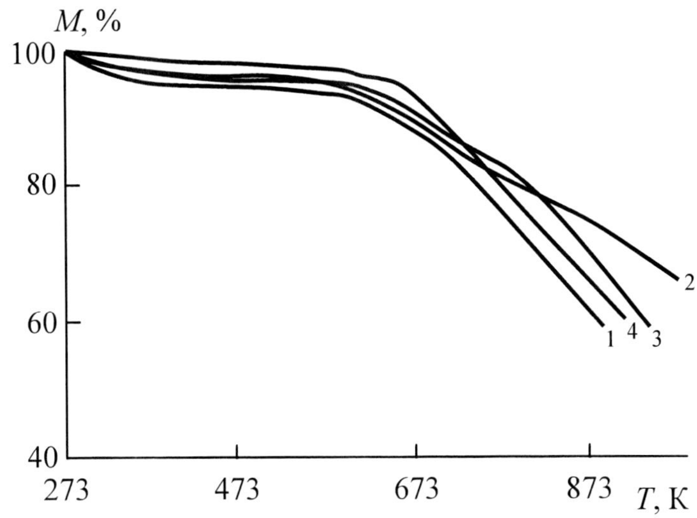

Figure 5. The TGA curves of phenylone (1) and nanocomposites phenylone/A with aerosil contents 0.3 (2), 0.5 (3) and 1.0 (4) mass. %.

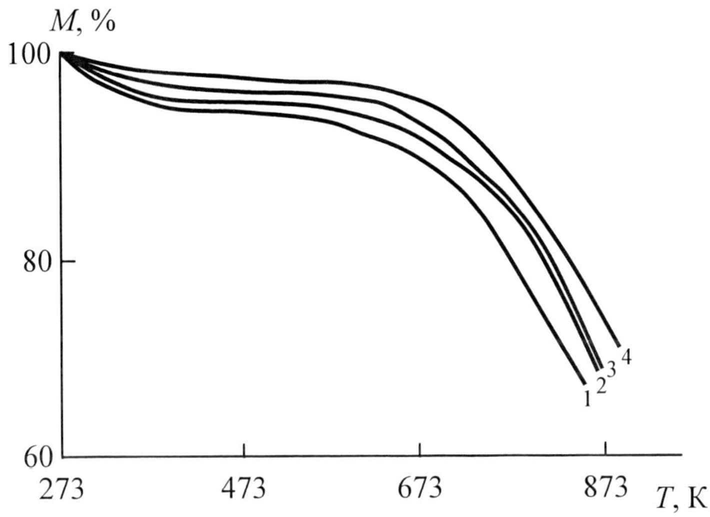

Figure 6. The TGA curves of phenylone (1) and nanocomposites phenylone/A-MF with aerosil contents 0.3 (2), 0.5 (3) and 1.0 (4) mass. %.

The experimentally determined temperatures of sample mass 20 %-th loss $T_{20\%}$ for all studied materials are listed in table 1. As one can see, the introduction of only one percent of aerosil into phenylone rises its thermal stability on 76-80 K. Let's consider the physical fundamentals of this strong enough effect.

As it was known [18], within the framework of a strange (anomalous) diffusion conception on fractal object it's possible to select two its main types: slow and fast diffusion. The dependence of mobile displacement s on time t is assumed as a basis of such division [18]:

Table 1. The comparison of experimental $T_{20\%}$ and calculated according to the equation (8) $T_{20\%}^{T}$ temperatures of sample 20 %-th mass loss for nanocomposites phenylone/aerosil

Material	The aerosil contents, mass. %	$T_{20\%}$, K	$T_{20\%}^{T}$, K, the equation (16)	$T_{20\%}^{T}$, K, the equation (14)
Phenylone S-2	-	773	799	799
Phenylone/A	0.3	775	803	805
	0.5	798	810	805
	1.0	849	825	804
Phenylone/A-MF	0.3	798	809	805
	0.5	818	823	805
	1.0	853	837	804

$$s \sim t^{\beta}, \tag{5}$$

where for classical case β=1/2, for slow diffusion β<1/2 and for the fast one - β>1/2.

The most precisely the macromolecular coil structure, which is fractal object, can be characterized with the aid of its fractal (Hausdorff) dimension Δ_f, describing the coil elements distribution in space [19]. Earlier within the framework of fractional derivatives theory the interrelation Δ_f and β was shown, which is expressed analytically as follows [18]:

$$\beta = \frac{\Delta_f - 1}{4} \tag{6}$$

for slow diffusion and

$$\beta = \frac{\Delta_f - 1}{\Delta_f} \tag{7}$$

for the fast one. The structural boundary between the mentioned kinds of diffusion one can consider the value Δ_f=2.5 at its general variation $2.0 \le \Delta_f < 3.0$: at $\Delta_f < 2.5$ (less compact macromolecular coils) the fast diffusion of oxidizer (oxygen) is realized, at $\Delta_f > 2.5$ – a slow one.

For theoretical estimation of value $T_{20\%}$ ($T_{20\%}^{T}$) the following equation was used [20]:

$$\Delta_f = c\left(T_{20\%}^{T} - T_g\right)^{\beta}, \tag{8}$$

where c is constant, accepted equal to 0.093 for fast diffusion and 0.305 – for slow one [20].

The equation (8) defines three factors, influencing on polymer materials thermal stability: chemical constitution of a polymer, characterized by T_g value, polymer melt structure, characterized by dimension Δ_f, and a type (intensity) of oxidizer diffusion, connected with the structure and characterized by exponent β [20].

To carry out the estimation of dimension Δ_f is possible with the aid of a number of methods. However, in the present case the most obvious one will be the following. As it is known [21, 22], the introduction of filler in polymer can change its effective fractal dimension Δ_f (let's remind, that for polymer materials melt the condition $\Delta_f = d_f$ was accepted), which can be calculated according to the equation [23]:

$$d_f = 3 - 6\left(\frac{\varphi_{den}}{SC_\infty}\right)^{1/2}, \tag{9}$$

where φ_{den} is relative fraction of nanocomposites densely-packed regions, in which relaxation processes are inhibited, S is a cross-sectional area of a macromolecule (for phenylone S=17.9 Å^2 [24]), C_∞ is characteristic ratio, which is equal to 3 for phenylone [24].

To densely-packed regions of nanocomposites clusters in polymeric matrix, nanofiller and interfacial regions with relative fractions φ_{cl}, φ_n and φ_{if}, respectively can be attribute. The value φ_{cl} for phenylone can be obtained according to the equation (9) at d_f=2.447 [24]. The relation between φ_{if} and φ_n can be obtained as follows. As it is known [7] both polymeric matrix and aerosil particles surface, interacting at interfacial regions formation, are fractal objects. At interaction of such objects there is the only linear scale l, defining their interpenetration distance [25]. As in polymer nanocomposites nanofiller elasticity modulus, as a rule, is larger than the corresponding parameter for polymeric matrix, then it is assumed [7], that in this case aerosil penetration in polymeric matrix occurs and then l is equal to interfacial layer thickness l_{if}. Further it can be written [25]:

$$l_{if} \sim a\left(\frac{r_n}{a}\right)^{2(d-d_{surf})/d}, \tag{10}$$

where a is a lower linear scale of polymeric matrix fractal behavior, r_n is a nanofiller particles radius, d is dimension of Euclidean space, in which fractal is considered (obviously, in our case d=3), d_{surf} is fractal dimension of nanofiller particles surface. The proportionality coefficient in the relationship (10) is accepted equal to 1.0 [22].

For polymers the value a is accepted equal to statistical segment length l_{st} [7], determined according to the formula [26]:

$$l_{st} = l_0 C_\infty, \tag{11}$$

where l_0 is the length of skeletal bond of the main chain, which is equal to 1.40 Å for phenylone [24].

The value d_{surf} can be determined according to the following equation [27]:

$$S_u = 400 r_n^{d_{surf}-d}, \tag{12}$$

where specific surface of aerosil particles S_u is given in m^2/g and r_n – in nm.

The equations (11) and (12) give l_{st}=0.42 nm, d_{surf}=2.95 and then l_{if}=2.74. Further the following equation can be used [22]:

$$\varphi_{if} = \varphi_n \left[\left(\frac{r_n + l_{if}}{r_n} \right)^3 - 1 \right], \tag{13}$$

which at r_n=7.5 nm and l_{if}=2.74 nm gives the following relationship between φ_{if} and φ_n for aerosil:

$$\varphi_{if} = 1.55\varphi_n. \tag{14}$$

Let's note, that for the layered silicates (organoclays) the equation (14) has a form [21, 22]:

$$\varphi_{if} = 1.91\varphi_n. \tag{15}$$

As nanocomposites reinforcement degree is defined by sum (φ_n+φ_{if}) (see the equation (18)), then the comparison of the equations (14) and (15) shows, that by its reinforcing action nonaggregated aerosil is close to exfoliated organoclays. The equations (14) and (15) were obtained for perfect adhesion polymeric matrix-nanofiller, i.e. for the condition b=1. At nanoadhesion effect availability (b>1) the value l_{if} is proportional to b [11] and at the first approximation the equation (14) is necessary to write as follows:

$$\varphi_{if} = 1.55\varphi_n b. \tag{16}$$

Then for the value φ_{den} in the equation (9) in case of nanocomposites phenylone/aerosil let's obtain finally:

$$\varphi_{den} = \varphi_{cl} + \varphi_n + 1.55\varphi_n b. \tag{17}$$

Further one can calculate theoretical values $T_{20\%}^T$, using the equations (7), (8), (9) and (17). The comparison of experimental $T_{20\%}$ and calculated by the indicated method $T_{20\%}^T$ values of sample 20 %-th mass loss temperatures in TGA tests is shown in table 1. As one can see, the good enough correspondence of theory and experiment is obtained (the average discrepancy values $T_{20\%}$ and $T_{20\%}^T$ makes up 2.2 %). In this table the similar calculation of

$T_{20\%}^{T}$ without accounting of nanoadhesion effect is given, i.e. at the calculation the equation (15) was used instead of the equation (17). In this case theory and experiment correspondence grows substantially worse. This means, that the indicated above $T_{20\%}$ enhancement at the introduction of particulate nanofiller (aerosil) small amounts by 76-80 K is due to nanoadhesion effect. At the indicated effect absence such enhancement would not exceed 32 K.

The equation (8) gives a strict and physically substantiated definition of temperature $T_{20\%}$. However, the thermal stability structural analysis can be made by empirical more obvious method. In Fig. 7 the dependence $T_{20\%}$ on value φ_{den} is shown, which demonstrates linear increase $T_{20\%}$ at φ_{den} enhancement. The extrapolation of linear plot $T_{20\%}(\varphi_{den})$ to limiting values φ_{den}=0 and φ_{den}=1.0 gives the values $T_{20\%}$=616 and 1040 K, respectively. Theoretically these values $T_{20\%}$ can be calculated according to the equation (8) at the condition Δ_f=3 (slow diffusion) and Δ_f=2 (fast diffusion), respectively. Such calculation gives the following values $T_{20\%}^{T}$: 645 and 1040 K, respectively, i.e. close enough to the one obtained by extrapolation. Hence, as follows from the equations (16) and (17), such empirical treatment assumes also nanoadhesion effect influence on the value $T_{20\%}$.

Let's consider briefly nanoadhesion influence on nanocomposites phenylone/aerosil mechanical properties, namely, on their reinforcement degree. Compression tests have shown, that the elasticity modulus E_m of initial matrix phenylone is equal to 3.05 GPa and nanocomposite phenylone/A-MF with aerosil content 0.3 mass. % - 3.29 GPa. As it is known [22], nanocomposites reinforcement degree is defined by the equation:

$$\frac{E_n}{E_m} = 1 + 11\left(\varphi_n + \varphi_{if}\right)^{1.7}, \qquad (18)$$

where E_n is nanocomposite elasticity modulus.

By using in the equation (18) the estimation φ_{if} according to the formula (14) let's obtain theoretical value E_n (E_n^T) equal to 3.06 GPa and by using formula (16) E_n^T=3.33 GPa. It is not difficult to see, that the second value E_n^T, obtained with nanoadhesion effect accounting, corresponds to the experiment much better (the discrepancy of E_n and E_n^T makes 1 %), than obtained without taking into consideration the indicated effect (the discrepancy is about 10 %). Therefore, nanoadhesion effect results to nanocomposites elasticity modulus increase by 0.263 GPa even at insignificant aerosil content of 0.3 mass. %.

Fig. 8 gives the obvious notion of nanoadhesion effectivity action, where the curves of stress-strain (σ-ε) for initial phenylone and nanocomposite phenylone/A-MF with aerosil content 0.3 mass. % are shown. As one can see, the introduction of aerosil indicated amount changes not only mechanical characteristics of matrix polymer, but its fracture type.

Let's consider in more detail nanoadhesion effect influence on nanocomposites reinforcement degree on the example of phenylone/OSY systems. For the indicated nanocomposites the equations (10)-(14) give the following relationship between φ_{if} and φ_n:

$$\varphi_{if} = 0.102\varphi_n. \tag{19}$$

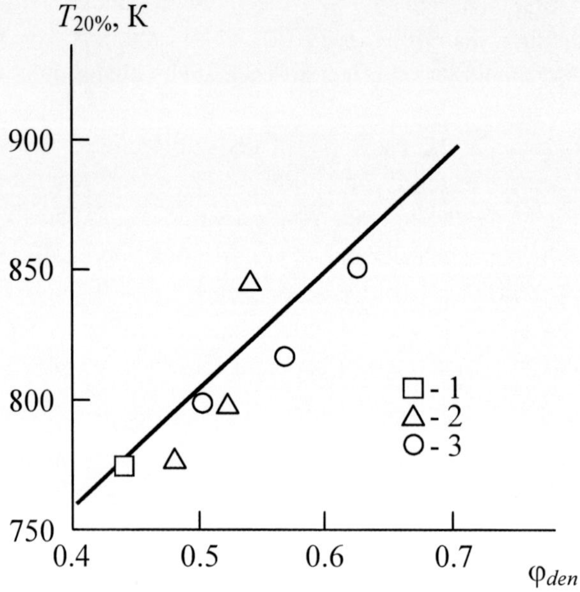

Figure 7. The dependence of sample 20 %-th mass loss temperature $T_{20\%}$ on relative fraction of densely-packed regions φ_{den} for phenylone (1) and nanocomposites phenylone/A (2), phenylone/A-MF (3).

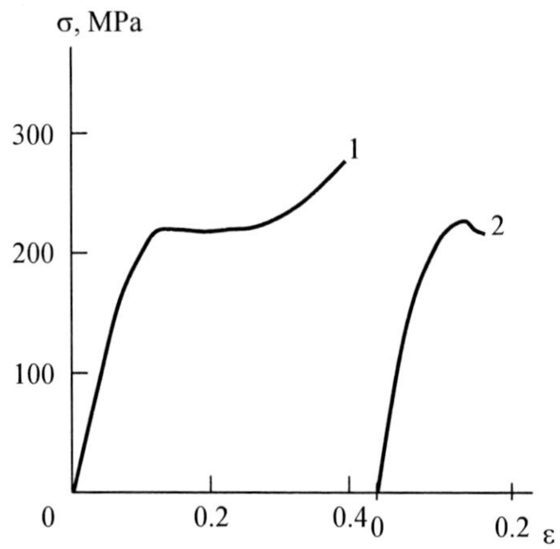

Figure 8. The diagrams stress-strain (σ-ε) for phenylone (1) and nanocomposite phenylone/A-MF with aerosil content 0.3 mass. % (2) in compression tests.

In Fig. 9 the comparison of dependences $E_n(\varphi_n)$ calculated according to the equations (18) and (19) is adduced and is also obtained experimentally. As one can see, the indicated curves $E_n(\varphi_n)$ correspondence leaves much to be desired. Therefore let's use the equation (19) modification similar to the formula (16):

$$\varphi_{if} = 0.102\varphi_n b. \tag{20}$$

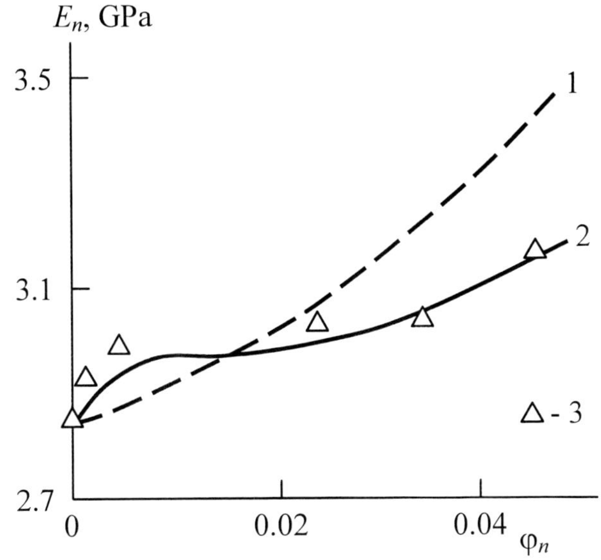

Figure 9. The dependences of elasticity modulus E_n on nanofiller volume contents φ_n for nanocomposites phenylone/OSY. 1 – the calculation according to the equations (18) and (19); 2 – according to the equations (18) and (20); 3 – experimental data.

In Fig. 9 the comparison of the dependences $E_n(\varphi_n)$, obtained experimentally and calculated according to the equations (18), (20) is also shown. As follows from Fig. 9 data, in this case much better correspondence of theory and experiment (their average discrepancy makes 1.9 %) is obtained. But the main merit of the equation (20) does not even consist in this (using of the equations (18) and (19) gives also acceptable average discrepancy of theory and experiment, namely, 4.2 %). Firstly, the calculation with due regard to the nanoadhesion effect (the equation (20)) allows to reproduce the main qualitative features of the experimental curve $E_n(\varphi_n)$, namely, to increase ("hump") of E_n at small φ_n ($b>1$, nanoadhesion effect) and slower enhancement of E_n at $\varphi_n \geq 0.023$ ($b<1$ – adhesion level reduction due to OSY particles aggregation). Secondly, the equations (16) and (20) allow quantitative accounting interfacial adhesion level on nanocomposites reinforcement degree.

Conclusions

Therefore, the present paper results have shown the existence in polymer particulate-filled nanocomposites nanoadhesion effect, which consists in the enhanced level of polymeric matrix-nanofiller interaction. This effect has purely dimensional origin and is realized for nanofiller particles with diameter less than 80 nm. The nanoadhesion effect influences strongly on macroscopic properties of those nanocomposites, in which it is realized. So, the essential thermal stability enhancement and the considerable thermal expansion coefficient reduction of nanocomposites on the base of phenylone at the introduction only 1 mass. %

aerosil are due to the nanoadhesion effect. Appreciation of the nanoadhesion effect allows qualitatively and quantitatively an exact description of nanocomposites reinforcement degree.

REFERENCES

[1] Pul C., Owens F. The Nanotechnologies. Moscow, *Technosfera,* 2005, 336 p.

[2] Sheng N., Boyce M.C., Parks D.M., Rutledge G.C., Abes J.I., Cohen R.E. *Polymer,* 2004, v. 45, № 2, p. 487-506.

[3] Buchachenko A.L. Uspekhi *Khimii*, 2003, v. 72, № 5, p. 419-437.

[4] Okhlopkova A.A., Vinogradov A.V., Pinchyk L.S. Plastics Filled by Ultradisperse Inorganic Particles. *Gomel,* IMMS NASB, 1999, 164 p.

[5] Lipatov Yu.S. The Interfacial Phenomena in Polymers. Kiev, *Naukova Dumka*, 1980, 260 p.

[6] Khunyants N.N., Lyapunova M.A., Manevich L.I., Oshmyan V.G., Shaulov A.Yu. *Mekhanika Kompozithykh Materialov*, 1986, v. 22, № 2, p. 231-234.

[7] Kozlov G.V., Yanovskii Yu.G., Lipatov Yu.S. *Mekhanika Kompozitsionnykh Materialov I Konstruktsii,* 2002, v. 8, № 1, p. 111-149.

[8] Kozlov G.V., Lipatov Yu.S. *Composite Interfaces*, 2002, v. 9, № 6, p. 509-527.

[9] Kubat J., Rigdahl M., Welander M. *J. Appl. Polymer Sci.,* 1990, v. 39, № 5, p. 1527-1539.

[10] Holliday L., Robinson J. In book: *Polymer Engineering Composites.* Ed. Richardson M.O.W. London, Applied Science Publishers LTD, 1978, p. 241-283.

[11] Tanaka K., Nagamura T. *Mater. Intern. Conf. on Polymer and Advanced Materials. POLYMEX*-2006, Huatulko, Mexico, November 5-9, 2006, Session 1, p. 51.

[12] Andrievskii R.A. Rossiiskii *Khimichesk. Zhurnal*, 2002, v. 46, № 5, p. 50-56.

[13] Wu S. Polymer *Interface and Adhesion.* New York, Marcel Dekker, 1982, 492 p.

[14] Wendlandt U. *Thermal Methods of Analysis*. Moscow, Mir, 1978, 526 p.

[15] Zuru A.A., Whitehead R., Griffiths D.L. *Thermochim. Acta*, 1990, v. 164, № 3, p. 285-305.

[16] Kozlov G.V., Zaikov G.E. *The Structural Stabilization of Polymers: Fractal Models. Leiden-Boston,* Brill Academic Publishers, 2006, 345 p.

[17] Bershtein V.A., Egorov V.M. Differential Scanning Calorymetry in Physics-Chemistry of Polymers. Leningrad, *Khimiya,* 1990, 256 p.

[18] Shogenov V.Kh., Akhkubekov A.A., Akhkubekov R.A. Izvestiya VUZov, Severo-Kavkazsk. region, *estestv. nauki*, 2004, № 1, p. 46-50.

[19] Vilgis T.A. *Physica* A, 1988, v. 153, № 2, p. 341-354.

[20] Dolbin I.V., Burya A.I., Kozlov G.V. *Teplofizika Vysokikh Temperatur*, 2007, v. 45, № 3, p. 355-358.

[21] Kozlov G.V., Malamatov A.Kh., Antipov E.M., Karnet Yu.N., Yanovskii Yu.G. *Mekhanika Kompozitsionnykh Materialov I Konstruktsii*, 2006, v. 12, № 1, p. 99-140.

[22] Malamatov A.Kh., Kozlov G.V., Mikitaev M.A. *The reinforcement Mechanisms of Polymer Nanocomposites.* Moscow, Publishers RKhTU, 2006, 240 p.

[23] Kozlov G.V., Novikov V.U. Uspekhi Fizichesk. *Nauk,* 2001, v. 171, № 7, p. 717-764.

[24] Kozlov G.V., Burya A.I., Dolbin I.V. *Prikladnaya Fizika*, 2006, № 1, p. 14-18.

[25] Hentschel H.G.E., Deutch J.M. *Phys. Rev.* A, 1986, v. 29, № 12, p. 1609-1611.
[26] Wu S. J. *Polymer Sci.: Part B: Polymer Phys.*, 1989, v. 27, № 4, p. 723-741.
[27] Kozlov G.V., Shustov G.B. *Khimicheskaya Technologiya*, 2006, № 1, p. 24-26.

In: Progress in Monomers, Oligomers, Polymers, Composites ... ISBN 978-1-60456-878-3
Editors: R. A. Pethrick, G.E. Zaikov et al.

Chapter 31

PHARMACOLOGICAL PREMISES OF THE CREATION NEW ANTITUMOR PREPARATIONS OF THE CLASS OF NITROSOALKYLUREA

***J.A. Djamanbaev*[1,*], *Ch. Kamchybekova*[1], *J.A. Abdurashitova*[1] and *G. E. Zaikov*[2,†]**

[1]Institute of Chemistry and Chemical Technology, Kyrgyz National Academy of Sciences, Bishkek, Kyrgyzstan

[2]N.M. Emanuel Institute of Biochemical Physics, Russian Academy of Sciences, Moscow, Russian Federation

ABSTRACT

Perspectives in the field of creation highly effective anticancerogenic preparations have been evaluated. For their creation it is offered a new regio-selective method of glycosylation alkylurea in conditions nucleophilic catalysis with the following nitrozing of glycosyl carbamides of the D- and L-rows.

This method opens principal new possibilities of modification of compounds by means of glycosylamides bond allowing to get the preparations, possessing small toxicity and selectivity of the action.

Key words: monosaccharides, glycosylureas, N-glycosyds.

Interest to nitrosoalkylurea, as potentially antitumor agents was shown at the beginning of 60-th years, after opening high antileukemic activity N-methyl-N-nitrozoguanidin and N-methyl-N-nitrosourea [1-2]. These compounds have soon increased the essential arsenal of antitumor preparations of alkylation action [3-4]. However, a number of side effects of the

* Correspondence to: Institute of Chemistry and Chemical Technology, Kyrgyz National Academy of Sciences, 720071, Kyrgyzstan, Bishkek, 267 Chui Prospect, e-mail: Djamanbaev-J@mail.ru

compounds of this type, first of all high myelotoxic , restrained thus introduction to medical practice.

Modification of preparations by introduction to the molecular structure of different substituent though has increased the range of potentially active compounds, however has not brought to eliminating undesirable influence of preparations on organism. The success reached in pharmacology of the nitrosoalkylureas is described in particular in works [1, 5]. First natural carbohydrates analogue N-nitrosomethylurea the antibiotic - "streptosotocin" got from cultural liquid - a streptomycin was found in the same years [3, 6]. It was realized that "streptosotocin" possesses the broad spectrum of actions - an antidiabetic, diabetogenic, mutagenic and antitumor and presents itself carbohydrate derivative of nitrosomethylurea from secondary carbon atom of glucoses.

The multiple pharmacological studies having been carried out, showed that though "streptosotocin" has above mentioned side effects, its general toxicity much below the most citotoxic fragment, connected with atom C_2 D-glucopiranosyd ring. These observations stimulated studies on syntheses and test on antitumor activity carbohydrate derivatives of N-methyl-N-nitrosourea. Repeatedly spoken out suggestions that carbohydrates molecule can enter as special transport carrier, relieving carrying the function active groups (in this instance - an citostatic of the groups) in the tumors fabric promoted the development of this work.

In this work we tried to motivate searching and elaboration of new methods of the syntheses more active antitumor preparations among derivatives of nitrosomethylurea.

The study of the reaction with the participation of glycosyl bonds is important not only for the theory of the carbohydrates structure and reactional ability of carbohydrates.

They present also the significant interest for solving a number of actual problems of bioorganic chemistry as the glycosyl bond is one of the most important structural elements of many biologically active compounds.

Studying of the reactions of glycosyl centre has allowed determining some typical peculiarities of this type of nucleophilic catalysis.

In this connection, the attention was concentrated on synthetic aspects of the homogeneous catalysis and elaboration of methods of directing synthesis of carbohydrates physiologically active substances. They attack great importance to the creation of medical preparations, in particular antitumor preparations, possessing the low of toxic activity, the high of solubility in water and selectivity of actions.

These investigations are of great interest for solution of practical important problems for ennoble physiologically active substances and medical preparations with the known therapeutic action. Given direction provides working out of the recommendations on decrease of toxicity, change of water and lipid solubility preparations as well as obtaining derivatives with selective permeability through cell membranes [7].

In the process of pharmacological investigations there was ascertained that glycosylation leads to the sharp decrease of toxicity of medicinal preparations (LD_{50} falls, as a rule, on two orders).

In clinical oncology there is of great importance not only high antitumor activity of preparations of the group of nitrosourea but also their ability to run through hematoencephalitic barrier that opens the real possibilities for chemotherapy metastasis and

† Correspondence to: *N. M. Emanuel Institute of Biochemical Physics, Russian Academy of Sciences, 119334, Russia, Moscow, 4 Kosygin Str.,e-mail: Chembio@sky.chph.ras.ru

primary tumors of the cerebrum. The majority of known preparation do not get through hematoencephalitic barrier and so do not warn spreading a metastasis in cerebrum, and they aren't practically efficient at defeat of the central nervous system.

As to selectivity of the action, it is known, that permeability of some cell membranes of animals regarding to monosaccharides is different, and the degree of permeability correlates with nucleophilic reactional ability of monosaccharides (tabl. 1) [8].

Thereof comes the principle possibility of the processes regulation of the passing carbohydrates preparation through cell barriers by changing the nature of carbohydrates carrier, as well as their biological activity. It drams attention the fact that a number of therapeutic characteristic greatly depends on place of joining of a physiological active fragment to the carbohydrates ring. Especially it reveals in the process of study preparations of antitumor action.

Table 1. Comparative absorption (B_1) and comparative reactional ability (B_2) of monosaccharides on C_1

Monosaccharides	B_1	B_2
glucose	1,0	0
galactose	1,1	0,64
fructose	0,4	-
mannose	0,2	0,37
ksilose	0,15	0,81
arabinose	0,09	1,03
ribose	-	1,19
licsose	-	1.25

Rather perspective for practical application, in our opinion, is usage as "blending links" more acid stable amide bonds. What is the value of the amid bond forming? What are the prospects of their use in applied aspects of the chemistry carbohydrates? It is enough to point to broad circulation of urea fragment amongst natural material - the derivatives of pyrimidin of the row: nucleosides, riboflavin, teobromin, caffeine and others, and development of the direction of heterocycles syntheses, including analogues of the nucleosides on the base of carbohydrates derivatives of urea.

The development of the work on specified directions greatly depends on elaboration technically acceptable and economic profitable methods of putting the carbamides fragment in to mono-, oligo- and polysaccharides.

As a result of searching conditions of the synthesis of N-nitroso derivatives of monosaccharides unprotected hydrocsyl groups, providing the simplicity, effectivity of the process, it is perfected the general methods of getting N-alkil-N-(β-D-glykopiranozyl)nitrosourea in which there were combined reactions: a) the interaction of monosaccharides with alkylurea in condition of nucleophilic catalysis with addition aryl amines and b) nitrozilation N-alkylglycosylureas [9, 10, 11].

Thus, rather easy and mobile method of syntheses of big group of casrbohydrates derivatives of nitrosoalkylureas was worked out. On this base are it seems to be perspective investigations of biological activity of modified medical preparations with the help of glycosylamides bond.

References

[1] Nitrozoalkilureas - a new class of antitumor preparations. Emanuel N. M., Corman D. B., Ostrovskaya L.A., Gorbacheva L.B, Dementieva N.P.-M., 1978. S. 290.

[2] Grenn M. O., Greenderg J. The activity of nitrozoguanidines against ascites tumors in mice// *Cancer. Res*.-1960.-V.-20.-N.-8.-P.1166-1171.

[3] Ross U. Biological alkylation substanses.-*M.: Medicine*, 1964.

[4] Chernov V.A. Citostatic substances in chemotherapies of oncomas.- *M.: Medicine*, 1964.

[5] The system of the creation of antitumor preparations in USSR and USA. Under N.N. Blohina, C.H. Zubroda. - *M.: Medicine*, 1977.

[6] Herr R.R., Jahnke H. K., Argoudelis A. D. The structure of streptozotocin // *J. Am. Chem. Soc*.-1967.-V.-89.-P. 4808-4809.

[7] Nikoliskiy N.N., Troshin A.S. The Transport sugar through celluar membrane. *L.: Science,* 1973.

[8] Afanasiev V.A., Strelicova I.F. and others. Construction and reactional ability N-glycosydes. Frunze: Ilim, 1976: 221.

[9] Djamanbaev J. A., Afanasiev V. A. The method of the synthesis of N-alkyl-(N-aril)-N-(β-D-glykosyl)urea. *Avt sv*. N772102 USSR.

[10] Afanaciev V.A., Djamanbaev J.A. *Patent USA*. 4: 656.259. 1987.

[11] Afanasiev V.A., Dzhamanbaev J. A., Zaikov G.E. The derivatives of carbohydrate with carbamides fragments // *Successes of chemistry* -V.-51.- 4.- 1982.-S.661.

In: Progress in Monomers, Oligomers, Polymers, Composites … ISBN 978-1-60456-878-3
Editors: R. A. Pethrick, G.E. Zaikov et al.

Chapter 32

INFLUENCE OF 1.1.5-TRIHYDROPERFLUOROPENTANOL ON THE STRUCTURE AND PROPERTIES OF POLYCAPROAMIDE SYSTEMS

I.A. Novakov[1,*]***, N.A. Rakhimova***[1]***, A.P. Krasnov***[2]***,***
Y.V. Zubavichus[2] ***and V.B. Ivanov***[3]
[1]Volgograd State Technical University of Russia, Russian Federation
[2]A.N.Nesmeyanov Institute of Organoelement Compounds of the Russian Academy of Sciences, Russian Federation
[3]Russian Academy of Sciences N.N. Semenov Institute of Chemical Physics, Russian Academy of Sciences, Russian Federation

ABSTRACT

The introduction of 0.035 wt% 1.1.5-trihydroperfluoropentanol decreases the rate of the photochemical destruction of polycaproamide fibres. The simultaneous modification of PCA with microquantity of 1.1.5-trihydroperfluoropentanol and 1.3-benzendicarboxamide, N.N'-bis(2.2.6.6-tetramethyl-4-piperidinyl) leads to the synergetic photostabilizing effect.

Changes in X-ray diffraction patterns of unoriented films of poly-ε-caproamide cast from its solutions in formic acid due to the introduction of 1,1,5-trihydroperfluoropentanol are analyzed. The plastification increases the fraction of polymer molecules in the planar all-trans conformation at the costs of those in the gauche conformation that constitutes the amorphous phase. This promotes the formation of hydrogen-bonded layers oriented perpendicular to the film plane as structural elements.

[*] Correspondence to: Volgograd State Technical University of Russia, Organic@vstu.ru

The improvement of thermal and photostability of polyamides, in particular, polycaproamide (PCA), is an urgent problem associated with the widespread use of polyamide items [1]. Among issues under examination in this field is the relationship between the kinetics of degradation processes and the structure of polymers. Only a few studies indicate that the stability of PCA markedly changes when the structure of macromolecules is modified via the introduction of functional units containing stabilizer fragments [2, 3] or when the supramolecular structure (primarily, the degree of crystallinity and the density of the amorphous phase) is changed by varying the sample preparation conditions [4, 5].

For the purposeful modification of the polymer structure during fabrication of articles, a number of special additives—primarily plasticizers, surfactants, and nucleation agents—are used (see, e.g., [1],). Recently, it has been established that fluorinated alcohols are extremely efficient modifiers for PCA [5–8].

The aim of this study was to examine the effect of 1,1,5-trihydroperfluoropentanol (TPFP) on the oxidative degradation of PCA. The choice of this compound was determined by its technological prospects related to unlimited miscibility with PCA and accessibility. A lower molecular mass analog, 1,1,3-trihydroperfluoropropanol, is a low-boiling explosive liquid, while higher homologues show a poorer miscibility with PCA. This class of modifiers for PCA shows promise due to the possibility of polyfunctional interaction between proton-donor CHF_2 and OH groups and between perfluorocarbon chains and PCA macromolecules.

The choice of experimental conditions was primarily determined by the fact that the influence of structural defects in the polymer matrix manifests itself to the greatest extent at relatively low temperatures, when relaxation processes occur rather slowly. The use of PCA fibers as an object of investigation is related to the known fact that the structure of PCA strongly affects the stability of fibers. Moreover, the improvement on the stability of such products represents an important problem.

Results and Discussion

A large body of experimental data suggests that there is a correlation between chemical processes leading to a decrease in the molecular mass of PCA and a drop in its tensile strength during aging [10]. This is particularly true of PCA fibers prepared by drawing [1]. In the course of oxidative degradation, the elongation at break for PCA fibers varies only slightly. On the basis of these data, a change of tensile strength (this parameter is most important for practice) was chosen as a criterion for the stability of PCA.

Table 1. Changes in mechanical properties of PCA fibers modified by polyfluorinated TPFP upon storage

PCA fiber	Tensile strength, N	Elongation at break, %
Initial	16.0/14.6	16.2/22.5
After storage for 2 years	8.5/12.9	16.0/22.5

Note: The values for the initial PCA fiber and the PCA fiber with the polyfluorinated modifier are in the numerator and denominator, respectively.

The data listed in Table 1 show that the oxidation of the unmodified PCA sample for a long period (2 years) at room temperature results in a sharp (practically twofold) drop in strength. At the same time, for the modified sample aged under the same conditions, the strength decreased only by 12.4%.

The effect of TPFP on the strength of PCA during long-term storage is identical to that of reduced water content in the PCA system [11], a condition which is also required for the retention of strength properties of PCA. In order to rationalize this effect we have studied the associative interactions of PCA macromolecules with water and TPFP molecules using the AM1 quantum-chemical method [12]. In this case, the following types of interaction were considered: namely, between the end groups of PCA, associates of PCA macromolecules–water, associates of PCA macromolecules– TPFP, and associates of PCA macromolecules–TPFP– water. The minimum total electron energy was obtained for the PCA–TPFP system. This condition favors a decrease in the content of water in the amorphous phase and is primarily responsible for the retention of strength properties of polycaproamide fibers upon storage.

The introduction of TPFP is also accompanied by a significant deceleration of the oxidative photodegradation of PCA upon irradiation of fibers in the weatherometer (Fig. 1). The photooxidation of the modified PCA is characterized by a significant induction period for a start of changes in strength. After the induction period, the photooxidation proceeds at a rate almost equal to the rate of degradation of the unmodified polymer. Analogous results were obtained in the study of the photooxidation of PCA in the Xenotest apparatus. The irradiation in this machine much better simulates the natural sunlight. It has been shown that, within 760 h, the strength of the modified fiber comprises 2.3 N, whereas that of the unmodified fiber is as low as 0.46 N.

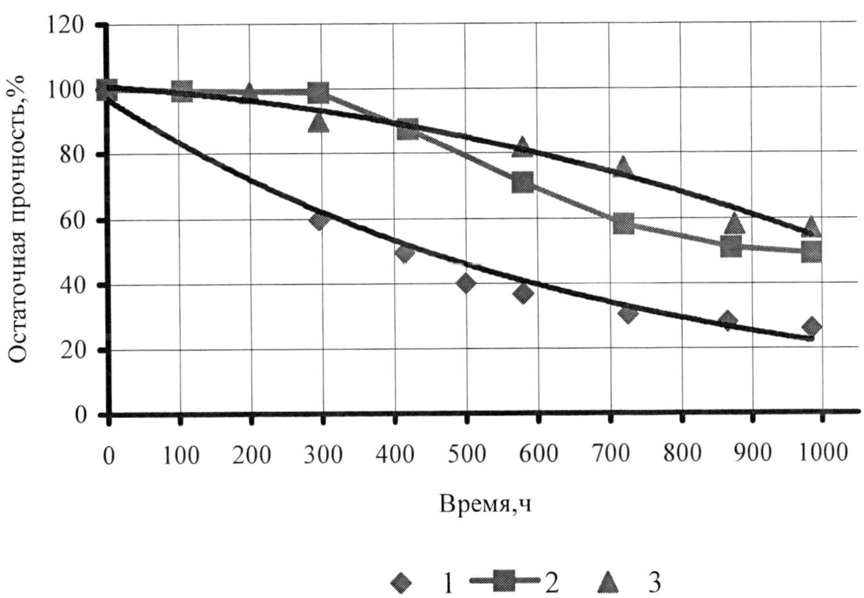

Figure 1. Variation of relative strength in the course of the photooxidative degradation during irradiation in the weatherometer of (*1*) PCA fibers without additives, (*2*) fibers modified by $3.5 \cdot 10^{-3}$ wt % TPFP, or (*3*) the mixture of TPFP ($5 \cdot 10^{-3}$ wt %) and TMBDC ($5 \cdot 10^{-2}$ wt %).

The influence of TPFP on the structure of PCA at the stage of formation and/or during the orientational drawing of fibers is among probable reasons for these phenomena and thus it is addressed in some detail below.

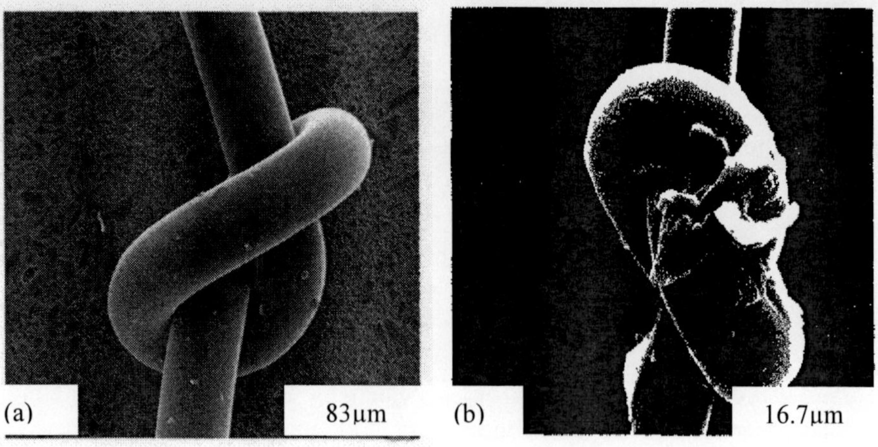

Figure 2. Electron micrographs of knots deliberately made on (a) PCA fibers modified by TPFP and (b) pristine PCA fibers.

Actually, a comparison of the electron micrographs of fiber knots presented in Fig. 2 indicates that the modified fiber is by far more perfect structurally and contains a significantly smaller number of ruptures and other defects than the unmodified one. Naturally, this strongly affects the photooxidative degradation of the polymer. Moreover, it is inconceivable that electrophilic polyfluorinated radicals arise from the perfluoroalcohol

$$H(CF_2CF_2)_2CH_2OH \xrightarrow[-RH]{R\cdot} \cdot(CF_2CF_2)_2CH_2OH$$

The radical center being formed differs from nonfluorinated analogs in that the tetrahedral configuration is retained. It is well known [13] that this electrophilic radical center reacts with any low- and high-molecular-mass compounds containing radical nucleophilic centers and thus terminates free-radical chains.

We also studied the combined effect of TPFP and TMBDC (the latter compound is one of the most efficient industrial stabilizers of aliphatic polyamides) on the photostability of modified PCA fibers. Figure 1 shows that, under these conditions, even at a low concentration of TMBDC (5·10–2wt %), this mixture markedly enhances the photostability of PCA fibers, whereas the addition of TMBDC alone increases the photostability of PCA only at much higher (nearly by one order of magnitude) concentrations [1]. This phenomenon may be explained in terms of well-known concepts assuming that there is a distinct stage of activation in the case of sterically hindered amines. This stage results from oxidative transformations induced by singlet oxygen and peroxide radicals and is accompanied by the formation of corresponding nitroxyl radicals, hydroxylamines, and hydroxylamine esters [14]. At higher conversion levels, when the inhibiting effect of the aforementioned products of amine transformations is rather pronounced, the oxidative degradation of PCA occurs at a significantly lower rate than the degradation of PCA in the absence of any composite additives.

The photostabilizing effect of the TPFP–TMBDC mixture allows one to conserve the strength properties of PCA fibers under UV irradiation: the residual strength was higher by a factor of 4.4 as compared to the strength of the unmodified fibers. The data on the photostabilizing effect during oxidative UV degradation of the PCA fiber modified by microquantities of TPFP, TMBDC, and their composition are presented in Table 2. (The photostabilizing properties of nitroxyl radicals were studied in [15].)

Table 2. The effect of the additives on the breaking strength after irradiation by light with a wavelength > 300 nm

Additive (content, wt %)	Density, tex	Residual strength, %*
None	14.34	37/30
TPFP (3.5·10–3)	14.21	75/60
TMBDC (5·10–1)	14.26	70/52
TPFP–TMBDC (5·10–3+ 5·10–2)	14.16	80/80

* The expositions are 575 and 710 h in the numerator and denominator, respectively.

The maximum photostabilizing effect in the oxidative degradation of the PCA fiber is observed in the case of the composite microadditive of TPFP (5·10–3wt %) and TMBDC (5·10-2wt %). The synergetic effect is probably related to both a decrease in the number of structural defects in the PCA fibers due to the presence of TPFP and the photostabilizing influence of nitroxyl radicals [15] and other products arising from transformations of sterically hindered amines generated from a tetramethylpiperidinyl derivative according to the following tentative scheme [8].

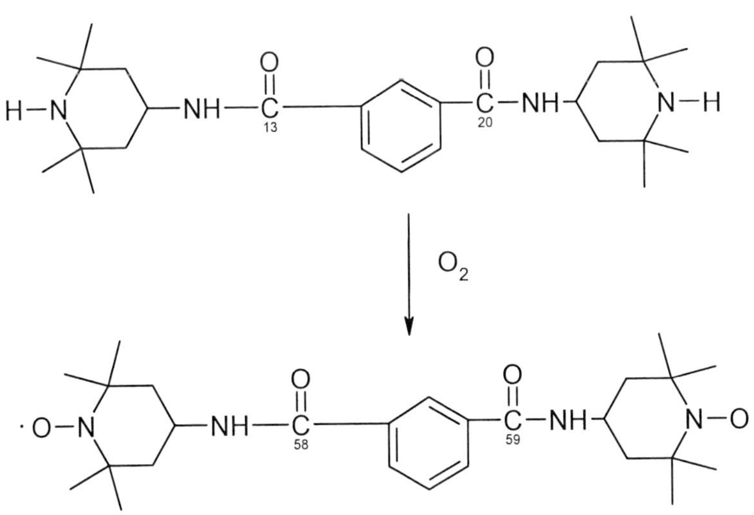

Thus, the modification of PCA by a small amount of TPFP (3.5·10–2wt %) improves the stability of fibers upon storage. In the course of the oxidative UV degradation in the presence of TPFP, the induction period was observed. The occurrence of this induction period is related to changes in the structure of the amorphous phase, a lower number of defects in the polycaproamide fiber, and the electrophilic character of generated polyfluorinated radicals. When microquantities of TPFP and TMBDC were simultaneously added, the synergetic photostabilizing effect is apparent.

Since above results strongly suggest that the introduction of TPFP into PCA somehow modify its structure, we performed X-ray diffraction structural study of modified polymers. Numerous publications are devoted to the study of the molecular and crystalline structure of PCA [16–24]. It was shown that PCA samples are mostly semicrystalline; there are three crystalline modifications of PCA— α, β, and γ [16, 17, 20]. The degree of crystallinity of PCA samples is primarily determined by their preparation technique and conditions. For PCA fibers, the contribution of the amorphous phase is, as a rule, insignificant, while during film formation of PCA solutions in formic acid, this fraction becomes prevailing [17].

The incorporation of various proton donors (water, alcohols, phenols, *etc.*) into PCA shows a strong effect on the supramolecular structure, mechanical characteristics, and thermal behavior of the as-modified polymer [22, 24]. The influence of water on supramolecular amorphous and crystalline structures derived from PCA under various preparation conditions has been studied in detail [17, 21]. If the content of water in PCA is small, the plasticization takes place, namely, water molecules form hydrogen bonds with the amide groups of PCA and function as short cross-linking bridges. In this case, water largely penetrates unoriented amorphous regions of the polymer. Furthermore, even small amounts of water strongly affect the shear deformation of crystallites in PCA films [23]. In the case of aliphatic alcohols, an increase in the size of plasticizer molecules reduces its activity. For example, ethanol is two times less active than water [24]. As regards oxyaromatic compounds, both reversible and irreversible bonding with PCA macromolecules incorporated into the amorphous phase was observed [22]. The irreversible character of bonding is related to the incorporation of modifying molecules into densely packed regions of the amorphous phase inaccessible to water molecules used for the extraction of phenols from PCA.

Recent studies showed that polyfluorinated alcohols (PFAs) of the general formula $H(CF_2CF_2)_nCH_2OH$ readily interact with PCA macromolecules owing to the presence of proton-donor groups OH and HCF_2 (the latter appear to be more acidic because of the strong electron-acceptor effect of the fluorine atoms). When the polyfluorinated alcohol is added in the 5-fold excess, the complete dissolution of PCA occurs, which is accompanied by a marked exothermic effect [25, 7, 8].

Quantum-chemical calculations of intermolecular interactions for the PFA-modified PCA revealed that the order of bonding between associated groups increases, indicating strengthening of the hydrogen bonds –NH ···O=C between neighboring macromolecules under the action of PFA even at a rather low concentration of the plasticizer (one PFA molecule per several PCA macromolecules) [8, 12]. The goal of this study was to gain insight into the effect of 1,1,5-trihydroperfluoropentanol (TPFP)—a representative of the family of polyfluorinated alcohols— on the structure of unoriented PCA films cast from formic acid solutions by means of X-ray diffraction.

The experimental diffraction patterns of test PCA films recorded in the reflection (Bragg-Brentano) mode are shown in Fig.3. The patterns of all three PCA films studied show similar features, namely, two narrow peaks at ~20.1° and 24.5° superimposed on a wide amorphous halo. Another two peaks of much lower intensity are observed at scattering angles of 38.7° and 40.9°. Narrow diffraction peaks are assigned to the α-crystalline phase of PCA. In accordance with the X-ray data reported by Holmes et al. [7], α -PCA has a monoclinic structure with unit cell parameters a= 9.56, b = 17.24, and c =8.01 Å and β = 67.5°.

Taking into account these crystallographic data, the following indices may be assigned to the above-described diffraction features (the theoretical values of 2θ calculated from these unit cell parameters are shown in parentheses): 200 (20.11°), 002/202 (24.05/24.78°), 402 (38.67°) , and 400 (40.87°) . The center of gravity of the amorphous halo is seen at a scattering angle of ~21.1°. This result is consistent with the data by Murthy *et al.* [28] who studied temperature-induced phase transitions in PCA. Despite similar features, the patterns of the plasticized PCA films somewhat differ from that of the film cast from the pristine polymer. Specifically, the height of the amorphous halo (for the diffraction patterns normalized to the peak intensity of the 200 line corresponding to the α– phase) is maximum for the starting commercial PCA. At the same time, the ratio between the intensity of 200 and 002/202 lines significantly increases in the case of TPFP-modified PCA films. To obtain a more detailed quantitative description of the above phenomenon, we performed the X-ray line profile analysis of the patterns at scattering angles in the range from 6° to 36°.

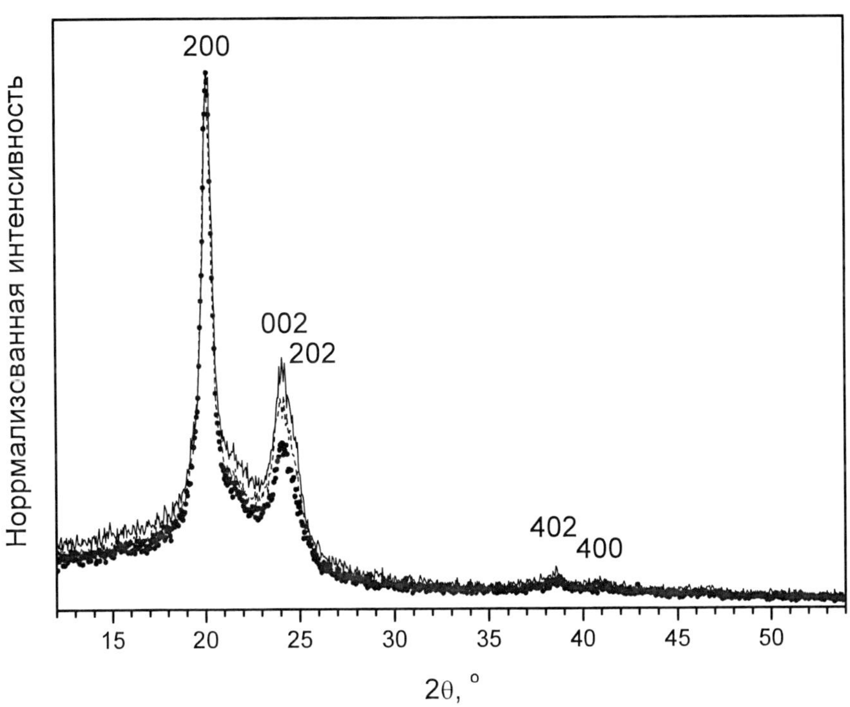

Figure 3. Experimental diffraction patterns of TPFP-modified PCA films: the solid line refers to the commercial PCA: the dashed line refers to PCA + 0.02% TPFP; and the data points refer to PCA + 0.25% TPFP; the reflection mode.

The linear background of incoherent scattering was subtracted from the experimental patterns, and the observed specific features were simulated by four functions of the Pearson VII type. The relevant data are shown in Fig. 4 and in the table. As is seen, formation of the PCA film in the presence of the polyfluorinated alcohol leads to a change in the film structure, as evidenced by an increase in the integral intensity of diffraction peaks due to the crystalline α- phase (the degree of crystallinity grows from ~33% for the starting commercial PCA to 45% for the sample containing 0.25 wt % TPFP) and a certain narrowing of diffraction lines (full width at half maximum of the peak decreases from 0.63° to 0.61°), suggesting an increase in the size of ordered domains (coherent scattering regions). It should be noted that the redistribution of the intensities of lines corresponding to the α- phase takes place, namely, the height of the 002 line and, especially of the 202 line, decreases relative to that of the 200 line with a rise in the concentration of TPFP.

Such changes in intensities may reflect the evolution of the film morphology resulting from modification of the mean orientation of polymer chains relative to the film plane. As demonstrated by the preliminary optical microscopy data, the incorporation of TPFP into PCA leads to a change in the type of morphology the films from globular to lamellar spherulite.

Table 3. Positions (2θ), relative integral intensities (*I*), and widths (*w*) of diffraction lines from profile analysis data

Sample	α-PCA			amorphous PCA		
	2θ, °	I, %	w, °	2θ, °	I, %	w, °
Commercial PCA	20.11 (200) 24.05 (002) 24.78 (202)	20 8 4.5	0.63	21.13	67.5	6.40
PCA + 0.02% PPFP	20.11 (200) 24.05 (002) 24.78 (202)	26 9 4	0.62	21.14	61.0	7.11
PCA + 0.25% PPFP	20.11 (200) 24.05 (002) 24.78 (202)	33 8 4	0.61	21.09	55.0	6.80

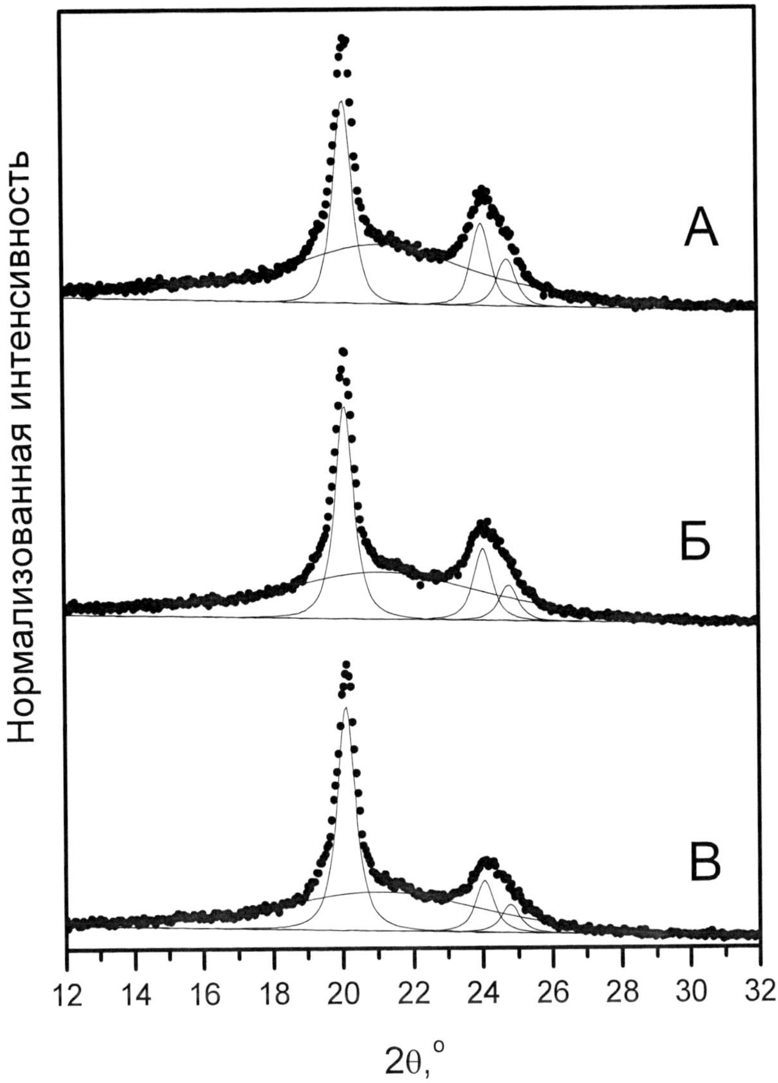

Figure 4. Profile analysis of experimental reflection diffraction patterns of TPFP-modified PCA films: (*1*) α- PCA, (*2*) the amorphous PCA, and (*3*) the linear background; (*a*) the commercial PCA, (*b*) PCA + 0.02 TPFP, and (*c*) PCA + 0.25% TPFP.

Additional knowledge about the preferable orientation of PCA molecules in unoriented films was derived from additional X-ray diffraction studies in the Debye–Scherrer geometry (the transmission mode). These experiments allowed one to provide information concerning structural elements oriented parallel to the plane of the sample (in the case of the above Bragg–Brentano diffraction geometry, the structural elements preferably oriented perpendicular to the surface of the test sample are most prominent). Figure 5 shows the diffraction patterns obtained in the transmission mode for the same three samples.

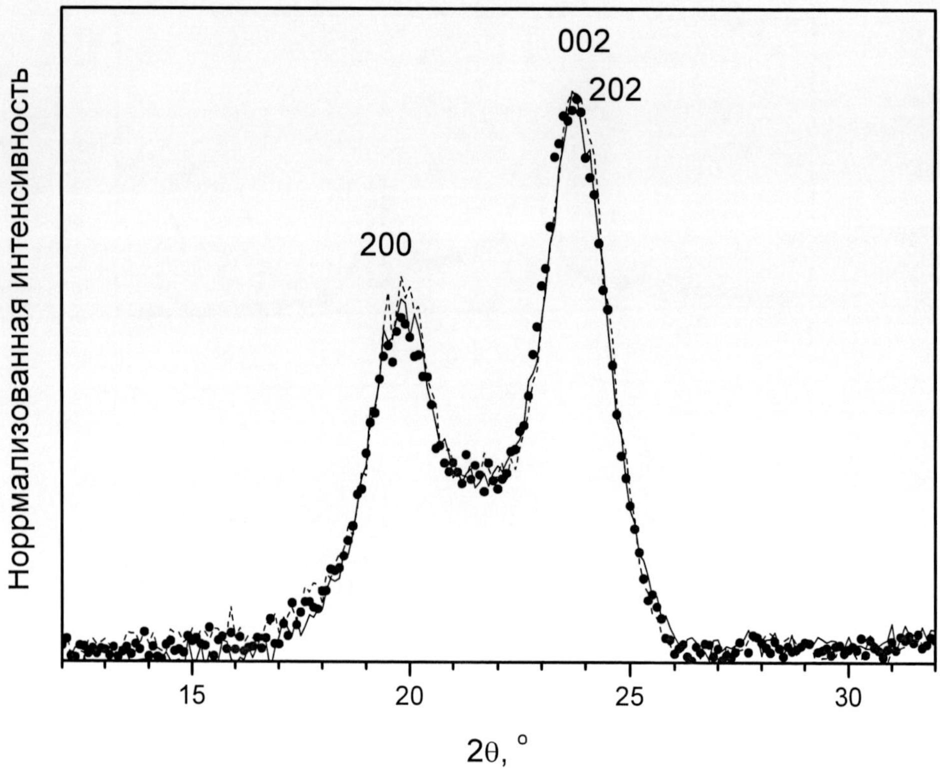

Figure 5. Experimental transmission diffraction patterns of TPFP-modified PCA films: the solid line refers to the commercial PCA; the dashed line refers to PCA + 0.02 TPFP; and the data points refer to PCA + 0.25% TPFP.

Along with a marked widening of diffraction lines compared to the reflection diffractograms (Fig. 3), a noticeable redistribution of relative intensities of diffraction lines occurs since the angular focusing conditions of the divergent X-ray beam are not met in the Debye–Scherrer geometry. Additionally, the 002/202 doublet appears to be much more intense than the 200 line. This result in combination with the above diffraction measurements in the Bragg–Brentano diffraction geometry confirms the occurrence of texturing in the unoriented PCA films. However, it should be emphasized that the diffractograms obtained for the three test samples in the transmission mode are indistinguishable within the experimental error. Most likely, this finding indicates a lower sensitivity of the procedure under discussion to small changes in the structure and morphology of polymer samples compared to the reflection diffractograms.

On the basis of Holmes *et al.* studies [20], the structure of the α-phase of PCA may be described as follows: the molecular axis of the polymer coincides with the crystallographic axis *b*; the polymer molecules adopt the planar zigzag conformation (ribbons) united into layers (approximately in the crystallographic plane *ab*) via the hydrogen bonds C=O ··· H–N . Neighboring H-bonded ribbons in the layer are shifted by 3/14 of a translation relative to each other in direction *b*. This ensures mutual arrangement of carbonyl and amide groups that is optimal for hydrogen bonding. Molecules of neighboring layers superimposed on each other in direction *c* are arranged according to the head-to-tail type. The unit cell is composed of four elementary units - HN–$(CH_2)_5$C(O)– .

Thus, the diffraction line 200 corresponds to the ordering along the crystallographic axis *a*, that is, along the direction of hydrogen bonds between the polymer chains in layers; the diffraction line 002 is determined by the parallel stacking of H-bonded layers; the mixed line 202 is more sensitive to mutual orientations of neighboring H bonded layers. A comparison of reflection and transmission diffractograms leads us to conclusion that, in PCA films cast from formic acid solutions, crystallites of the α-modification of PCA are arranged in such a mode that hydrogen-bonded layers (the crystallographic axis *a*) are predominantly oriented perpendicular to the plane of the film. The effect of TPFP on the formation of PCA films from formic acid solutions includes an increase in the proportion of polymer molecules occurring in the planar *trans* conformation (as follows from [28], the amorphous PCA consists of molecules with a large proportion of *gauche* defects) and facilitation of formation of extended hydrogen-bonded layers oriented perpendicular to the surface of the film as structural elements.

Acknowledgments

We are grateful to T.E. Sukhanova (Institute of Macromolecular Compounds, Russian Academy of Sciences) for providing us with the electron micrographs.

References

[1] *Plastics Additives Handbook* , Ed. by H. Zweifel (Hanser, Munich, 2002) pp. 80, 352.

[2] V. B. Ivanov, V. V. Selikhov, N. N. Barashkov, *et al.* , Vysokomol. Soedin., Ser. B 32 , 140 (1991).

[3] V. B. Ivanov, I. I. Barashkova, V. V. Selikhov, *et al.* Polym. Degrad. Stab. 35 , 267 (1992).

[4] P. Gijsman, D. Tummer, and K. Jansen, Polym. Degrad. Stab. 49 , 121 (1995).

[5] G. A. George and M. S. O'Shea, Polym. Degrad. Stab. 28 , 289 (1990).

[6] N. A. Storozhakova, S. A. Strel'chenko, and V. S. Kuz'min, RF Patent No. 2187517, Byull. Izobret., No. 23 (2002).

[7] N. Storozhakova, Russ. Polym. News **7** (4), 57 (2002).

[8] N. A. Storozhakova, A. B. Golovanchikov, M. K. Tatarnikov, *et al.*,Khim. Tekhnol. Khim. Volokon, No. 4, 41 (2002).

[9] B. N. Maksimov, V. G. Barabanov, I. L. Serushkin, V. S. Zotikov, I. A. Semerikova, V. P. Stepanov, N. G. agaidakova, and G. I. Kaurova, *Commercial Organofluoric Products* (Khimiya, Leningrad, 1990) [in Russian].

[10] V. Ya. Shlyapintokh, *Photochemical Conversion and Stabilization of Polymers* (Hanser, Munich, 1984).

[11] A. K. Kuznetsov, Yu. M. Bazarov, A. S. Belikov, and N. V. Kuleshova, Khim. Volokna, No. 1, 58 (2004).

[12] N. A. Storozhakova, R. G. Fedunov, and V. A. Babkin, J. Balkan Tribol. Assoc. 10 , 318 (2004).

[13] A. I. Rakhimov, *Chemistry and Technology of Fluoroorganic Compounds* (Khimiya, Moscow, 1986) [in Russian].

[14] M. Dagonneau, V. B. Ivanov, E. G. Rozantsev, *et al.*, J. Macromol. Sci., Rev. Macromol. Chem. Phys. 22, 169 (1983).

[15] S. M. Kavun and A. L. Buchachenko, Vysokomol. Soedin

[16] F. Danusso, Usp. Khim. 39, 304 (1970).

[17] T. L. Lebedeva, S. V. Shapovalov, and N. A. Platé, Polymer Science, Ser. A 38, (1996) [Vysokomol. Soedin., Ser. A 38, 1986 (1996)].

[18] R. Houwink and A. J. Staverman, *Polymer Chemistry and Technology* (Khimiya, Moscow, 1996) [in Russian].

[19] V. A. Kargin and G. L. Slonimskii, *Short Essays on Physical Chemistry of Polymers* (Khimiya, Moscow, 1967) [in Russian].

[20] D. R. Holmes, C. W. Bunn, and D. J. Smith, J. Polym. Sci. 17, 159 (1955).

[21] I. I. Perepechko and S. S. Yakovenko, Vysokomol. Soedin., Ser. A 23, 1166 (1981).

[22] M. S. Arzhakov, E. V. Nechaeva, N. V. Red'ko, et al., Polymer Science, Ser. A 38, (1996) [Vysokomol. Soedin., Ser. A 38, 66 (1996)].

[23] B. M. Ginzburg, N. Sultonov, and A. A. Shepelevskii, Polymer Science, Ser. A 46, (2004) [Vysokomol. Soedin., Ser. A 46, 285 (2004)].

[24] L. P. Razumovskii, V. S. Markin, and G. E. Zaikov, Vysokomol. Soedin., Ser. A 27, 863 (1985).

[25] N. A. Storozhakova, V. F. Zheltobryukhov, E. A. Zauer, and A. I. Rakhimov, Russ. Polym. News **5** (3), 38 (2000).

[26] W. R. Sorenson and T. W. Campbell, *Preparative Methods of Polymer Chemistry* (Wiley, New York, 1961; Inostrannaya Literatura, Moscow, 1963).

[27] D. Midgley and K. Torrance, *Potentiometric Water Analysis* (Wiley, Chichester, 1978; Mir, Moscow, 1980).

[28] N. S. Murthy, S. A. Curran, S. M. Aharoni, and H. Minor, Macromolecules 24, 3215 (1991).

INDEX

A

B

C

D

E

F

G

H

I

J

K

L

M

P

Q

R

S

T

U

V

W

X

Y

Z